P9-AFN-629

Intermediate Algebra

EIGHTH EDITION

Charles P. McKeague
Cuesta College

Prepared by

Ross Rueger
College of the Sequoias

THOMSON
™
BROOKS/COLE

Australia • Brazil • Canada • Mexico • Singapore • Spain • United Kingdom • United States

Printed in the United States of America
1 2 3 4 5 6 7   11 10 09 08 07

Printer: Thomson/West

ISBN-13: 978-0-495-38267-6
ISBN-10: 0-495-38267-1

Thomson Higher Education
10 Davis Drive
Belmont, CA 94002-3098
USA

For more information about our products, contact us at:
Thomson Learning Academic Resource Center
1-800-423-0563

For permission to use material from this text or product, submit a request online at
http://www.thomsonrights.com.
Any additional questions about permissions can be submitted by email to **thomsonrights@thomson.com.**

Contents

Preface

This *Student Solutions Manual* contains complete solutions to all odd-numbered exercises and all chapter review/test exercises of *Intermediate Algebra* by Charles P. McKeague. I have attempted to format solutions for readability and accuracy, and apologize to you for any errors that you may encounter. If you have any comments, suggestions, error corrections, or alternative solutions please feel free to drop me a note or send an email (address below).

Please use this manual with some degree of caution. Be sure that you have attempted a solution, and re-attempted it, before you look it up in this manual. Algebra can only be learned by *doing*, and not by observing! As you use this manual, do not just read the solution but work it along with the manual, using my solution to check your work. If you use this manual in that fashion then it should be helpful to you as you do homework and study for tests.

I would like to thank a number of people for their assistance in preparing this manual. Thanks go to Laura Localio at Thomson Brooks/Cole Publishing for her valuable assistance and support. Special thanks go to Matt Bourez of College of the Sequoias for his meticulous error-checking of my solutions, and prompt return of my manuscript under tight deadlines. Also thanks go to Patrick McKeague for keeping me updated to manuscript corrections and prompt response to my emails.

I wish to express my appreciation to Pat McKeague for asking me to be involved with this textbook. This book provides a complete course in intermediate algebra, and you will find the text very easy to read and understand. Good luck!

Ross Rueger
College of the Sequoias
matmanross@aol.com

February, 2007

Chapter 1
Basic Properties and Definitions

1.1 Fundamental Definitions and Notation

1. Written in symbols, the expression is: $x + 5 = 2$

3. Written in symbols, the expression is: $6 - x = y$

5. Written in symbols, the expression is: $2t < y$

7. Written in symbols, the expression is: $x + y < x - y$

9. Written in symbols, the expression is: $3(x - 5) > y$

11. Multiplying: $6^2 = 6 \cdot 6 = 36$

13. Multiplying: $10^2 = 10 \cdot 10 = 100$

15. Multiplying: $2^3 = 2 \cdot 2 \cdot 2 = 8$

17. Multiplying: $2^4 = 2 \cdot 2 \cdot 2 \cdot 2 = 16$

19. Multiplying: $10^4 = 10 \cdot 10 \cdot 10 \cdot 10 = 10,000$

21. Multiplying: $11^2 = 11 \cdot 11 = 121$

23.
  a. Using the rule for order of operations: $3 \cdot 5 + 4 = 15 + 4 = 19$
  b. Using the rule for order of operations: $3(5 + 4) = 3 \cdot 9 = 27$
  c. Using the rule for order of operations: $3 \cdot 5 + 3 \cdot 4 = 15 + 12 = 27$

25.
  a. Using the rule for order of operations: $6 + 3 \cdot 4 - 2 = 6 + 12 - 2 = 16$
  b. Using the rule for order of operations: $6 + 3(4 - 2) = 6 + 3 \cdot 2 = 6 + 6 = 12$
  c. Using the rule for order of operations: $(6 + 3)(4 - 2) = 9 \cdot 2 = 18$

27.
  a. Using the rule for order of operations: $(7 - 4)(7 + 4) = 3 \cdot 11 = 33$
  b. Using the rule for order of operations: $7^2 - 4^2 = 49 - 16 = 33$

29.
  a. Using the rule for order of operations: $(5 + 7)^2 = 12^2 = 144$
  b. Using the rule for order of operations: $5^2 + 7^2 = 25 + 49 = 74$
  c. Using the rule for order of operations: $5^2 + 2 \cdot 5 \cdot 7 + 7^2 = 25 + 70 + 49 = 144$

31.
  a. Using the rule for order of operations: $2 + 3 \cdot 2^2 + 3^2 = 2 + 3 \cdot 4 + 9 = 2 + 12 + 9 = 23$
  b. Using the rule for order of operations: $2 + 3\left(2^2 + 3^2\right) = 2 + 3(4 + 9) = 2 + 3 \cdot 13 = 2 + 39 = 41$
  c. Using the rule for order of operations: $(2 + 3)\left(2^2 + 3^2\right) = (5)(4 + 9) = 5 \cdot 13 = 65$

33.
  a. Using the rule for order of operations: $40 - 10 \div 5 + 1 = 40 - 2 + 1 = 38 + 1 = 39$
  b. Using the rule for order of operations: $(40 - 10) \div 5 + 1 = 30 \div 5 + 1 = 6 + 1 = 7$
  c. Using the rule for order of operations: $(40 - 10) \div (5 + 1) = 30 \div 6 = 5$

35.
  a. Using the rule for order of operations: $40 + \left[10 - (4 - 2)\right] = 40 + (10 - 2) = 40 + 8 = 48$
  b. Using the rule for order of operations: $40 - 10 - 4 - 2 = 30 - 4 - 2 = 26 - 2 = 24$

37.   **a.**   Using the rule for order of operations:

$$3+2\left(2\bullet 3^2+1\right)=3+2\left(2\bullet 9+1\right)=3+2\left(18+1\right)=3+2\bullet 19=3+38=41$$

    b.   Using the rule for order of operations: $\left(3+2\right)\left(2\bullet 3^2+1\right)=5\left(2\bullet 9+1\right)=5\left(18+1\right)=5\bullet 19=95$

39.   Using the rule for order of operations: $5\bullet 10^3+4\bullet 10^2+3\bullet 10+1=5,431$

41.   Using the rule for order of operations:

$$3\left[2+4\left(5+2\bullet 3\right)\right]=3\left[2+4\left(5+6\right)\right]=3\left(2+4\bullet 11\right)=3\left(2+44\right)=3\bullet 46=138$$

43.   Using the rule for order of operations:

$$6\left[3+2\left(5\bullet 3-10\right)\right]=6\left[3+2\left(15-10\right)\right]=6\left(3+2\bullet 5\right)=6\left(3+10\right)=6\bullet 13=78$$

45.   Using the rule for order of operations:

$$5\left(7\bullet 4-3\bullet 4\right)+8\left(5\bullet 9-4\bullet 9\right)=5\left(28-12\right)+8\left(45-36\right)=5\bullet 16+8\bullet 9=80+72=152$$

47.   Using the rule for order of operations: $25-17+3=8+3=11$

49.   Using the rule for order of operations: $109-36+14=73+14=87$

51.   Using the rule for order of operations: $20-13-3=7-3=4$

53.   Using the rule for order of operations: $63-37-4=26-4=22$

55.   Using the rule for order of operations: $36\div 9\bullet 4=4\bullet 4=16$

57.   Using the rule for order of operations: $75\div 3\bullet 25=25\bullet 25=625$

59.   Using the rule for order of operations: $64\div 16\div 4=4\div 4=1$

61.   Using the rule for order of operations: $75\div 25\div 5=3\div 5=0.6$

63.   Calculating the value: $18,000-9,300=8,700$

65.   Calculating the value: $3.45+2.6-1.004=6.05-1.004=5.046$

67.   Calculating the value: $275\div 55=5$

69.   Calculating the value: $4\left(2\right)\left(4\right)^2=4\left(2\right)\left(16\right)=8\bullet 16=128$

71.   Calculating the value: $250\left(5\right)-25\left(5\right)^2=250\left(5\right)-25\left(25\right)=1,250-625=625$

73.   Calculating the value: $5\bullet 2^3-3\bullet 2^2+4\bullet 2-5=5\bullet 8-3\bullet 4+4\bullet 2-5=40-12+8-5=31$

75.   Calculating the value: $125\bullet 2^3=125\bullet 8=1,000$ **77.**   Calculating the value: $500\left(1.5\right)=750$

79.   Calculating the value: $5\left(0.10\right)=0.5$ **81.**   Calculating the value: $0.20\left(8\right)=1.6$

83.   Calculating the value: $0.08\left(4,000\right)=320$ **85.**   Using a calculator: $0.6931+1.0986=1.7917$

87.   Using a calculator: $3\left(0.6931\right)=2.0793$ **89.**   Using a calculator: $250\left(3.14\right)=785$

91.   Using a calculator: $4,628\div 25=185.12$ **93.**   Using a calculator: $65,000\div 5,280\approx 12.3106$

95.   Using a calculator: $0.8413-1=-0.1587$ **97.**   Using a calculator: $16\left(3.5\right)^2=16\left(12.25\right)=196$

99.   Using a calculator: $11.5\left(130\right)-0.05\left(130\right)^2=11.5\left(130\right)-0.05\left(16,900\right)=1,495-845=650$

101.   **a.**   Evaluating when $x=5$: $x+2=5+2=7$ **b.**   Evaluating when $x=5$: $2x=2\left(5\right)=10$

    c.   Evaluating when $x=5$: $x^2=5^2=5\bullet 5=25$ **d.**   Evaluating when $x=5$: $2^x=2^5=2\bullet 2\bullet 2\bullet 2\bullet 2=32$

103.   **a.**   Evaluating when $x=10$: $x^2+2x+1=\left(10\right)^2+2\left(10\right)+1=100+20+1=121$

    b.   Evaluating when $x=10$: $\left(x+1\right)^2=\left(10+1\right)^2=11^2=121$

    c.   Evaluating when $x=10$: $x^2+1=10^2+1=100+1=101$

    d.   Evaluating when $x=10$: $\left(x-1\right)^2=\left(10-1\right)^2=9^2=81$

105.   **a.**   Evaluating when $a=2$, $b=5$, and $c=3$: $b^2-4ac=\left(5\right)^2-4\left(2\right)\left(3\right)=25-24=1$

    b.   Evaluating when $a=10$, $b=60$, and $c=30$: $b^2-4ac=\left(60\right)^2-4\left(10\right)\left(30\right)=3,600-1,200=2,400$

    c.   Evaluating when $a=0.4$, $b=1$, and $c=0.3$: $b^2-4ac=\left(1\right)^2-4\left(0.4\right)\left(0.3\right)=1-0.48=0.52$

107. The set is $A \cup B = \{0,1,2,3,4,5,6\}$.

109. The set is $A \cap B = \{2,4\}$.

111. The set is $B \cap C = \{1,3,5\}$.

113. The set is $A \cup (B \cap C) = \{0,1,2,3,4,5,6\}$.

115. The set is $\{0,2\}$.

117. The set is $\{0,6\}$.

119. The set is $\{0,1,2,3,4,5,6,7\}$.

121. The set is $\{1,2,4,5\}$.

123.   **a.**   Subtracting: $3:47.83 - 3:47.80 = 0.03$ seconds

     b.   Subtracting: $3:48.20 - 3:47.83 = 0.37$ seconds

125.   **a.**   Substituting $n = 1$, $x = 5$, $y = 8$, and $z = 13$: $x^n + y^n = 5^1 + 8^1 = 5 + 8 = 13 = 13^1 = z^n$

     b.   Substituting $n = 1$, $x = 2$, $y = 3$, and $z = 5$: $x^n + y^n = 2^1 + 3^1 = 2 + 3 = 5 = 5^1 = z^n$

     c.   Substituting $n = 2$, $x = 3$, $y = 4$, and $z = 5$: $x^n + y^n = 3^2 + 4^2 = 9 + 16 = 25 = 5^2 = z^n$

     d.   Substituting $n = 2$, $x = 7$, $y = 24$, and $z = 25$: $x^n + y^n = 7^2 + 24^2 = 49 + 576 = 625 = 25^2 = z^n$

1.2 The Real Numbers

1.   Graphing the inequality:

3.   Graphing the inequality:

5.   Graphing the inequality:

7.   Graphing the inequality:

9.   Graphing the inequality:

11.   Graphing the inequality:

13.   Graphing the inequality:

15.   Graphing the inequality:

17.   Graphing the inequality:

19.   Graphing the inequality:

21.   Graphing the inequality (no solution):

23.   Graphing the inequality (no solution):

25.   Graphing the inequality:

27.   Graphing the inequality:

29.   Graphing the inequality:

31.   Graphing the inequality:

33.   Graphing the inequality:

35.   Graphing the inequality:

37.   The inequality is: $x \geq 5$

39.   The inequality is: $x \leq -3$

41.   The inequality is: $x \leq 4$

43.   The inequality is: $-4 < x < 4$

45.   The inequality is: $-4 \leq x \leq 4$

47.   The counting numbers are: $1, 2$

49.   The rational numbers are: $-6, -5.2, 0, 1, 2, 2.3, \frac{9}{2}$

51.   The irrational numbers are: $-\sqrt{7}, -\pi, \sqrt{17}$

53.   The nonnegative integers are: $0, 1, 2$

55.   Factoring: $60 = 4 \cdot 15 = 2^2 \cdot 3 \cdot 5$

57.   Factoring: $266 = 14 \cdot 19 = 2 \cdot 7 \cdot 19$

59.   Factoring: $111 = 3 \cdot 37$

61.   Factoring: $369 = 9 \cdot 41 = 3^2 \cdot 41$

63. Reducing the fraction: $\frac{165}{385} = \frac{3 \cdot 55}{7 \cdot 55} = \frac{3}{7}$

65. Reducing the fraction: $\frac{385}{735} = \frac{35 \cdot 11}{35 \cdot 21} = \frac{11}{21}$

67. Reducing the fraction: $\frac{111}{185} = \frac{37 \cdot 3}{37 \cdot 5} = \frac{3}{5}$

69. Reducing the fraction: $\frac{525}{630} = \frac{105 \cdot 5}{105 \cdot 6} = \frac{5}{6}$

71. Reducing the fraction: $\frac{75}{135} = \frac{15 \cdot 5}{15 \cdot 9} = \frac{5}{9}$

73. Reducing the fraction: $\frac{6}{8} = \frac{2 \cdot 3}{2 \cdot 4} = \frac{3}{4}$

75. Reducing the fraction: $\frac{200}{5} = \frac{40 \cdot 5}{1 \cdot 5} = 40$

77. Reducing the fraction: $\frac{10}{22} = \frac{2 \cdot 5}{2 \cdot 11} = \frac{5}{11}$

79. The numbers are either $2 - 5 = -3$ or $2 + 5 = 7$.

81. The numbers are between 3 and 13, which is the inequality $3 < x < 13$.

83. The numbers are either less than 3 or more than 13, which is the inequality $x < 3$ or $x > 13$.

85.
  a. $10 = 3 + 7$ or $10 = 5 + 5$
  b. $16 = 11 + 5$ or $16 = 13 + 3$
  c. $24 = 19 + 5$ or $24 = 17 + 7$ or $24 = 13 + 11$
  d. $36 = 19 + 17$ or $36 = 23 + 13$ or $36 = 29 + 7$ or $36 = 31 + 5$

87. The inequality is $50 \leq F \leq 270$.

89.
  a. Adding the percents: $15\% + 29\% + 26\% + 16\% + 12\% = 98\%$
  b. Adding the percents: $2\% + 15\% + 29\% + 26\% = 72\%$
  c. Adding the percents: $26\% + 16\% + 12\% = 54\%$

91. Answers will vary.

93. Calculating the value: $5! = 5 \cdot 4 \cdot 3 \cdot 2 \cdot 1 = 120$

95. Calculating the value: $6! = 6 \cdot 5 \cdot 4 \cdot 3 \cdot 2 \cdot 1 = 6 \cdot 5!$. So the statement is true.

1.3 Properties of Real Numbers

1-11. Completing the table:

	Number	Opposite	Reciprocal
1.	4	−4	$\frac{1}{4}$
3.	$-\frac{1}{2}$	2	−2
5.	5	−5	$\frac{1}{5}$
7.	$\frac{3}{8}$	$-\frac{3}{8}$	$\frac{8}{3}$
9.	$-\frac{1}{6}$	$\frac{1}{6}$	−6
11.	3	−3	$\frac{1}{3}$

13. −1 and 1 are their own reciprocals.

15. 0 is its own opposite.

17. Writing without absolute values: $\left|-2\right| = 2$

19. Writing without absolute values: $\left|-\frac{3}{4}\right| = \frac{3}{4}$

21. Writing without absolute values: $\left|\pi\right| = \pi$

23. Writing without absolute values: $-\left|4\right| = -4$

25. Writing without absolute values: $-\left|-2\right| = -2$

27. Writing without absolute values: $-\left|-\frac{3}{4}\right| = -\frac{3}{4}$

29. Multiplying the fractions: $\frac{3}{5} \cdot \frac{7}{8} = \frac{21}{40}$

31. Multiplying the fractions: $\frac{1}{3} \cdot 6 = \frac{6}{3} = 2$

33. Multiplying the fractions: $\left(\frac{2}{3}\right)^3 = \frac{2}{3} \cdot \frac{2}{3} \cdot \frac{2}{3} = \frac{8}{27}$

35. Multiplying the fractions: $\left(\frac{1}{10}\right)^4 = \frac{1}{10} \cdot \frac{1}{10} \cdot \frac{1}{10} \cdot \frac{1}{10} = \frac{1}{10,000}$

37. Multiplying the fractions: $\frac{3}{5} \cdot \frac{4}{7} \cdot \frac{6}{11} = \frac{72}{385}$

39. Multiplying the fractions: $\frac{4}{3} \cdot \frac{3}{4} = \frac{12}{12} = 1$

41. Using the associative property: $4 + (2 + x) = (4 + 2) + x = 6 + x$

43. Using the associative property: $(a + 3) + 5 = a + (3 + 5) = a + 8$

45. Using the associative property: $5(3y) = (5 \cdot 3) y = 15y$ **47.**   Using the associative property: $\frac{1}{3}(3x) = \left(\frac{1}{3} \cdot 3\right) x = x$

49. Using the associative property: $4\left(\frac{1}{4} a\right) = \left(4 \cdot \frac{1}{4}\right) a = a$ **51.**   Using the associative property: $\frac{2}{3}\left(\frac{3}{2} x\right) = \left(\frac{2}{3} \cdot \frac{3}{2}\right) x = x$

53. Applying the distributive property: $3(x + 6) = 3 \cdot x + 3 \cdot 6 = 3x + 18$

55. Applying the distributive property: $2(6x + 4) = 2 \cdot 6x + 2 \cdot 4 = 12x + 8$

57. Applying the distributive property: $5(3a + 2b) = 5 \cdot 3a + 5 \cdot 2b = 15a + 10b$

59. Applying the distributive property: $\frac{1}{3}(4x + 6) = \frac{1}{3} \cdot 4x + \frac{1}{3} \cdot 6 = \frac{4}{3} x + 2$

61. Applying the distributive property: $\frac{1}{5}(10 + 5y) = \frac{1}{5} \cdot 10 + \frac{1}{5} \cdot 5y = 2 + y$

63. Applying the distributive property: $(5t + 1) 8 = 5t \cdot 8 + 1 \cdot 8 = 40t + 8$

65. Applying the distributive property: $\frac{3}{4}(8x - 4) = \frac{3}{4} \cdot 8x - \frac{3}{4} \cdot 4 = 6x - 3$

67. Applying the distributive property: $\frac{5}{6}(12x - 18) = \frac{5}{6} \cdot 12x - \frac{5}{6} \cdot 18 = 10x - 15$

69. Applying the distributive property: $3(3x + y - 2z) = 3 \cdot 3x + 3 \cdot y - 3 \cdot 2z = 9x + 3y - 6z$

71. Applying the distributive property: $10(0.3x + 0.7y) = 10 \cdot 0.3x + 10 \cdot 0.7y = 3x + 7y$

73. Applying the distributive property: $100(0.06x + 0.07y) = 100 \cdot 0.06x + 100 \cdot 0.07y = 6x + 7y$

75. Applying the distributive property: $3\left(x + \frac{1}{3}\right) = 3 \cdot x + 3 \cdot \frac{1}{3} = 3x + 1$

77. Applying the distributive property: $2\left(x - \frac{1}{2}\right) = 2 \cdot x - 2 \cdot \frac{1}{2} = 2x - 1$

79. Applying the distributive property: $x\left(1 + \frac{2}{x}\right) = x \cdot 1 + x \cdot \frac{2}{x} = x + 2$

81. Applying the distributive property: $a\left(1 - \frac{3}{a}\right) = a \cdot 1 - a \cdot \frac{3}{a} = a - 3$

83. Applying the distributive property: $8\left(\frac{1}{8} x + 3\right) = 8 \cdot \frac{1}{8} x + 8 \cdot 3 = x + 24$

85. Applying the distributive property: $6\left(\frac{1}{2} x - \frac{1}{3} y\right) = 6 \cdot \frac{1}{2} x - 6 \cdot \frac{1}{3} y = 3x - 2y$

87. Applying the distributive property: $12\left(\frac{1}{4} x + \frac{2}{3} y\right) = 12 \cdot \frac{1}{4} x + 12 \cdot \frac{2}{3} y = 3x + 8y$

89. Applying the distributive property: $20\left(\frac{2}{5} x + \frac{1}{4} y\right) = 20 \cdot \frac{2}{5} x + 20 \cdot \frac{1}{4} y = 8x + 5y$

91. Applying the distributive property: $24\left(\frac{2}{3} x + \frac{1}{2}\right) = 24 \cdot \frac{2}{3} x + 24 \cdot \frac{1}{2} = 16x + 12$

93. Applying the distributive property: $3(5x + 2) + 4 = 15x + 6 + 4 = 15x + 10$

95. Applying the distributive property: $4(2y + 6) + 8 = 8y + 24 + 8 = 8y + 32$

97. Applying the distributive property: $5(1 + 3t) + 4 = 5 + 15t + 4 = 15t + 9$

99. Applying the distributive property: $3 + (2 + 7x) 4 = 3 + 8 + 28x = 28x + 11$

101. Adding the fractions: $\frac{2}{5} + \frac{1}{15} = \frac{2}{5} \cdot \frac{3}{3} + \frac{1}{15} = \frac{6}{15} + \frac{1}{15} = \frac{7}{15}$

103. Adding the fractions: $\frac{17}{30} + \frac{11}{42} = \frac{17}{30} \cdot \frac{7}{7} + \frac{11}{42} \cdot \frac{5}{5} = \frac{119}{210} + \frac{55}{210} = \frac{174}{210} = \frac{29}{35}$

105. Adding the fractions: $\frac{9}{48} + \frac{3}{54} = \frac{9}{48} \cdot \frac{9}{9} + \frac{3}{54} \cdot \frac{8}{8} = \frac{81}{432} + \frac{24}{432} = \frac{105}{432} = \frac{35}{144}$

107. Adding the fractions: $\frac{25}{84} + \frac{41}{90} = \frac{25}{84} \cdot \frac{15}{15} + \frac{41}{90} \cdot \frac{14}{14} = \frac{375}{1260} + \frac{574}{1260} = \frac{949}{1260}$

109. Simplifying: $\frac{3}{14} + \frac{7}{30} = \frac{3}{14} \cdot \frac{15}{15} + \frac{7}{30} \cdot \frac{7}{7} = \frac{45}{210} + \frac{49}{210} = \frac{94}{210} = \frac{47}{105}$

111. Simplifying: $32\left(\frac{3}{4}\right) - 16\left(\frac{3}{4}\right)^2 = 32\left(\frac{3}{4}\right) - 16\left(\frac{9}{16}\right) = 24 - 9 = 15$

113. Simplifying the expression: $5a + 7 + 8a + a = (5a + 8a + a) + 7 = 14a + 7$

115. Simplifying the expression: $3y + y + 5 + 2y + 1 = (3y + y + 2y) + (5 + 1) = 6y + 6$

117. Simplifying the expression: $2(5x + 1) + 2x = 10x + 2 + 2x = 12x + 2$

119. Simplifying the expression: $7 + 2(4y + 2) = 7 + 8y + 4 = 8y + 11$

121. Simplifying the expression: $3 + 4(5a + 3) + 4a = 3 + 20a + 12 + 4a = 24a + 15$

123. Simplifying the expression: $5x + 2(3x + 8) + 4 = 5x + 6x + 16 + 4 = 11x + 20$

125. commutative property of addition   127. commutative property of multiplication

129. additive inverse property   131. commutative property of addition

133. associative and commutative properties of multiplication

135. commutative and associative properties of addition   137. distributive property

139. Computing the expression: $12 + \frac{1}{4} \cdot 12 = 15$

141. a.   Though the number of people incarcerated is increasing, the rate of increase is slowing down.

     b.   The year 2001 showed a decline in the number of people incarcerated.

     c.   The rate of increase was a negative number.

1.4 Arithmetic with Real Numbers

1. Finding the sum: $6 + (-2) = 4$ 3. Finding the sum: $-6 + 2 = -4$

5. Finding the difference: $-7 - 3 = -7 + (-3) = -10$ 7. Finding the difference: $-7 - (-3) = -7 + 3 = -4$

9. Finding the difference: $\frac{3}{4} - \left(-\frac{5}{6}\right) = \frac{3}{4} + \frac{5}{6} = \frac{3}{4} \cdot \frac{3}{3} + \frac{5}{6} \cdot \frac{2}{2} = \frac{9}{12} + \frac{10}{12} = \frac{19}{12}$

11. Finding the difference: $\frac{11}{42} - \frac{17}{30} = \frac{11}{42} \cdot \frac{5}{5} - \frac{17}{30} \cdot \frac{7}{7} = \frac{55}{210} - \frac{119}{210} = -\frac{64}{210} = -\frac{32}{105}$

13. Subtracting: $-3 - 5 = -8$ 15. Subtracting: $-4 - 8 = -12$

17. Subtracting: $-3x - 4x = -7x$ 19. The number is 13, since $5 - 13 = -8$.

21. Computing the value: $-7 + (2 - 9) = -7 + (-7) = -14$ 23. Simplifying the value: $(8a + a) - 3a = 9a - 3a = 6a$

25. Finding the product: $3(-5) = -15$ 27. Finding the product: $-3(-5) = 15$

29. Finding the product: $2(-3)(4) = -6(4) = -24$ 31. Finding the product: $-2(5x) = -10x$

33. Finding the product: $-\frac{1}{3}(-3x) = \frac{3}{3}x = x$ 35. Finding the product: $-\frac{2}{3}\left(-\frac{3}{2}y\right) = \frac{6}{6}y = y$

37. Finding the product: $-2(4x - 3) = -8x + 6$ 39. Finding the product: $-\frac{1}{2}(6a - 8) = -\frac{6}{2}a + \frac{8}{2} = -3a + 4$

41. Simplifying: $3(-4) - 2 = -12 - 2 = -14$ 43. Simplifying: $4(-3) - 6(-5) = -12 + 30 = 18$

45. Simplifying: $2 - 5(-4) - 6 = 2 + 20 - 6 = 22 - 6 = 16$

47. Simplifying: $4 - 3(7 - 1) - 5 = 4 - 3(6) - 5 = 4 - 18 - 5 = -19$

49. Simplifying: $2(-3)^2 - 4(-2)^3 = 2(9) - 4(-8) = 18 + 32 = 50$

51. Simplifying: $7(3 - 5)^3 - 2(4 - 7)^3 = 7(-2)^3 - 2(-3)^3 = 7(-8) - 2(-27) = -56 + 54 = -2$

53. Simplifying: $1(-2) - 2(-16) + 1(9) = -2 + 32 + 9 = 39$

55. Simplifying: $1(1) - 3(-2) + (-2)(-2) = 1 + 6 + 4 = 11$

57. Simplifying: $-4(0)(-2) - (-1)(1)(1) - 1(2)(3) = 0 + 1 - 6 = -5$

59. Simplifying: $1[0 - (-1)] - 3(2 - 4) + (-2)(-2 - 0) = 1(1) - 3(-2) - 2(-2) = 1 + 6 + 4 = 11$

61. Simplifying: $3(-2)^2 + 2(-2) - 1 = 3(4) + 2(-2) - 1 = 12 - 4 - 1 = 7$

63. Simplifying: $2(-2)^3 - 3(-2)^2 + 4(-2) - 8 = 2(-8) - 3(4) + 4(-2) - 8 = -16 - 12 - 8 - 8 = -44$

65. Simplifying: $-24\left(\frac{3}{8}\right) = -\frac{24}{1} \cdot \frac{3}{8} = -9$ 67. Simplifying: $24\left(-\frac{3}{8}\right) = -\frac{24}{1} \cdot \frac{3}{8} = -9$

69. Simplifying: $-15\left(\dfrac{x}{5}\right) = -3x$

71. Simplifying: $-15\left(\dfrac{y}{-3}\right) = 5y$

73. Applying the distributive property: $-1(5-x) = (-1) \bullet 5 - (-1) \bullet x = -5 + x = x - 5$

75. Applying the distributive property: $-1(7-x) = (-1) \bullet 7 - (-1) \bullet x = -7 + x = x - 7$

77. Applying the distributive property: $-3(2x - 3y) = (-3) \bullet 2x - (-3) \bullet 3y = -6x + 9y$

79. Applying the distributive property: $6\left(\dfrac{x}{2} - 3\right) = 6 \bullet \dfrac{x}{2} - 6 \bullet 3 = 3x - 18$

81. Applying the distributive property: $12\left(\dfrac{a}{4} + \dfrac{1}{2}\right) = 12 \bullet \dfrac{a}{4} + 12 \bullet \dfrac{1}{2} = 3a + 6$

83. Applying the distributive property: $15\left(\dfrac{x}{5} + 4\right) = 15 \bullet \dfrac{x}{5} + 15 \bullet 4 = 3x + 60$

85. Applying the distributive property: $8\left(\dfrac{x}{8} + \dfrac{y}{2}\right) = 8 \bullet \dfrac{x}{8} + 8 \bullet \dfrac{y}{2} = x + 4y$

87. Applying the distributive property: $-15\left(\dfrac{x}{5} + \dfrac{y}{-3}\right) = -15 \bullet \dfrac{x}{5} - 15 \bullet \dfrac{y}{-3} = -3x + 5y$

89. Applying the distributive property: $12\left(\dfrac{y}{2} + \dfrac{y}{4} + \dfrac{y}{6}\right) = 12 \bullet \dfrac{y}{2} + 12 \bullet \dfrac{y}{4} + 12 \bullet \dfrac{y}{6} = 6y + 3y + 2y = 11y$

91. Simplifying: $3(5x + 4) - x = 15x + 12 - x = 14x + 12$

93. Simplifying: $6 - 7(3 - m) = 6 - 21 + 7m = 7m - 15$

95. Simplifying: $7 - 2(3x - 1) + 4x = 7 - 6x + 2 + 4x = -2x + 9$

97. Simplifying: $5(3y + 1) - (8y - 5) = 15y + 5 - 8y + 5 = 7y + 10$

99. Simplifying: $4(2 - 6x) - (3 - 4x) = 8 - 24x - 3 + 4x = -20x + 5$

101. Simplifying: $10 - 4(2x + 1) - (3x - 4) = 10 - 8x - 4 - 3x + 4 = -11x + 10$

103. Simplifying: $0.06x + 0.05(10,000 - x) = 0.06x + 500 - 0.05x = 0.01x + 500$

105. Simplifying: $0.12x + 0.10(15,000 - x) = 0.12x + 1,500 - 0.10x = 0.02x + 1,500$

107. Simplifying: $-(a + 1) - 4a = -a - 1 - 4a = -5a - 1$

109. Dividing: $-\dfrac{3}{4} \div \dfrac{9}{8} = -\dfrac{3}{4} \bullet \dfrac{8}{9} = -\dfrac{2}{3}$

111. Dividing: $-8 \div \left(-\dfrac{1}{4}\right) = -8 \bullet \left(-\dfrac{4}{1}\right) = 32$

113. Dividing: $\dfrac{4}{9} \div (-8) = \dfrac{4}{9} \bullet \left(-\dfrac{1}{8}\right) = -\dfrac{1}{18}$

115. Simplifying: $\dfrac{0 - 4}{0 - 2} = \dfrac{-4}{-2} = 2$

117. Simplifying: $\dfrac{-4 - 4}{-4 - 2} = \dfrac{-8}{-6} = \dfrac{4}{3}$

119. Simplifying: $\dfrac{-6 + 6}{-6 - 3} = \dfrac{0}{-9} = 0$

121. Simplifying: $\dfrac{2 - 4}{2 - 2} = \dfrac{-2}{0}$, which is undefined

123. Simplifying: $\dfrac{3 - (-1)}{-3 - 3} = \dfrac{4}{-6} = -\dfrac{2}{3}$

125. Simplifying: $\dfrac{3(-1) - 4(-2)}{8 - 5} = \dfrac{-3 + 8}{3} = \dfrac{5}{3}$

127. Simplifying: $8 - (-6)\left[\dfrac{2(-3) - 5(4)}{-8(6) - 4}\right] = 8 - (-6)\left(\dfrac{-6 - 20}{-48 - 4}\right) = 8 - (-6)\left(\dfrac{-26}{-52}\right) = 8 - (-6)\left(\dfrac{1}{2}\right) = 8 + 3 = 11$

129. Simplifying: $6 - (-3)\left[\dfrac{2 - 4(3 - 8)}{1 - 5(1 - 3)}\right] = 6 - (-3)\left[\dfrac{2 - 4(-5)}{1 - 5(-2)}\right] = 6 - (-3)\left(\dfrac{2 + 20}{1 + 10}\right) = 6 - (-3)\left(\dfrac{22}{11}\right) = 6 + 6 = 12$

a	b	SUM $a+b$	DIFFERENCE $a-b$	PRODUCT ab	QUOTIENT a/b
3	12	15	-9	36	$\frac{1}{4}$
-3	12	9	-15	-36	$-\frac{1}{4}$
3	-12	-9	15	-36	$-\frac{1}{4}$
-3	-12	-15	9	36	$\frac{1}{4}$

131. Completing the table:

x	$3(5x-2)$	$15x-6$	$15x-2$
-2	-36	-36	-32
-1	-21	-21	-17
0	-6	-6	-2
1	9	9	13
2	24	24	28

133. Completing the table:

135. **a.** Evaluating when $a = 3$ and $b = -6$: $-\dfrac{b}{2a} = -\dfrac{-6}{2(3)} = -\dfrac{-6}{6} = 1$

b. Evaluating when $a = -2$ and $b = 6$: $-\dfrac{b}{2a} = -\dfrac{6}{2(-2)} = -\dfrac{6}{-4} = \dfrac{3}{2}$

c. Evaluating when $a = -1$ and $b = -2$: $-\dfrac{b}{2a} = -\dfrac{-2}{2(-1)} = -\dfrac{-2}{-2} = -1$

d. Evaluating when $a = -0.1$ and $b = 27$: $-\dfrac{b}{2a} = -\dfrac{27}{2(-0.1)} = -\dfrac{27}{-0.2} = 135$

137. **a.** Adding: $-2.25 + 7.5 = 5.25$ **b.** Subtracting: $-2.25 - 7.5 = -9.75$

c. Multiplying: $-2.25(7.5) = -16.875$ **d.** Dividing: $\dfrac{-2.25}{7.5} = -0.3 = -\dfrac{3}{10}$

139. Using a calculator: $\dfrac{1.3802}{0.9031} \approx 1.5283$ **141.** Using a calculator: $\frac{1}{2}(-0.1587) \approx -0.0794$

143. Using a calculator: $\dfrac{1}{2}\left(\dfrac{1.2}{1.4} - 1\right) \approx -0.0714$ **145.** Using a calculator: $\dfrac{(6.8)(3.9)}{7.8} = 3.4$

147. Using a calculator: $\dfrac{0.0005(200)}{(0.25)^2} = 1.6$

149. Using a calculator: $-500 + 27(100) - 0.1(100)^2 = -500 + 2,700 - 1,000 = 1,200$

151. Using a calculator: $-0.05(130)^2 + 9.5(130) - 200 = -845 + 1,235 - 200 = 190$

153. For Santa Fe, the final time would be: $6:55 + 2:15 + 1 = 9:70 = 10:10$ P.M.
For Detroit, the final time would be: $10:10 + 3:20 + 2 = 15:30 = 3:30$ A.M.

155. **a.** The difference is: $8:46:15 - 8:46:05 = 0:00:10$ (10 seconds)
b. The difference is: $8:39:19 - 8:36:59 = 0:02:20$ (2 minutes 20 seconds)
c. The difference is: $8:46:05 - 8:23:12 = 0:22:53$ (22 minutes 53 seconds)

1.5 Recognizing Patterns

1. The pattern is to add 1, so the next term is 5.   **3.** The pattern is to add 2, so the next term is 10.
5. These numbers are squares, so the next term is $5^2 = 25$.
7. The pattern is to add 7, so the next term is 29.
9. The pattern is to add 7, then add 6, then add 5, so the next term is $19 + 4 = 23$.
11. One possibility is: Δ   **13.** One possibility is: ☉
15. The pattern is to add 4, so the next two numbers are 17 and 21.
17. The pattern is to add –1, so the next two numbers are –2 and –3.
19. The pattern is to add –3, so the next two numbers are –4 and –7.
21. The pattern is to add $-\frac{1}{4}$, so the next two numbers are $-\frac{1}{2}$ and $-\frac{3}{4}$.
23. The pattern is to add $\frac{1}{2}$, so the next two numbers are $\frac{5}{2}$ and 3.
25. The pattern is to multiply by 3, so the next number is 27.
27. The pattern is to multiply by –3, so the next number is –270.
29. The pattern is to multiply by $\frac{1}{2}$, so the next number is $\frac{1}{8}$.
31. The pattern is to multiply by $\frac{1}{2}$, so the next number is $\frac{5}{2}$.
33. The pattern is to multiply by –5, so the next number is –625.
35. The pattern is to multiply by $-\frac{1}{5}$, so the next number is $-\frac{1}{125}$.
37. **a**. The pattern is to add 4, so the next number is 12.   **b**. The pattern is to multiply by 2, so the next number is 16.
39. The sequence is 1,1,2,3,5,8,13,21,34,55,89,144,233,..., so the 12th term is 144.
41. Three Fibonacci numbers that are prime numbers are 2, 3, and 5 (among others).
43. The even numbers are 2, 8, and 34.
45. Completing the table:

Two Numbers a and b	Their Product ab	Their Sum $a+b$
1, –24	–24	–23
–1, 24	–24	23
2, –12	–24	–10
–2, 12	–24	10
3, –8	–24	–5
–3, 8	–24	5
4, –6	–24	–2
–4, 6	–24	2

47. The sequence of air temperatures is: 41°, 37.5°, 34°, 30.5°, 27°, 23.5°. This is an arithmetic sequence.
49. The sequence of air temperatures is: 41°, 45.5°, 50°, 54.5°, 59°, 63.5°. This is an arithmetic sequence.
51. The patient on antidepressant 1 will have less medication in the body, approximately half as much. The patient on antidepressant 2 will still have more of the medication in the body, since a full half-life has not passed by the second day.
53. Completing the table:

Hours Since Discontinuing	Concentration (ng/ml)
0	60
4	30
8	15
12	7.5
16	3.75

55. Completing the table:

Elevation (ft)	Boiling Point (°F)
–2,000	215.6
–1,000	213.8
0	212
1,000	210.2
2,000	208.4
3,000	206.6

Sketching the graph:

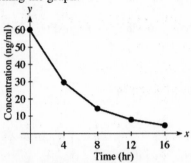

57.   **a.** The steep straight line segments show when the patient takes his medication.
b. The graph is falling off because the patient stops taking his medication.
c. The maximum medication is approximately 50 ng/ml.
d. Since the patient takes his medication every 4 hours, the values are $A = 4$ hours, $B = 8$ hours, and $C = 12$ hours.

59.   Completing the table:

Temperature (Fahrenheit)	Shelf Life (days)
32°	24
40°	10
50°	2
60°	1
70°	$\frac{1}{2}$

Constructing a line graph:

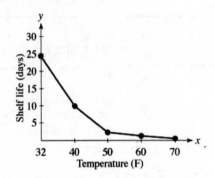

Chapter 1 Review/Test

1.   Written in symbols, the expression is: $x + 2$

2.   Written in symbols, the expression is: $x - 2$

3.   Written in symbols, the expression is: $\dfrac{x}{2}$

4.   Written in symbols, the expression is: $2x$

5.   Written in symbols, the expression is: $2(x + y)$

6.   Written in symbols, the expression is: $2x + y$

7.   Multiplying: $3^3 = 3 \cdot 3 \cdot 3 = 27$

8.   Multiplying: $5^3 = 5 \cdot 5 \cdot 5 = 125$

9.   Multiplying: $8^2 = 8 \cdot 8 = 64$

10.   Multiplying: $1^8 = 1 \cdot 1 \cdot 1 \cdot 1 \cdot 1 \cdot 1 \cdot 1 \cdot 1 = 1$

11.   Multiplying: $2^5 = 2 \cdot 2 \cdot 2 \cdot 2 \cdot 2 = 32$

12.   Multiplying: $3^4 = 3 \cdot 3 \cdot 3 \cdot 3 = 81$

13.   Using the rule for order of operations: $2 + 3 \cdot 5 = 2 + 15 = 17$

14.   Using the rule for order of operations: $10 - 2 \cdot 3 = 10 - 6 = 4$

15.   Using the rule for order of operations: $20 \div 2 + 3 = 10 + 3 = 13$

16.   Using the rule for order of operations: $30 \div 6 + 4 \div 2 = 5 + 2 = 7$

17.   Using the rule for order of operations: $3 + 2(5 - 2) = 3 + 2(3) = 3 + 6 = 9$

18.   Using the rule for order of operations: $(10 - 2)(7 - 3) = 8 \cdot 4 = 32$

19. Using the rule for order of operations: $3 \cdot 4^2 - 2 \cdot 3^2 = 3 \cdot 16 - 2 \cdot 9 = 48 - 18 = 30$

20. Using the rule for order of operations: $3 + 5(2 \cdot 3^2 - 10) = 3 + 5(2 \cdot 9 - 10) = 3 + 5(18 - 10) = 3 + 5(8) = 3 + 40 = 43$

21. The set is $A \cup B = \{1, 2, 3, 4, 5, 6\}$.

22. The set is $A \cap C = \{1, 3\}$.

23. The set is $\{5\}$.

24. The set is $\{6\}$.

25. Sketching a number line:

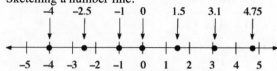

26. Sketching a number line:

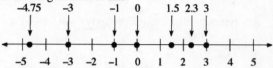

27. The opposite is -2 and the reciprocal is $\frac{1}{2}$.

28. The opposite is $\frac{2}{5}$ and the reciprocal is $-\frac{5}{2}$.

29. Writing without absolute values: $|-3| = 3$

30. Writing without absolute values: $-|-5| = -5$

31. Writing without absolute values: $|-4| = 4$

32. Writing without absolute values: $|10 - 16| = |-6| = 6$

33. The integers are: $-7, 0, 5$

34. The rational numbers are: $-7, -4.2, 0, \frac{3}{4}, 5$

35. The irrational numbers are: $-\sqrt{3}, \pi$

36. Factoring into primes: $4{,}356 = 36 \cdot 121 = 4 \cdot 9 \cdot 121 = 2^2 \cdot 3^2 \cdot 11^2$

37. Reducing the fraction: $\frac{4{,}356}{5{,}148} = \frac{396 \cdot 11}{396 \cdot 13} = \frac{11}{13}$

38. Multiplying the fractions: $\frac{3}{4} \cdot \frac{8}{5} \cdot \frac{5}{6} = 1$

39. Multiplying the fractions: $\left(\frac{3}{4}\right)^3 = \frac{3}{4} \cdot \frac{3}{4} \cdot \frac{3}{4} = \frac{27}{64}$

40. Multiplying the fractions: $\frac{1}{4} \cdot 8 = \frac{8}{4} = 2$

41. Graphing the inequality:

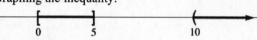

42. Graphing the inequality:

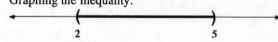

43. Graphing the inequality:

44. Graphing the inequality:

45. The inequality is: $x \geq 4$

46. The inequality is: $x \leq 5$

47. The inequality is: $0 < x < 8$

48. The inequality is: $0 \leq x \leq 8$

49. Simplifying: $-2y + 4y = 2y$

50. Simplifying: $-3x - x + 7x = 3x$

51. Simplifying: $3x - 2 + 5x + 7 = (3x + 5x) + (-2 + 7) = 8x + 5$

52. Simplifying: $2y + 4 - y - 2 = (2y - y) + (4 - 2) = y + 2$

53. Commutative property of addition (a)

54. Associative property of addition (c)

55. Commutative property of addition (a)

56. Commutative and associative properties of multiplication (b,d)

57. Commutative and associative properties of addition (a,c)

58. Multiplicative identity property (f)

59. Additive identity property (e)

60. Additive inverse property (g)

61. Finding the difference: $5 - 3 = 2$

62. Finding the difference: $-5 - (-3) = -5 + 3 = -2$

63. Computing: $7 + (-2) - 4 = 5 - 4 = 1$

64. Computing: $6 - (-3) + 8 = 6 + 3 + 8 = 17$

65. Computing: $|-4| - |-3| + |-2| = 4 - 3 + 2 = 3$

66. Finding the difference: $|7 - 9| - |-3 - 5| = |-2| - |-8| = 2 - 8 = -6$

67. Finding the differences: $6 - (-3) - 2 - 5 = 6 + 3 - 2 - 5 = 9 - 2 - 5 = 7 - 5 = 2$

68. Computing: $2 \cdot 3^2 - 4 \cdot 2^3 + 5 \cdot 4^2 = 2 \cdot 9 - 4 \cdot 8 + 5 \cdot 16 = 18 - 32 + 80 = 66$

69. Finding the differences: $-\frac{1}{12} - \frac{1}{6} - \frac{1}{4} - \frac{1}{3} = -\frac{1}{12} - \frac{2}{12} - \frac{3}{12} - \frac{4}{12} = -\frac{10}{12} = -\frac{5}{6}$

70. Finding the differences: $-\frac{1}{3} - \frac{1}{4} - \frac{1}{6} - \frac{1}{12} = -\frac{4}{12} - \frac{3}{12} - \frac{2}{12} - \frac{1}{12} = -\frac{10}{12} = -\frac{5}{6}$

71.   Finding the product: $6(-7) = -42$

72.   Finding the product: $-3(5)(-2) = -15(-2) = 30$

73.   Finding the product: $7(3x) = 21x$

74.   Finding the product: $-3(2x) = -6x$

75.   Applying the distributive property: $-2(3x - 5) = -2(3x) - (-2)(5) = -6x + 10$

76.   Applying the distributive property: $-3(2x - 7) = -3(2x) - (-3)(7) = -6x + 21$

77.   Applying the distributive property: $-\frac{1}{2}(2x - 6) = -\frac{1}{2}(2x) - \left(-\frac{1}{2}\right)(6) = -x + 3$

78.   Applying the distributive property: $-3(5x - 1) = -3(5x) - (-3)(1) = -15x + 3$

79.   Dividing: $-\frac{5}{8} \div \frac{3}{4} = -\frac{5}{8} \cdot \frac{4}{3} = -\frac{5}{6}$

80.   Dividing: $-12 \div \frac{1}{3} = -12 \cdot \frac{3}{1} = -36$

81.   Dividing: $\frac{3}{5} \div 6 = \frac{3}{5} \cdot \frac{1}{6} = \frac{1}{10}$

82.   Dividing: $\frac{4}{7} \div (-2) = \frac{4}{7} \cdot \left(-\frac{1}{2}\right) = -\frac{2}{7}$

83.   Simplifying: $2(-5) - 3 = -10 - 3 = -13$

84.   Simplifying: $3(-4) - 5 = -12 - 5 = -17$

85.   Simplifying: $6 + 3(-2) = 6 + (-6) = 0$

86.   Simplifying: $7 + 2(-4) = 7 + (-8) = -1$

87.   Simplifying: $-3(2) - 5(6) = -6 - 30 = -36$

88.   Simplifying: $-4(3)^2 - 2(-1)^3 = -4(9) - 2(-1) = -36 + 2 = -34$

89.   Simplifying: $8 - 2(6 - 10) = 8 - 2(-4) = 8 + 8 = 16$

90.   Simplifying: $(8 - 2)(6 - 10) = (6)(-4) = -24$

91.   Simplifying: $\dfrac{3(-4) - 8}{-5 - 5} = \dfrac{-12 - 8}{-5 - 5} = \dfrac{-20}{-10} = 2$

92.   Simplifying: $\dfrac{9(-1)^3 - 3(-6)^2}{6 - 9} = \dfrac{9(-1) - 3(36)}{6 - 9} = \dfrac{-9 - 108}{-3} = \dfrac{-117}{-3} = 39$

93.   Simplifying: $4 - (-2)\left[\dfrac{6 - 3(-4)}{1 + 5(-2)}\right] = 4 - (-2)\left(\dfrac{6 + 12}{1 - 10}\right) = 4 - (-2)\left(\dfrac{18}{-9}\right) = 4 - (-2)(-2) = 4 - 4 = 0$

94.   Simplifying: $7 - 2(3y - 1) + 4y = 7 - 6y + 2 + 4y = -2y + 9$

95.   Simplifying: $4(3x - 1) - 5(6x + 2) = 12x - 4 - 30x - 10 = -18x - 14$

96.   Simplifying: $4(2a - 5) - (3a + 2) = 8a - 20 - 3a - 2 = 5a - 22$

97.   The pattern is to add -3, so the next term is $2 + (-3) = -1$. This is an arithmetic sequence.

98.   The pattern is to add the two previous terms, so the next term is $3 + 5 = 8$.

99.   The pattern is to add $-\frac{1}{2}$, so the next term is $-\frac{1}{2} - \frac{1}{2} = -1$. This is an arithmetic sequence.

100.   The pattern is to multiply by $-\frac{1}{2}$, so the next term is $-\frac{1}{8}\left(-\frac{1}{2}\right) = \frac{1}{16}$. This is a geometric sequence.

Chapter 2
Equations and Inequalities in One Variable

2.1 Linear Equations in One Variable

1. Solving the equation:
$$x - 5 = 3$$
$$x = 8$$

3. Solving the equation:
$$2x - 4 = 6$$
$$2x = 10$$
$$x = \frac{10}{2} = 5$$

5. Solving the equation:
$$7 = 4a - 1$$
$$8 = 4a$$
$$a = \frac{8}{4} = 2$$

7. Solving the equation:
$$3 - y = 10$$
$$-y = 7$$
$$y = -7$$

9. Solving the equation:
$$-3 - 4x = 15$$
$$-4x = 18$$
$$x = -\frac{9}{2}$$

11. Solving the equation:
$$-3 = 5 + 2x$$
$$-8 = 2x$$
$$x = -4$$

13. Solving the equation:
$$-300y + 100 = 500$$
$$-300y = 400$$
$$y = -\frac{4}{3}$$

15. Solving the equation:
$$160 = -50x - 40$$
$$200 = -50x$$
$$x = -4$$

17. Solving the equation:
$$-x = 2$$
$$x = -1 \bullet 2 = -2$$

19. Solving the equation:
$$-a = -\frac{3}{4}$$
$$a = -1 \bullet \left(-\frac{3}{4}\right) = \frac{3}{4}$$

21. Solving the equation:
$$\frac{2}{3}x = 8$$
$$x = \frac{3}{2} \bullet 8 = 12$$

23. Solving the equation:
$$-\frac{3}{5}a + 2 = 8$$
$$-\frac{3}{5}a = 6$$
$$a = -\frac{5}{3} \bullet 6 = -10$$

25. Solving the equation:
$$8 = 6 + \frac{2}{7}y$$
$$2 = \frac{2}{7}y$$
$$y = \frac{7}{2} \bullet 2 = 7$$

27. Solving the equation:
$$2x - 5 = 3x + 2$$
$$-x - 5 = 2$$
$$-x = 7$$
$$x = -7$$

29.   Solving the equation:

$$-3a + 2 = -2a - 1$$
$$-a + 2 = -1$$
$$-a = -3$$
$$a = 3$$

31.   Solving the equation:

$$5 - 2x = 3x + 1$$
$$5 - 5x = 1$$
$$-5x = -4$$
$$x = \frac{4}{5}$$

33.   Solving the equation:

$$11x - 5 + 4x - 2 = 8x$$
$$15x - 7 = 8x$$
$$7x = 7$$
$$x = 1$$

35.   Solving the equation:

$$6 - 7(m - 3) = -1$$
$$6 - 7m + 21 = -1$$
$$-7m + 27 = -1$$
$$-7m = -28$$
$$m = 4$$

37.   Solving the equation:

$$7 + 3(x + 2) = 4(x - 1)$$
$$7 + 3x + 6 = 4x - 4$$
$$3x + 13 = 4x - 4$$
$$-x + 13 = -4$$
$$-x = -17$$
$$x = 17$$

39.   Solving the equation:

$$5 = 7 - 2(3x - 1) + 4x$$
$$5 = 7 - 6x + 2 + 4x$$
$$5 = -2x + 9$$
$$-2x = -4$$
$$x = 2$$

41.   Solving the equation:

$$\tfrac{1}{2}x + \tfrac{1}{4} = \tfrac{1}{3}x + \tfrac{5}{4}$$
$$12\left(\tfrac{1}{2}x + \tfrac{1}{4}\right) = 12\left(\tfrac{1}{3}x + \tfrac{5}{4}\right)$$
$$6x + 3 = 4x + 15$$
$$2x + 3 = 15$$
$$2x = 12$$
$$x = 6$$

43.   Solving the equation:

$$-\tfrac{2}{5}x + \tfrac{2}{15} = \tfrac{2}{3}$$
$$15\left(-\tfrac{2}{5}x + \tfrac{2}{15}\right) = 15\left(\tfrac{2}{3}\right)$$
$$-6x + 2 = 10$$
$$-6x = 8$$
$$x = -\tfrac{4}{3}$$

45.   Solving the equation:

$$\tfrac{3}{4}(8x - 4) = \tfrac{2}{3}(6x - 9)$$
$$6x - 3 = 4x - 6$$
$$2x - 3 = -6$$
$$2x = -3$$
$$x = -\tfrac{3}{2}$$

47.   Solving the equation:

$$\tfrac{1}{4}(12a + 1) - \tfrac{1}{4} = 5$$
$$3a + \tfrac{1}{4} - \tfrac{1}{4} = 5$$
$$3a = 5$$
$$a = \tfrac{5}{3}$$

49.   Solving the equation:

$$0.35x - 0.2 = 0.15x + 0.1$$
$$0.2x - 0.2 = 0.1$$
$$0.2x = 0.3$$
$$2x = 3$$
$$x = \tfrac{3}{2}$$

51.   Solving the equation:

$$0.42 - 0.18x = 0.48x - 0.24$$
$$0.42 - 0.66x = -0.24$$
$$-0.66x = -0.66$$
$$x = 1$$

53.   Solving the equation:

$$3x - 6 = 3(x + 4)$$
$$3x - 6 = 3x + 12$$
$$-6 = 12$$

Since this statement is false, there is no solution $(\varnothing)$.

55.   Solving the equation:

$$4y + 2 - 3y + 5 = 3 + y + 4$$
$$y + 7 = y + 7$$
$$7 = 7$$

Since this statement is true, the solution is all real numbers.

57.   Solving the equation:
$$2(4t-1)+3=5t+4+3t$$
$$8t-2+3=8t+4$$
$$8t+1=8t+4$$
$$1=4$$

Since this statement is false, there is no solution $(\varnothing)$.

59.   Solving the equation:
$$3x+2=0$$
$$3x=-2$$
$$x=-\frac{2}{3}$$

61.   Solving the equation:
$$0=6,400a+70$$
$$-70=6,400a$$
$$a=-\frac{70}{6,400}=-\frac{7}{640}$$

63.   Solving the equation:
$$x+2=2x$$
$$2=2x-x$$
$$x=2$$

65.   Solving the equation:
$$0.07x=1.4$$
$$x=\frac{1.4}{0.07}=20$$

67.   Solving the equation:
$$5(2x+1)=12$$
$$10x+5=12$$
$$10x=7$$
$$x=\frac{7}{10}=0.7$$

69.   Solving the equation:
$$50=\frac{K}{48}$$
$$50\bullet 48=\frac{K}{48}\bullet 48$$
$$K=2,400$$

71.   Solving the equation:
$$100P=2,400$$
$$P=\frac{2,400}{100}=24$$

73.   Solving the equation:
$$x+(3x+2)=26$$
$$4x+2=26$$
$$4x=24$$
$$x=6$$

75.   Solving the equation:
$$2x-3(3x-5)=-6$$
$$2x-9x+15=-6$$
$$-7x+15=-6$$
$$-7x=-21$$
$$x=3$$

77.   Solving the equation:
$$3x+(x-2)\bullet 2=6$$
$$3x+2x-4=6$$
$$5x-4=6$$
$$5x=10$$
$$x=2$$

79.   Solving the equation:
$$15-3(x-1)=x-2$$
$$15-3x+3=x-2$$
$$-3x+18=x-2$$
$$-4x+18=-2$$
$$-4x=-20$$
$$x=5$$

81.   Solving the equation:
$$2(2x-3)+2x=45$$
$$4x-6+2x=45$$
$$6x-6=45$$
$$6x=51$$
$$x=\frac{51}{6}=\frac{17}{2}$$

83.   Solving the equation:
$$2(x+3)+x=4(x-3)$$
$$2x+6+x=4x-12$$
$$3x+6=4x-12$$
$$-x+6=-12$$
$$-x=-18$$
$$x=18$$

85.   Solving the equation:
$$6(y-3)-5(y+2)=8$$
$$6y-18-5y-10=8$$
$$y-28=8$$
$$y=36$$

87.   Solving the equation:
$$2(20+x)=3(20-x)$$
$$40+2x=60-3x$$
$$40+5x=60$$
$$5x=20$$
$$x=4$$

89.   Solving the equation:
$$2x+1.5(75-x)=127.5$$
$$2x+112.5-1.5x=127.5$$
$$0.5x+112.5=127.5$$
$$0.5x=15$$
$$x=30$$

91.   Solving the equation:
$$0.08x + 0.09(9,000 - x) = 750$$
$$0.08x + 810 - 0.09x = 750$$
$$-0.01x + 810 = 750$$
$$-0.01x = -60$$
$$x = 6,000$$

93.   Solving the equation:
$$0.12x + 0.10(15,000 - x) = 1,600$$
$$0.12x + 1,500 - 0.10x = 1,600$$
$$0.02x + 1,500 = 1,600$$
$$0.02x = 100$$
$$x = 5,000$$

95.   Solving the equation:
$$5\left(\frac{19}{15}\right) + 5y = 9$$
$$\frac{19}{3} + 5y = 9$$
$$5y = \frac{8}{3}$$
$$y = \frac{8}{15}$$

97.   Solving the equation:
$$2\left(\frac{29}{22}\right) - 3y = 4$$
$$\frac{29}{11} - 3y = 4$$
$$-3y = \frac{15}{11}$$
$$y = -\frac{5}{11}$$

99.   **a.**   Solving the equation:
$$8x - 5 = 0$$
$$8x = 5$$
$$x = \frac{5}{8}$$

b.   Solving the equation:
$$8x - 5 = -5$$
$$8x = 0$$
$$x = 0$$

c.   Adding: $(8x - 5) + (2x - 5) = 10x - 10$

d.   Solving the equation:
$$8x - 5 = 2x - 5$$
$$6x - 5 = -5$$
$$6x = 0$$
$$x = 0$$

e.   Multiplying: $8(x - 5) = 8x - 40$

f.   Solving the equation:
$$8(x - 5) = 2(x - 5)$$
$$8x - 40 = 2x - 10$$
$$6x - 40 = -10$$
$$6x = 30$$
$$x = 5$$

101.   **a.**   The equation is $6.60 = 0.4n + 1.80$

b.   Solving the equation:
$$6.60 = 0.4n + 1.80$$
$$4.80 = 0.4n$$
$$n = 12$$
The woman traveled 12 miles.

103.   **a.**   The equation is $1,125A = 3,937,000$.

b.   Solving the equation:
$$1,125A = 3,937,000$$
$$A \approx 3,500$$

The area of Puerto Rico is approximately 3,500 square miles.

105.   This is the commutative property of multiplication.   **107.**   This is the associative property of addition.

109.   These are the commutative and associative properties of addition.

111.   This is the multiplicative identity property.   **113.**   This is the commutative property of multiplication.

115.   This is the additive identity property.

117.   Solving the equation:
$$x \bullet 42 = 21$$
$$x = \frac{21}{42} = 0.5$$

119.   Solving the equation:
$$25 = 0.4x$$
$$x = \frac{25}{0.4} = 62.5$$

121. Solving the equation:
$$12 - 4y = 12$$
$$-4y = 0$$
$$y = 0$$

123. Solving the equation:
$$525 = 900 - 300p$$
$$-375 = -300p$$
$$p = \frac{-375}{-300} = 1.25$$

125. Solving the equation:
$$486.7 = 78.5 + 31.4h$$
$$408.2 = 31.4h$$
$$h = \frac{408.2}{31.4} = 13$$

127. **a.** Evaluating when $x = 2$: $2x - 1 = 2(2) - 1 = 4 - 1 = 3$

b. Evaluating when $x = 3$: $2x - 1 = 2(3) - 1 = 6 - 1 = 5$

c. Evaluating when $x = 5$: $2x - 1 = 2(5) - 1 = 10 - 1 = 9$

129. Solving the equation:
$$\frac{x+4}{5} - \frac{x+3}{3} = -\frac{7}{15}$$
$$15\left(\frac{x+4}{5} - \frac{x+3}{3}\right) = 15\left(-\frac{7}{15}\right)$$
$$3(x+4) - 5(x+3) = -7$$
$$3x + 12 - 5x - 15 = -7$$
$$-2x - 3 = -7$$
$$-2x = -4$$
$$x = 2$$

131. Solving the equation:
$$\frac{1}{x} - \frac{2}{3} = \frac{2}{x}$$
$$3x\left(\frac{1}{x} - \frac{2}{3}\right) = 3x\left(\frac{2}{x}\right)$$
$$3 - 2x = 6$$
$$-2x = 3$$
$$x = -\frac{3}{2}$$

133. Solving the equation:
$$\frac{x-3}{5} - \frac{x+1}{10} = -\frac{1}{10}$$
$$10\left(\frac{x-3}{5} - \frac{x+1}{10}\right) = 10\left(-\frac{1}{10}\right)$$
$$2(x-3) - 1(x+1) = -1$$
$$2x - 6 - x - 1 = -1$$
$$x - 7 = -1$$
$$x = 6$$

135. Solving the equation:
$$\frac{x+2}{4} - \frac{x-1}{3} = -\frac{x+2}{6}$$
$$12\left(\frac{x+2}{4} - \frac{x-1}{3}\right) = 12\left(-\frac{x+2}{6}\right)$$
$$3(x+2) - 4(x-1) = -2(x+2)$$
$$3x + 6 - 4x + 4 = -2x - 4$$
$$-x + 10 = -2x - 4$$
$$x + 10 = -4$$
$$x = -14$$

2.2 Formulas

1. Substituting $x = 0$:
$$3(0) - 4y = 12$$
$$-4y = 12$$
$$y = -3$$

3. Substituting $x = 4$:
$$3(4) - 4y = 12$$
$$12 - 4y = 12$$
$$-4y = 0$$
$$y = 0$$

5. Substituting $y = 0$:
$$2x - 3 = 0$$
$$2x = 3$$
$$x = \frac{3}{2}$$

7. Substituting $y = 5$:
$$2x - 3 = 5$$
$$2x = 8$$
$$x = 4$$

9. Substituting $y = -\frac{6}{5}$:

$$x - 2\left(-\frac{6}{5}\right) = 4$$
$$x + \frac{12}{5} = 4$$
$$x = \frac{8}{5}$$

11. Substituting $x = 0$:

$$2(0) + 3y = 6$$
$$3y = 6$$
$$y = 2$$

13. Substituting $x = 160$ and $y = 0$:

$$0 = a(160 - 80)^2 + 70$$
$$-70 = a(80)^2$$
$$6,400a = -70$$
$$a = -\frac{70}{6,400} = -\frac{7}{640}$$

15. Substituting $p = 1.5$: $R = (900 - 300 \bullet 1.5)(1.5) = (450)(1.5) = 675$

17.
   a. Substituting $x = 100$: $P = -0.1(100)^2 + 27(100) - 1,700 = -1,000 + 2,700 - 1,700 = 0$

   b. Substituting $x = 170$: $P = -0.1(170)^2 + 27(170) - 1,700 = -2,890 + 4,590 - 1,700 = 0$

19.
   a. Substituting $t = \frac{1}{4}$: $h = 16 + 32\left(\frac{1}{4}\right) - 16\left(\frac{1}{4}\right)^2 = 16 + 8 - 1 = 23$

   b. Substituting $t = \frac{7}{4}$: $h = 16 + 32\left(\frac{7}{4}\right) - 16\left(\frac{7}{4}\right)^2 = 16 + 56 - 49 = 23$

21. Substituting $x = \frac{3}{2}$: $y = -2\left(\frac{3}{2}\right)^2 + 6\left(\frac{3}{2}\right) - 5 = -\frac{9}{2} + 9 - 5 = -\frac{1}{2}$

23. Substituting $x = 5$ and $y = 15$:

$$15 = K(5)$$
$$K = 3$$

25. Substituting $P = 48$ and $V = 50$:

$$50 = \frac{K}{48}$$
$$K = 50 \bullet 48 = 2,400$$

27. Solving for l:
$$A = lw$$
$$l = \frac{A}{w}$$

29. Solving for t:
$$I = prt$$
$$t = \frac{I}{pr}$$

31. Solving for T:
$$PV = nRT$$
$$T = \frac{PV}{nR}$$

33. Solving for x:
$$y = mx + b$$
$$y - b = mx$$
$$x = \frac{y - b}{m}$$

35. Solving for F:
$$C = \frac{5}{9}(F - 32)$$
$$\frac{9}{5}C = F - 32$$
$$F = \frac{9}{5}C + 32$$

37. Solving for v:
$$h = vt + 16t^2$$
$$h - 16t^2 = vt$$
$$v = \frac{h - 16t^2}{t}$$

39. Solving for d:

$$A = a + (n - 1)d$$
$$A - a = (n - 1)d$$
$$d = \frac{A - a}{n - 1}$$

41. Solving for y:
$$2x + 3y = 6$$
$$3y = -2x + 6$$
$$y = \frac{-2x + 6}{3}$$
$$y = -\frac{2}{3}x + 2$$

43. Solving for y:

$$-3x + 5y = 15$$
$$5y = 3x + 15$$
$$y = \frac{3x + 15}{5}$$
$$y = \frac{3}{5}x + 3$$

45. Solving for y:

$$2x - 6y + 12 = 0$$
$$-6y = -2x - 12$$
$$y = \frac{-2x - 12}{-6}$$
$$y = \frac{1}{3}x + 2$$

47. Solving for x:

$$ax + 4 = bx + 9$$
$$ax - bx + 4 = 9$$
$$ax - bx = 5$$
$$x(a - b) = 5$$
$$x = \frac{5}{a - b}$$

49. Solving for P:

$$A = P + \mathrm{Pr}\,t$$
$$A = P(1 + rt)$$
$$P = \frac{A}{1 + rt}$$

51. Solving for y:

$$\frac{x}{8} + \frac{y}{2} = 1$$
$$8\left(\frac{x}{8} + \frac{y}{2}\right) = 8(1)$$
$$x + 4y = 8$$
$$4y = -x + 8$$
$$y = -\frac{1}{4}x + 2$$

53. Solving for y:

$$\frac{x}{5} + \frac{y}{-3} = 1$$
$$15\left(\frac{x}{5} + \frac{y}{-3}\right) = 15(1)$$
$$3x - 5y = 15$$
$$-5y = -3x + 15$$
$$y = \frac{3}{5}x - 3$$

55. Solving for y:

$$x = 2y - 3$$
$$2y = x + 3$$
$$y = \frac{x + 3}{2} = \frac{1}{2}x + \frac{3}{2}$$

57. Solving for y:

$$y - 3 = -2(x + 4)$$
$$y - 3 = -2x - 8$$
$$y = -2x - 5$$

59. Solving for y:

$$y - 3 = -\frac{2}{3}(x + 3)$$
$$y - 3 = -\frac{2}{3}x - 2$$
$$y = -\frac{2}{3}x + 1$$

61. Solving for y:

$$y - 4 = -\frac{1}{2}(x + 1)$$
$$y - 4 = -\frac{1}{2}x - \frac{1}{2}$$
$$y = -\frac{1}{2}x + \frac{7}{2}$$

63. a. Solving for y:

$$\frac{y + 1}{x - 0} = 4$$
$$y + 1 = 4(x - 0)$$
$$y + 1 = 4x$$
$$y = 4x - 1$$

b. Solving for y:

$$\frac{y + 2}{x - 4} = -\frac{1}{2}$$
$$y + 2 = -\frac{1}{2}(x - 4)$$
$$y + 2 = -\frac{1}{2}x + 2$$
$$y = -\frac{1}{2}x$$

c. Solving for y:

$$\frac{y + 3}{x - 7} = 0$$
$$y + 3 = 0(x - 7)$$
$$y + 3 = 0$$
$$y = -3$$

65. Writing a linear equation and solving:

$$x = 0.54 \cdot 38$$
$$x = 20.52$$

67. Writing a linear equation and solving:

$$x \cdot 36 = 9$$
$$x = \frac{1}{4} = 25\%$$

69. Writing a linear equation and solving:

$0.04 \cdot x = 37$

$x = \dfrac{37}{0.04} = 925$

71. **a.** Solving the equation:

$-4x + 5 = 20$
$-4x = 15$
$x = -\dfrac{15}{4} = -3.75$

b. Substituting $x = 3$: $-4x + 5 = -4(3) + 5 = -7$

c. Solving for y:
$-4x + 5y = 20$
$5y = 4x + 20$
$y = \dfrac{4}{5}x + 4$

d. Solving for x:
$-4x + 5y = 20$
$-4x = -5y + 20$
$x = \dfrac{5}{4}y - 5$

73. Let p represent the percent. The equation is:

$p \cdot 25 = 4$

$p = \dfrac{4}{25} = 0.16$

The tip represented 16% of the cost.

75. Computing the sales tax: $0.065(50) = 3.25$. The sales tax was $3.25.

77. The discount was: $\$94.00 - \$56.40 = \$37.60$. Finding the percent discount:

$p \cdot 94 = 37.60$

$p = \dfrac{37.6}{94} = 0.40$

The percent discount was 40%.

79. Substituting $x = 800$:
$800 = 1,300 - 100p$
$-500 = -100p$
$p = 5$

They should charge $5.00 per cartridge.

81. Substituting $x = 300$:
$300 = 1,300 - 100p$
$-1,000 = -100p$
$p = 10$

They should charge $10.00 per cartridge.

83. Let c represent the rate of the current. The equation is:
$2(15 - c) = 18$
$30 - 2c = 18$
$-2c = -12$
$c = 6$

The speed of the current is 6 mph.

85. Let w represent the rate of the wind. The equation is:
$4(258 - w) = 864$
$1032 - 4w = 864$
$-4w = -168$
$w = 42$

The speed of the wind is 42 mph.

87. Finding the rate: $\dfrac{220 \text{ miles}}{4 \text{ hours}} = 55$ miles per hour

89. Finding the rate: $\dfrac{252 \text{ kilometers}}{3 \text{ hours}} = 84$ kilometers per hour

91. The distance traveled by the rider is the circumference: $C = \pi(65) \approx (3.14)(65) \approx 204.1$ feet

Finding the rate: $\dfrac{204.1 \text{ feet}}{30 \text{ seconds}} \approx 6.8$ feet per second

93. Substituting $n = 1$, $y = 7$, and $z = 15$:

$x^1 + 7^1 = 15^1$

$x + 7 = 15$

$x = 8$

95. For Shar, $M = 220 - 46 = 174$ and $R = 60$: $T = R + 0.6(M - R) = 60 + 0.6(174 - 60) = 128.4$ beats per minute

For Sara, $M = 220 - 26 = 194$ and $R = 60$: $T = R + 0.6(M - R) = 60 + 0.6(194 - 60) = 140.4$ beats per minute

97. Using the order of operations: $38 - 19 + 1 = 19 + 1 = 20$

99. Using the order of operations: $57 - 18 - 8 = 39 - 8 = 31$

101. Using the order of operations: $28 \div 7 \bullet 2 = 4 \bullet 2 = 8$

103. Using the order of operations: $125 \div 25 \div 5 = 5 \div 5 = 1$

105. Translating into symbols: $2x - 3$

107. Translating into symbols: $x + y = 180$

109. Solving the equation:

$$x + 2x = 90$$
$$3x = 90$$
$$x = 30$$

111. Solving the equation:

$$2(2x - 3) + 2x = 45$$
$$4x - 6 + 2x = 45$$
$$6x - 6 = 45$$
$$6x = 51$$
$$x = \frac{51}{6} = 8.5$$

113. Solving the equation:

$$6x + 5(10,000 - x) = 56,000$$
$$6x + 50,000 - 5x = 56,000$$
$$x + 50,000 = 56,000$$
$$x = 6,000$$

115. Solving the equation:

$$\frac{x}{a} + \frac{y}{b} = 1$$
$$ab\left(\frac{x}{a} + \frac{y}{b}\right) = ab(1)$$
$$bx + ay = ab$$
$$bx = -ay + ab$$
$$x = a - \frac{a}{b}y$$

117. Solving the equation for a:

$$\frac{1}{a} + \frac{1}{b} = \frac{1}{c}$$
$$abc\left(\frac{1}{a} + \frac{1}{b}\right) = abc\left(\frac{1}{c}\right)$$
$$ac + bc = ab$$
$$bc = ab - ac$$
$$bc = a(b - c)$$
$$a = \frac{bc}{b - c}$$

2.3 Applications

1. Let w represent the width and $2w$ represent the length. Using the perimeter formula:

$$2w + 2(2w) = 60$$
$$2w + 4w = 60$$
$$6w = 60$$
$$w = 10$$

The dimensions are 10 feet by 20 feet.

3. Let s represent the side of the square. Using the perimeter formula:

$$4s = 28$$
$$s = 7$$

The length of each side is 7 feet.

5. Let x represent the shortest side, $x + 3$ represent the medium side, and $2x$ represent the longest side. Using the perimeter formula:

$$x + x + 3 + 2x = 23$$
$$4x + 3 = 23$$
$$4x = 20$$
$$x = 5$$

The shortest side is 5 inches.

7. Let w represent the width and $2w - 3$ represent the length. Using the perimeter formula:
$$2w + 2(2w - 3) = 18$$
$$2w + 4w - 6 = 18$$
$$6w - 6 = 18$$
$$6w = 24$$
$$w = 4$$
The width is 4 meters.

9. Let w represent the width and $2w$ represent the length. Using the perimeter formula:
$$2w + 2(2w) = 48$$
$$2w + 4w = 48$$
$$6w = 48$$
$$w = 8$$
The width is 8 feet and the length is 16 feet. Finding the cost: $C = 1.75(32) + 2.25(16) = 56 + 36 = 92$
The cost to build the pen is $92.00.

11. Let b represent the amount of money Shane had at the beginning of the trip. Using the percent increase:
$$b + 0.50b = 300$$
$$1.5b = 300$$
$$b = 200$$
Shane had $200.00 at the beginning of the trip.

13. Let I represent the accountant's monthly income before the raise. The equation is:
$$I + 0.055I = 3440$$
$$1.055I = 3440$$
$$I \approx 3260.66$$
The accountant's monthly income before the raise was approximately $3,260.66.

15. Let R represent the total box office receipts. The equation is:
$$0.53R = 52.8$$
$$R \approx 99.6$$
The total receipts were approximately $99.6 million.

17. Let x represent one angle and $8x$ represent the other angle. Since the angles are supplementary:
$$x + 8x = 180$$
$$9x = 180$$
$$x = 20$$
The two angles are 20° and 160°.

19. **a.** Let x represent one angle and $4x - 12$ represent the other angle. Since the angles are complementary:
$$x + 4x - 12 = 90$$
$$5x - 12 = 90$$
$$5x = 102$$
$$x = 20.4$$
$$4x - 12 = 4(20.4) - 12 = 69.6$$
The two angles are 20.4° and 69.6°.

    b. Let x represent one angle and $4x - 12$ represent the other angle. Since the angles are supplementary:
$$x + 4x - 12 = 180$$
$$5x - 12 = 180$$
$$5x = 192$$
$$x = 38.4$$
$$4x - 12 = 4(38.4) - 12 = 141.6$$
The two angles are 38.4° and 141.6°.

21. Let x represent the smallest angle, $3x$ represent the largest angle, and $3x - 9$ represent the third angle. The equation is:
$$x + 3x + 3x - 9 = 180$$
$$7x - 9 = 180$$
$$7x = 189$$
$$x = 27$$
The three angles are 27°, 72° and 81°.

23. Let x represent the largest angle, $\frac{1}{3}x$ represent the smallest angle, and $\frac{1}{3}x+10$ represent the third angle.
   The equation is:

$$\frac{1}{3}x+x+\frac{1}{3}x+10 = 180$$
$$\frac{5}{3}x+10 = 180$$
$$\frac{5}{3}x = 170$$
$$x = 102$$

   The three angles are 34°, 44° and 102°.

25. Let x represent the measure of the two base angles, and $2x + 8$ represent the third angle. The equation is:

$$x+x+2x+8 = 180$$
$$4x+8 = 180$$
$$4x = 172$$
$$x = 43$$

   The three angles are 43°, 43° and 94°.

27. Let x represent the amount invested at 8% and $9,000 - x$ represent the amount invested at 9%. The equation is:

$$0.08x+0.09(9,000-x) = 750$$
$$0.08x+810-0.09x = 750$$
$$-0.01x+810 = 750$$
$$-0.01x = -60$$
$$x = 6,000$$

   She invested $6,000 at 8% and $3,000 at 9%.

29. Let x represent the amount invested at 12% and $15,000 - x$ represent the amount invested at 10%. The equation is:

$$0.12x+0.10(15,000-x) = 1,600$$
$$0.12x+1,500-0.10x = 1,600$$
$$0.02x+1,500 = 1,600$$
$$0.02x = 100$$
$$x = 5,000$$

   The investment was $5,000 at 12% and $10,000 at 10%.

31. Let x represent the amount invested at 8% and $6,000 - x$ represent the amount invested at 9%. The equation is:

$$0.08x+0.09(6,000-x) = 500$$
$$0.08x+540-0.09x = 500$$
$$-0.01x+540 = 500$$
$$-0.01x = -40$$
$$x = 4,000$$

   Stacy invested $4,000 at 8% and $2,000 at 9%.

33. Let x represent the number of adult tickets sold and $x + 6$ the number of children's tickets sold. The equation is:

$$6(x)+4(x+6) = 184$$
$$6x+4x+24 = 184$$
$$10x+24 = 184$$
$$10x = 160$$
$$x = 16$$

   Miguel sold 16 adult tickets and 22 children's tickets.

35. The total money collected is: $1204 - $250 = $954

   Let x represent the amount of her sales (not including tax). Since this amount includes the tax collected, the equation is:

$$x+0.06x = 954$$
$$1.06x = 954$$
$$x = 900$$

   Her sales were $900, so the sales tax is: $0.06(900) = \$54$

37. Completing the table:

t	0	$\frac{1}{4}$	1	$\frac{7}{4}$	2
h	0	7	16	7	0

39. Completing the table:

Year	Sales (billions of dollars)
2005	7
2006	7.5
2007	8.0
2008	8.6
2009	9.2

41. Completing the table:

Speed (miles per hour)	Distance (miles)
20	10
30	15
40	20
50	25
60	30
70	35

43. Completing the table:

Time (hours)	Distance Upstream (miles)	Distance Downstream (miles)
1	6	14
2	12	28
3	18	42
4	24	56
5	30	70
6	36	84

45. Completing the table:

Age (years)	Maximum Heart Rate (beats per minute)
18	202
19	201
20	200
21	199
22	198
23	197

47. Completing the table:

w	l	A
2	22	44
4	20	80
6	18	108
8	16	128
10	14	140
12	12	144

49. Graphing the inequality:

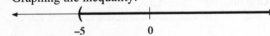

51. Graphing the inequality:

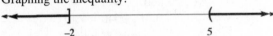

53. Graphing the inequality:

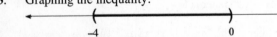

55. Graphing the inequality:

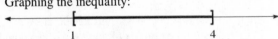

57. Graphing the inequality:

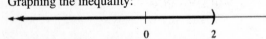

59. Graphing the inequality:

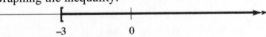

61. Solving the equation:

$$-2x - 3 = 7$$
$$-2x = 10$$
$$x = -5$$

63. Solving the equation:

$$3(2x - 4) - 7x = -3x$$
$$6x - 12 - 7x = -3x$$
$$-x - 12 = -3x$$
$$-12 = -2x$$
$$x = 6$$

2.4 Linear Inequalities in One Variable

1. Solving the inequality:

$$2x \leq 3$$
$$x \leq \frac{3}{2}$$

Graphing the solution set:

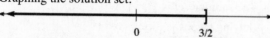

3. Solving the inequality:

$$\frac{1}{2}x > 2$$
$$x > 4$$

Graphing the solution set:

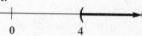

5. Solving the inequality:

$$-5x \leq 25$$
$$x \geq -5$$

Graphing the solution set:

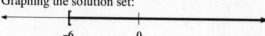

7. Solving the inequality:

$$-\frac{3}{2}x > -6$$
$$-3x > -12$$
$$x < 4$$

Graphing the solution set:

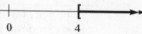

9. Solving the inequality:

$$-12 \leq 2x$$
$$x \geq -6$$

Graphing the solution set:

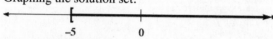

11. Solving the inequality:

$$-1 \geq -\frac{1}{4}x$$
$$x \geq 4$$

Graphing the solution set:

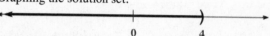

13. Solving the inequality:

$$-3x + 1 > 10$$
$$-3x > 9$$
$$x < -3$$

Graphing the solution set:

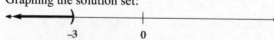

15. Solving the inequality:

$$\frac{1}{2} - \frac{m}{12} \leq \frac{7}{12}$$
$$12\left(\frac{1}{2} - \frac{m}{12}\right) \leq 12\left(\frac{7}{12}\right)$$
$$6 - m \leq 7$$
$$-m \leq 1$$
$$m \geq -1$$

Graphing the solution set:

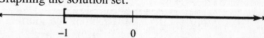

17. Solving the inequality:

$$\frac{1}{2} \geq -\frac{1}{6} - \frac{2}{9}x$$
$$18\left(\frac{1}{2}\right) \geq 18\left(-\frac{1}{6} - \frac{2}{9}x\right)$$
$$9 \geq -3 - 4x$$
$$12 \geq -4x$$
$$x \geq -3$$

Graphing the solution set:

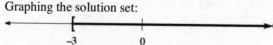

19. Solving the inequality:

$$-40 \leq 30 - 20y$$
$$-70 \leq -20y$$
$$y \leq \frac{7}{2}$$

Graphing the solution set:

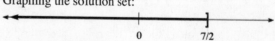

21. Solving the inequality:

$$\frac{2}{3}x - 3 < 1$$
$$\frac{2}{3}x < 4$$
$$2x < 12$$
$$x < 6$$

Graphing the solution set:

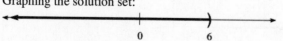

23. Solving the inequality:

$$10 - \frac{1}{2}y \leq 36$$
$$-\frac{1}{2}y \leq 26$$
$$y \geq -52$$

Graphing the solution set:

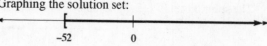

25. Solving the inequality:

$$2(3y+1) \le -10$$
$$6y+2 \le -10$$
$$6y \le -12$$
$$y \le -2$$

The solution set is $(-\infty, -2]$.

27. Solving the inequality:

$$-(a+1)-4a \le 2a-8$$
$$-a-1-4a \le 2a-8$$
$$-5a-1 \le 2a-8$$
$$-7a \le -7$$
$$a \ge 1$$

The solution set is $[1, \infty)$.

29. Solving the inequality:

$$\tfrac{1}{3}t - \tfrac{1}{2}(5-t) < 0$$
$$6\left(\tfrac{1}{3}t - \tfrac{1}{2}(5-t)\right) < 6(0)$$
$$2t - 3(5-t) < 0$$
$$2t - 15 + 3t < 0$$
$$5t - 15 < 0$$
$$5t < 15$$
$$t < 3$$

The solution set is $(-\infty, 3)$.

31. Solving the inequality:

$$-2 \le 5 - 7(2a+3)$$
$$-2 \le 5 - 14a - 21$$
$$-2 \le -16 - 14a$$
$$14 \le -14a$$
$$a \le -1$$

The solution set is $(-\infty, -1]$.

33. Solving the inequality:

$$-\tfrac{1}{3}(x+5) \le -\tfrac{2}{9}(x-1)$$
$$9\left[-\tfrac{1}{3}(x+5)\right] \le 9\left[-\tfrac{2}{9}(x-1)\right]$$
$$-3(x+5) \le -2(x-1)$$
$$-3x - 15 \le -2x + 2$$
$$-x - 15 \le 2$$
$$-x \le 17$$
$$x \ge -17$$

The solution set is $[-17, \infty)$.

35. Solving the inequality:

$$20x + 9,300 > 18,000$$
$$20x > 8,700$$
$$x > 435$$

37. Solving the inequality:

$$-2 \le m - 5 \le 2$$
$$3 \le m \le 7$$

The solution set is $[3, 7]$. Graphing the solution set:

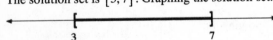

39. Solving the inequality:

$$-60 < 20a + 20 < 60$$
$$-80 < 20a < 40$$
$$-4 < a < 2$$

The solution set is $(-4, 2)$. Graphing the solution set:

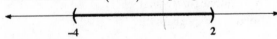

41. Solving the inequality:

$$0.5 \le 0.3a - 0.7 \le 1.1$$
$$1.2 \le 0.3a \le 1.8$$
$$4 \le a \le 6$$

The solution set is $[4, 6]$. Graphing the solution set:

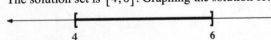

43. Solving the inequality:

$$3 < \tfrac{1}{2}x + 5 < 6$$
$$-2 < \tfrac{1}{2}x < 1$$
$$-4 < x < 2$$

The solution set is $(-4, 2)$. Graphing the solution set:

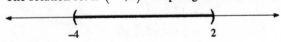

45. Solving the inequality:

$$4 < 6 + \tfrac{2}{3}x < 8$$
$$-2 < \tfrac{2}{3}x < 2$$
$$-6 < 2x < 6$$
$$-3 < x < 3$$

The solution set is $(-3, 3)$. Graphing the solution set:

47.   Solving the inequality:

$$x+5 \le -2 \qquad \text{or} \qquad x+5 \ge 2$$
$$x \le -7 \qquad \text{or} \qquad x \ge -3$$

The solution set is $(-\infty, -7] \cup [-3, \infty)$. Graphing the solution set:

-7   -3

49.   Solving the inequality:

$$5y+1 \le -4 \qquad \text{or} \qquad 5y+1 \ge 4$$
$$5y \le -5 \qquad \text{or} \qquad 5y \ge 3$$
$$y \le -1 \qquad \text{or} \qquad y \ge \frac{3}{5}$$

The solution set is $(-\infty, -1] \cup \left[\frac{3}{5}, \infty\right)$. Graphing the solution set:

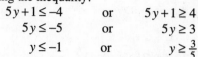

-1   3/5

51.   Solving the inequality:

$$2x+5 < 3x-1 \qquad \text{or} \qquad x-4 > 2x+6$$
$$-x+5 < -1 \qquad \text{or} \qquad -x-4 > 6$$
$$-x < -6 \qquad \text{or} \qquad -x > 10$$
$$x > 6 \qquad \text{or} \qquad x < -10$$

The solution set is $(-\infty, -10) \cup (6, \infty)$. Graphing the solution set:

-10   6

53.   Writing as an inequality: $-2 < x \le 4$

55.   Writing as an inequality: $x < -4$ or $x \ge 1$

57.   **a.**   Writing as an inequality: $x > 0$

b.   Writing as an inequality: $x \ge 0$

c.   Writing as an inequality: $x \ge 0$

59.   There is no solution, since $x^2 \ge 0$ for all values of x.

61.   This inequality is true for all values of x, since $x^2 \ge 0$ is always true.

63.   There is no solution, since $\dfrac{1}{x^2} > 0$ for all values of x.

65.   **a.**   Evaluating when $x = 0$: $-\frac{1}{2}x+1 = -\frac{1}{2}(0)+1 = 1$

b.   Solving the equation:

$$-\frac{1}{2}x+1 = -7$$
$$-\frac{1}{2}x = -8$$
$$x = 16$$

c.   Substituting $x = 0$: $-\frac{1}{2}x+1 = -\frac{1}{2}(0)+1 = 1$

No, 0 is not a solution to the inequality.

d.   Solving the inequality:

$$-\frac{1}{2}x+1 < -7$$
$$-\frac{1}{2}x < -8$$
$$x > 16$$

67.   Solving the inequality:

$$900 - 300p \ge 300$$
$$-300p \ge -600$$
$$p \le 2$$

They should charge $2.00 per pad or less.

69.   Solving the inequality:

$$900 - 300p < 525$$
$$-300p < -375$$
$$p > 1.25$$

They should charge more than $1.25 per pad.

71.   Solving the inequality:
$$22149 - 399x > 20500$$
$$-399x > -1649$$
$$x < 4.13$$
In the years 1990 through 1994 Amtrak had more than 20,500 million passengers.

73.   For adults, the inequality is $0.72 - 0.11 \le r \le 0.72 + 0.11$, or $0.61 \le r \le 0.83$. The survival rate for adults is between 61% and 83%. For juveniles, the inequality is $0.13 - 0.07 \le r \le 0.13 + 0.07$, or $0.06 \le r \le 0.20$. The survival rate for juveniles is between 6% and 20%.

75. **a.**   Solving the inequality:
$$95 \le \tfrac{9}{5}C + 32 \le 113$$
$$63 \le \tfrac{9}{5}C \le 81$$
$$315 \le 9C \le 405$$
$$35° \le C \le 45°$$

b.   Solving the inequality:
$$68 \le \tfrac{9}{5}C + 32 \le 86$$
$$36 \le \tfrac{9}{5}C \le 54$$
$$180 \le 9C \le 270$$
$$20° \le C \le 30°$$

c.   Solving the inequality:
$$-13 \le \tfrac{9}{5}C + 32 \le 14$$
$$-45 \le \tfrac{9}{5}C \le -18$$
$$-225 \le 9C \le -90$$
$$-25° \le C \le -10°$$

d.   Solving the inequality:
$$-4 \le \tfrac{9}{5}C + 32 \le 23$$
$$-36 \le \tfrac{9}{5}C \le -9$$
$$-180 \le 9C \le -45$$
$$-20° \le C \le -5°$$

77.   Simplifying: $|-3| = 3$

79.   Simplifying: $-|-3| = -3$

81. $|x|$ represents the distance from x to 0 on the number line.

83.   Solving the equation:
$$2a - 1 = -7$$
$$2a = -6$$
$$a = -3$$

85.   Solving the equation:
$$\tfrac{2}{3}x - 3 = 7$$
$$\tfrac{2}{3}x = 10$$
$$x = 15$$

87.   Solving the equation:
$$x - 5 = x - 7$$
$$-5 = -7$$

The equation has no solution $(\varnothing)$.

89.   Solving the equation:
$$x - 5 = -x - 7$$
$$2x - 5 = -7$$
$$2x = -2$$
$$x = -1$$

91.   Solving the inequality:
$$ax + b < c$$
$$ax < c - b$$
$$x < \frac{c - b}{a}$$

93.   Solving the inequality:
$$-c < ax + b < c$$
$$-c - b < ax < c - b$$
$$\frac{-c - b}{a} < x < \frac{c - b}{a}$$

2.5 Equations with Absolute Value

1.   Solving the equation:
$$|x| = 4$$
$$x = -4, 4$$

3.   Solving the equation:
$$2 = |a|$$
$$a = -2, 2$$

5.   The equation $|x| = -3$ has no solution, or $\varnothing$.

7.   Solving the equation:
$$|a| + 2 = 3$$
$$|a| = 1$$
$$a = -1, 1$$

9.   Solving the equation:
$$|y| + 4 = 3$$
$$|y| = -1$$

The equation $|y| = -1$ has no solution, or $\varnothing$.

11. Solving the equation:
$$4 = |x| - 2$$
$$|x| = 6$$
$$x = -6, 6$$

13. Solving the equation:
$$|x - 2| = 5$$
$$x - 2 = -5, 5$$
$$x = -3, 7$$

15. Solving the equation:
$$|a - 4| = \frac{5}{3}$$
$$a - 4 = -\frac{5}{3}, \frac{5}{3}$$
$$a = \frac{7}{3}, \frac{17}{3}$$

17. Solving the equation:
$$1 = |3 - x|$$
$$3 - x = -1, 1$$
$$-x = -4, -2$$
$$x = 2, 4$$

19. Solving the equation:
$$\left|\frac{3}{5}a + \frac{1}{2}\right| = 1$$
$$\frac{3}{5}a + \frac{1}{2} = -1, 1$$
$$\frac{3}{5}a = -\frac{3}{2}, \frac{1}{2}$$
$$a = -\frac{5}{2}, \frac{5}{6}$$

21. Solving the equation:
$$60 = |20x - 40|$$
$$20x - 40 = -60, 60$$
$$20x = -20, 100$$
$$x = -1, 5$$

23. Since $|2x + 1| = -3$ is impossible, there is no solution, or $\varnothing$.

25. Solving the equation:
$$\left|\frac{3}{4}x - 6\right| = 9$$
$$\frac{3}{4}x - 6 = -9, 9$$
$$\frac{3}{4}x = -3, 15$$
$$3x = -12, 60$$
$$x = -4, 20$$

27. Solving the equation:
$$\left|1 - \frac{1}{2}a\right| = 3$$
$$1 - \frac{1}{2}a = -3, 3$$
$$-\frac{1}{2}a = -4, 2$$
$$a = -4, 8$$

29. Solving the equation:
$$|3x + 4| + 1 = 7$$
$$|3x + 4| = 6$$
$$3x + 4 = -6, 6$$
$$3x = -10, 2$$
$$x = -\frac{10}{3}, \frac{2}{3}$$

31. Solving the equation:
$$|3 - 2y| + 4 = 3$$
$$|3 - 2y| = -1$$

Since this equation is impossible, there is no solution, or $\varnothing$.

33. Solving the equation:
$$3 + |4t - 1| = 8$$
$$|4t - 1| = 5$$
$$4t - 1 = -5, 5$$
$$4t = -4, 6$$
$$t = -1, \frac{3}{2}$$

35. Solving the equation:
$$\left|9 - \frac{3}{5}x\right| + 6 = 12$$
$$\left|9 - \frac{3}{5}x\right| = 6$$
$$9 - \frac{3}{5}x = -6, 6$$
$$-\frac{3}{5}x = -15, -3$$
$$-3x = -75, -15$$
$$x = 5, 25$$

37.   Solving the equation:

$$5 = \left| \frac{2x}{7} + \frac{4}{7} \right| - 3$$

$$\left| \frac{2x}{7} + \frac{4}{7} \right| = 8$$

$$\frac{2x}{7} + \frac{4}{7} = -8, 8$$

$$2x + 4 = -56, 56$$

$$2x = -60, 52$$

$$x = -30, 26$$

39.   Solving the equation:

$$2 = -8 + \left| 4 - \frac{1}{2} y \right|$$

$$\left| 4 - \frac{1}{2} y \right| = 10$$

$$4 - \frac{1}{2} y = -10, 10$$

$$-\frac{1}{2} y = -14, 6$$

$$y = -12, 28$$

41.   Solving the equation:

$$|3a + 1| = |2a - 4|$$

$$3a + 1 = 2a - 4 \qquad \text{or} \qquad 3a + 1 = -2a + 4$$
$$a + 1 = -4 \qquad\qquad\qquad\qquad 5a = 3$$
$$a = -5 \qquad\qquad\qquad\qquad\quad a = \frac{3}{5}$$

43.   Solving the equation:

$$\left| x - \frac{1}{3} \right| = \left| \frac{1}{2} x + \frac{1}{6} \right|$$

$$x - \frac{1}{3} = \frac{1}{2} x + \frac{1}{6} \qquad \text{or} \qquad x - \frac{1}{3} = -\frac{1}{2} x - \frac{1}{6}$$
$$6x - 2 = 3x + 1 \qquad\qquad\qquad 6x - 2 = -3x - 1$$
$$3x - 2 = 1 \qquad\qquad\qquad\qquad 9x - 2 = -1$$
$$3x = 3 \qquad\qquad\qquad\qquad\quad 9x = 1$$
$$x = 1 \qquad\qquad\qquad\qquad\qquad x = \frac{1}{9}$$

45.   Solving the equation:

$$|y - 2| = |y + 3|$$

$$y - 2 = y + 3 \qquad \text{or} \qquad y - 2 = -y - 3$$
$$-2 = -3 \qquad\qquad\qquad\qquad 2y = -1$$
$$y = \text{impossible} \qquad\qquad\qquad y = -\frac{1}{2}$$

47.   Solving the equation:

$$|3x - 1| = |3x + 1|$$

$$3x - 1 = 3x + 1 \qquad \text{or} \qquad 3x - 1 = -3x - 1$$
$$-1 = 1 \qquad\qquad\qquad\qquad 6x = 0$$
$$x = \text{impossible} \qquad\qquad\qquad x = 0$$

49.   Solving the equation:

$$|3 - m| = |m + 4|$$

$$3 - m = m + 4 \qquad \text{or} \qquad 3 - m = -m - 4$$
$$-2m = 1 \qquad\qquad\qquad\qquad 3 = -4$$
$$m = -\frac{1}{2} \qquad\qquad\qquad m = \text{impossible}$$

51.   Solving the equation:

$$|0.03 - 0.01x| = |0.04 + 0.05x|$$

$$0.03 - 0.01x = 0.04 + 0.05x \qquad \text{or} \qquad 0.03 - 0.01x = -0.04 - 0.05x$$
$$-0.06x = 0.01 \qquad\qquad\qquad\qquad\qquad 0.04x = -0.07$$
$$x = -\frac{1}{6} \qquad\qquad\qquad\qquad\qquad\qquad x = -\frac{7}{4}$$

53.   Since $|x - 2| = |2 - x|$ is always true, the solution set is all real numbers.

55.   Since $\left| \frac{x}{5} - 1 \right| = \left| 1 - \frac{x}{5} \right|$ is always true, the solution set is all real numbers.

57. **a.** Solving the equation:

$$4x - 5 = 0$$
$$4x = 5$$
$$x = \tfrac{5}{4} = 1.25$$

b. Solving the equation:

$$|4x - 5| = 0$$
$$4x - 5 = 0$$
$$4x = 5$$
$$x = \tfrac{5}{4} = 1.25$$

c. Solving the equation:

$$4x - 5 = 3$$
$$4x = 8$$
$$x = 2$$

d. Solving the equation:

$$|4x - 5| = 3$$
$$4x - 5 = -3, 3$$
$$4x = 2, 8$$
$$x = \tfrac{1}{2}, 2$$

e. Solving the equation:

$$|4x - 5| = |2x + 3|$$

$$4x - 5 = 2x + 3 \quad\text{or}\quad 4x - 5 = -2x - 3$$
$$2x - 5 = 3 \qquad\qquad 6x - 5 = -3$$
$$2x = 8 \qquad\qquad 6x = 2$$
$$x = 4 \qquad\qquad x = \tfrac{1}{3}$$

59. Setting $R = 722$:

$$-60|x - 11| + 962 = 722$$
$$-60|x - 11| = -240$$
$$|x - 11| = 4$$
$$x - 11 = -4, 4$$
$$x = 7, 15$$

The revenue was 722 million dollars in the years 1987 and 1995.

61. Graphing the inequality:

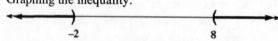

63. Graphing the inequality:

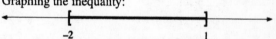

65. Simplifying: $\dfrac{38}{30} = \dfrac{2 \bullet 19}{2 \bullet 15} = \dfrac{19}{15}$

67. Simplifying: $\dfrac{240}{6} = \dfrac{6 \bullet 40}{6 \bullet 1} = 40$

69. Simplifying: $\dfrac{0+6}{0-3} = \dfrac{6}{-3} = -2$

71. Simplifying: $\dfrac{4-4}{4-2} = \dfrac{0}{2} = 0$

73. Solving the inequality:
$$2x - 5 < 3$$
$$2x < 8$$
$$x < 4$$

75. Solving the inequality:
$$-4 \le 3a + 7$$
$$-11 \le 3a$$
$$a \ge -\tfrac{11}{3}$$

77. Solving the inequality:
$$4t - 3 \le -9$$
$$4t \le -6$$
$$t \le -\tfrac{3}{2}$$

79. Solving the inequality:
$$|x - a| = b$$
$$x - a = -b, b$$
$$x = a - b \ \text{ or } \ x = a + b$$

81. Solving the equation:

$$|ax + b| = c$$
$$ax + b = -c, c$$
$$ax = -b - c, -b + c$$
$$x = \dfrac{-b-c}{a} \ \text{ or } \ x = \dfrac{-b+c}{a}$$

83. Solving the equation:

$$\left|\dfrac{x}{a} + \dfrac{y}{b}\right| = 1$$
$$\dfrac{x}{a} + \dfrac{y}{b} = -1, 1$$
$$\dfrac{x}{a} = -\dfrac{y}{b} - 1, -\dfrac{y}{b} + 1$$
$$x = -\dfrac{a}{b}y - a \ \text{ or } \ x = -\dfrac{a}{b}y + a$$

2.6 Inequalities Involving Absolute Value

1. Solving the inequality:

$|x| < 3$

$-3 < x < 3$

Graphing the solution set:

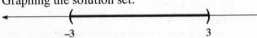

3. Solving the inequality:

$|x| \ge 2$

$x \le -2$ or $x \ge 2$

Graphing the solution set:

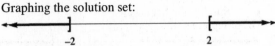

5. Solving the inequality:

$|x| + 2 < 5$

$|x| < 3$

$-3 < x < 3$

Graphing the solution set:

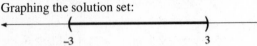

7. Solving the inequality:

$|t| - 3 > 4$

$|t| > 7$

$t < -7$ or $t > 7$

Graphing the solution set:

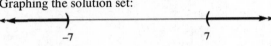

9. Since the inequality $|y| < -5$ is never true, there is no solution, or $\varnothing$. Graphing the solution set:

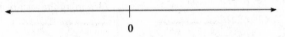

11. Since the inequality $|x| \ge -2$ is always true, the solution set is all real numbers.

Graphing the solution set:

13. Solving the inequality:

$|x - 3| < 7$

$-7 < x - 3 < 7$

$-4 < x < 10$

Graphing the solution set:

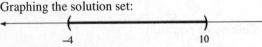

15. Solving the inequality:

$|a + 5| \ge 4$

$a + 5 \le -4$ or $a + 5 \ge 4$

$a \le -9$ or $a \ge -1$

Graphing the solution set:

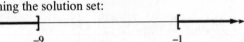

17. Since the inequality $|a - 1| < -3$ is never true, there is no solution, or $\varnothing$. Graphing the solution set:

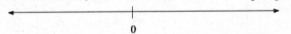

19. Solving the inequality:

$|2x - 4| < 6$

$-6 < 2x - 4 < 6$

$-2 < 2x < 10$

$-1 < x < 5$

Graphing the solution set:

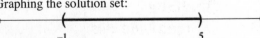

21. Solving the inequality:

$|3y + 9| \ge 6$

$3y + 9 \le -6$ or $3y + 9 \ge 6$

$3y \le -15$ $3y \ge -3$

$y \le -5$ $y \ge -1$

Graphing the solution set:

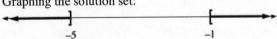

23. Solving the inequality:

$|2k + 3| \ge 7$

$2k + 3 \le -7$ or $2k + 3 \ge 7$

$2k \le -10$ $2k \ge 4$

$k \le -5$ $k \ge 2$

Graphing the solution set:

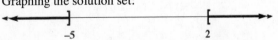

25. Solving the inequality:

$|x - 3| + 2 < 6$

$|x - 3| < 4$

$-4 < x - 3 < 4$

$-1 < x < 7$

Graphing the solution set:

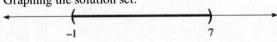

27. Solving the inequality:

$$|2a+1|+4 \geq 7$$
$$|2a+1| \geq 3$$

$2a+1 \leq -3$ or $2a+1 \geq 3$
$2a \leq -4$ $2a \geq 2$
$a \leq -2$ $a \geq 1$

Graphing the solution set:

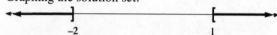
−2   1

29. Solving the inequality:

$$|3x+5|-8 < 5$$
$$|3x+5| < 13$$
$$-13 < 3x+5 < 13$$
$$-18 < 3x < 8$$
$$-6 < x < \frac{8}{3}$$

Graphing the solution set:

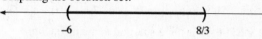
−6   8/3

31. Solving the inequality:

$$|5-x| > 3$$

$5-x < -3$ or $5-x > 3$
$-x < -8$ $-x > -2$
$x > 8$ $x < 2$

Graphing the solution set:

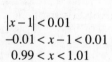
2   8

32. Solving the inequality:

$$\left|3-\frac{2}{3}x\right| \geq 5$$

$3-\frac{2}{3}x \leq -5$ or $3-\frac{2}{3}x \geq 5$
$-\frac{2}{3}x \leq -8$ $-\frac{2}{3}x \geq 2$
$-2x \leq -24$ $-2x \geq 6$
$x \geq 12$ $x \leq -3$

Graphing the solution set:

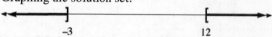
−3   12

35. Solving the inequality:

$$\left|2-\frac{1}{2}x\right| > 1$$

$2-\frac{1}{2}x < -1$ or $2-\frac{1}{2}x > 1$
$-\frac{1}{2}x < -3$ $-\frac{1}{2}x > -1$
$x > 6$ $x < 2$

Graphing the solution set:

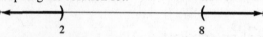
2   6

37. Solving the inequality:

$$|x-1| < 0.01$$
$$-0.01 < x-1 < 0.01$$
$$0.99 < x < 1.01$$

39. Solving the inequality:

$$|2x+1| \geq \frac{1}{5}$$

$2x+1 \leq -\frac{1}{5}$ or $2x+1 \geq \frac{1}{5}$
$2x \leq -\frac{6}{5}$ $2x \geq -\frac{4}{5}$
$x \leq -\frac{3}{5}$ $x \geq -\frac{2}{5}$

41. Solving the inequality:

$$\left|\frac{3x-2}{5}\right| \leq \frac{1}{2}$$
$$-\frac{1}{2} \leq \frac{3x-2}{5} \leq \frac{1}{2}$$
$$-\frac{5}{2} \leq 3x-2 \leq \frac{5}{2}$$
$$-\frac{1}{2} \leq 3x \leq \frac{9}{2}$$
$$-\frac{1}{6} \leq x \leq \frac{3}{2}$$

43. Solving the inequality:

$$\left|2x-\frac{1}{5}\right| < 0.3$$
$$-0.3 < 2x-0.2 < 0.3$$
$$-0.1 < 2x < 0.5$$
$$-0.05 < x < 0.25$$

45. Writing as an absolute value inequality: $|x| \leq 4$

47. Writing as an absolute value inequality: $|x-5| \leq 1$

49. **a.** Evaluating when $x = 0$: $|5x + 3| = |5(0) + 3| = |3| = 3$

   b. Solving the equation:
$$|5x + 3| = 7$$
$$5x + 3 = -7, 7$$
$$5x = -10, 4$$
$$x = -2, \frac{4}{5}$$

   c. Substituting $x = 0$: $|5x + 3| = |5(0) + 3| = |3| = 3$

   No, 0 is not a solution to the inequality.

   d. Solving the inequality:
$$|5x + 3| > 7$$

$$5x + 3 < -7 \qquad \text{or} \qquad 5x + 3 > 7$$
$$5x < -10 \qquad\qquad\qquad 5x > 4$$
$$x < -2 \qquad\qquad\qquad x > \frac{4}{5}$$

51. The absolute value inequality is: $|x - 65| \le 10$

53. Simplifying: $-9 \div \frac{3}{2} = -9 \cdot \frac{2}{3} = -6$ **55.** Simplifying: $3 - 7(-6 - 3) = 3 - 7(-9) = 3 + 63 = 66$

57. Simplifying: $-4(-2)^3 - 5(-3)^2 = -4(-8) - 5(9) = 32 - 45 = -13$

59. Simplifying: $-2(-3 + 8) - 7(-9 + 6) = -2(5) - 7(-3) = -10 + 21 = 11$

61. Simplifying: $\dfrac{2(-3) - 5(-6)}{-1 - 2 - 3} = \dfrac{-6 + 30}{-1 - 2 - 3} = \dfrac{24}{-6} = -4$ **63.** Simplifying: $6(1) - 1(-5) + 1(2) = 6 + 5 + 2 = 13$

65. Simplifying: $1(0)(1) + 3(1)(4) + (-2)(2)(-1) = 0 + 12 + 4 = 16$

67. Simplifying: $4(-1)^2 + 3(-1) - 2 = 4(1) + 3(-1) - 2 = 4 - 3 - 2 = -1$

69. Solving the inequality:

$$|x - a| < b$$
$$-b < x - a < b$$
$$a - b < x < a + b$$

71. Solving the inequality:
$$|ax - b| > c$$

$$ax - b < -c \qquad \text{or} \qquad ax - b > c$$
$$ax < b - c \qquad\qquad\qquad ax > b + c$$
$$x < \frac{b - c}{a} \qquad\qquad\qquad x > \frac{b + c}{a}$$

73. Solving the inequality:
$$|ax + b| \le c$$
$$-c \le ax + b \le c$$
$$-c - b \le ax \le c - b$$
$$\frac{-c - b}{a} \le x \le \frac{c - b}{a}$$

Chapter 2 Review/Test

1.   Solving the equation:

$$x - 3 = 7$$
$$x = 10$$

2.   Solving the equation:
$$5x - 2 = 8$$
$$5x = 10$$
$$x = 2$$

3.   Solving the equation:

$$400 - 100a = 200$$
$$-100a = -200$$
$$a = 2$$

4.   Solving the equation:
$$5 - \frac{2}{3}a = 7$$
$$-\frac{2}{3}a = 2$$
$$-2a = 6$$
$$a = -3$$

5.   Solving the equation:

$$4x - 2 = 7x + 7$$
$$-3x - 2 = 7$$
$$-3x = 9$$
$$x = -3$$

6.   Solving the equation:
$$\frac{3}{2}x - \frac{1}{6} = -\frac{7}{6}x - \frac{1}{6}$$
$$6\left(\frac{3}{2}x - \frac{1}{6}\right) = 6\left(-\frac{7}{6}x - \frac{1}{6}\right)$$
$$9x - 1 = -7x - 1$$
$$16x = 0$$
$$x = 0$$

7.   Solving the equation:

$$7y - 5 - 2y = 2y - 3$$
$$5y - 5 = 2y - 3$$
$$3y - 5 = -3$$
$$3y = 2$$
$$y = \frac{2}{3}$$

8.   Solving the equation:
$$\frac{3y}{4} - \frac{1}{2} + \frac{3y}{2} = 2 - y$$
$$8\left(\frac{3y}{4} - \frac{1}{2} + \frac{3y}{2}\right) = 8(2 - y)$$
$$6y - 4 + 12y = 16 - 8y$$
$$18y - 4 = -8y + 16$$
$$26y - 4 = 16$$
$$26y = 20$$
$$y = \frac{10}{13}$$

9.   Solving the equation:
$$3(2x + 1) = 18$$
$$6x + 3 = 18$$
$$6x = 15$$
$$x = \frac{5}{2}$$

10.   Solving the equation:
$$-\frac{1}{2}(4x - 2) = -x$$
$$-2x + 1 = -x$$
$$-x = -1$$
$$x = 1$$

11.   Solving the equation:
$$8 - 3(2t + 1) = 5(t + 2)$$
$$8 - 6t - 3 = 5t + 10$$
$$-6t + 5 = 5t + 10$$
$$-11t + 5 = 10$$
$$-11t = 5$$
$$t = -\frac{5}{11}$$

12.   Solving the equation:
$$8 + 4(1 - 3t) = -3(t - 4) + 2$$
$$8 + 4 - 12t = -3t + 12 + 2$$
$$-12t + 12 = -3t + 14$$
$$-9t + 12 = 14$$
$$-9t = 2$$
$$t = -\frac{2}{9}$$

13.   Solving for h:
$$40 = 2(3) + 2h$$
$$40 = 6 + 2h$$
$$2h = 34$$
$$h = 17$$

14.   Solving for t:
$$2000 = 1000 + 1000(0.05)t$$
$$2000 = 1000 + 50t$$
$$50t = 1000$$
$$t = 20$$

15.   Solving for p:
$$I = prt$$
$$p = \frac{I}{rt}$$

16.   Solving for x:
$$y = mx + b$$
$$y - b = mx$$
$$x = \frac{y - b}{m}$$

17.   Solving for y:

$$4x - 3y = 12$$
$$-3y = -4x + 12$$
$$y = \frac{4}{3}x - 4$$

18.   Solving for v:

$$d = vt + 16t^2$$
$$vt = d - 16t^2$$
$$v = \frac{d - 16t^2}{t}$$

19.   The first four terms are: 3,6,9,12. This sequence is increasing.

20.   The first four terms are: 4,5,6,7. This sequence is increasing.

21.   The first four terms are: $-2,4,-8,16$. This sequence is alternating.

22.   The first four terms are: $1, \frac{1}{2}, \frac{1}{3}, \frac{1}{4}$. This sequence is decreasing.

23.   Let w represent the width and $3w$ represent the length. Using the perimeter formula:

$$2w + 2(3w) = 32$$
$$2w + 6w = 32$$
$$8w = 32$$
$$w = 4$$

The dimensions are 4 feet by 12 feet.

24.   Let x, $x + 1$, and $x + 2$ represent the three sides. Using the perimeter formula:

$$x + x + 1 + x + 2 = 12$$
$$3x + 3 = 12$$
$$3x = 9$$
$$x = 3$$

The sides are 3 meters, 4 meters, and 5 meters.

25.   **a.**   Substituting $L = 45$ and $H = 12$: $N = 7(45)(12) = 3,780$ bricks

b.   Substituting $N = 35,000$ and $H = 8$:

$$35000 = 7L(8)$$
$$35000 = 56L$$
$$L = 625$$

The wall could be 625 feet long.

26.   Let s represent her first year salary. Using the percent increase:

$$s + 0.042s = 25,920$$
$$1.042s = 25,920$$
$$s \approx 24,875.24$$

Her first year salary was approximately $24,875.24.

27.   Solving the inequality:

$$-8a > -4$$
$$a < \frac{1}{2}$$

The solution set is $\left(-\infty, \frac{1}{2}\right)$.

28.   Solving the inequality:

$$6 - a \geq -2$$
$$-a \geq -8$$
$$a \leq 8$$

The solution set is $(-\infty, 8]$.

29.   Solving the inequality:

$$\frac{3}{4}x + 1 \leq 10$$
$$\frac{3}{4}x \leq 9$$
$$3x \leq 36$$
$$x \leq 12$$

The solution set is $(-\infty, 12]$.

30.   Solving the inequality:

$$800 - 200x < 1000$$
$$-200x < 200$$
$$x > -1$$

The solution set is $(-1, \infty)$.

31.   Solving the inequality:

$$\frac{1}{3} \leq \frac{1}{6}x \leq 1$$
$$2 \leq x \leq 6$$

The solution set is $[2, 6]$.

32.   Solving the inequality:

$$-0.01 \leq 0.02x - 0.01 \leq 0.01$$
$$0 \leq 0.02x \leq 0.02$$
$$0 \leq x \leq 1$$

The solution set is $[0, 1]$.

33. Solving the inequality:

$$5t + 1 \le 3t - 2 \quad \text{or} \quad -7t \le -21$$
$$2t \le -3 \qquad\qquad t \ge 3$$
$$t \le -\frac{3}{2} \qquad\qquad t \ge 3$$

The solution set is $\left(-\infty, -\frac{3}{2}\right] \cup [3, \infty)$.

34. Solving the inequality:

$$3(x + 1) < 2(x + 2) \quad \text{or} \quad 2(x - 1) \ge x + 2$$
$$3x + 3 < 2x + 4 \qquad\qquad 2x - 2 \ge x + 2$$
$$x + 3 < 4 \qquad\qquad\qquad x - 2 \ge 2$$
$$x < 1 \qquad\qquad\qquad\qquad x \ge 4$$

The solution set is $(-\infty, 1) \cup [4, \infty)$.

35. Solving the equation:

$$|x| = 2$$
$$x = -2, 2$$

36. Solving the equation:

$$|a| - 3 = 1$$
$$|a| = 4$$
$$a = -4, 4$$

37. Solving the equation:

$$|x - 3| = 1$$
$$x - 3 = -1, 1$$
$$x = 2, 4$$

38. Solving the equation:

$$|2y - 3| = 5$$
$$2y - 3 = -5, 5$$
$$2y = -2, 8$$
$$y = -1, 4$$

39. Solving the equation:

$$|4x - 3| + 2 = 11$$
$$|4x - 3| = 9$$
$$4x - 3 = -9, 9$$
$$4x = -6, 12$$
$$x = -\frac{3}{2}, 3$$

40. Solving the equation:

$$\left|\frac{7}{3} - \frac{x}{3}\right| + \frac{4}{3} = 2$$
$$\left|\frac{7}{3} - \frac{x}{3}\right| = \frac{2}{3}$$
$$\frac{7}{3} - \frac{x}{3} = -\frac{2}{3}, \frac{2}{3}$$
$$7 - x = -2, 2$$
$$-x = -9, -5$$
$$x = 5, 9$$

41. Solving the equation:

$$|5t - 3| = |3t - 5|$$
$$5t - 3 = 3t - 5 \quad \text{or} \quad 5t - 3 = -3t + 5$$
$$2t = -2 \qquad\qquad\qquad 8t = 8$$
$$t = -1 \qquad\qquad\qquad t = 1$$

42. Solving the equation:

$$\left|\frac{1}{2} - x\right| = \left|x + \frac{1}{2}\right|$$
$$\frac{1}{2} - x = x + \frac{1}{2} \quad \text{or} \quad \frac{1}{2} - x = -x - \frac{1}{2}$$
$$-2x = 0 \qquad\qquad\qquad 0 = -1$$
$$x = 0 \qquad\qquad\qquad x = \text{impossible}$$

43. Solving the inequality:

$$|x| < 5$$
$$-5 < x < 5$$

Graphing the solution set:

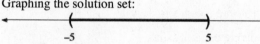

44. Solving the inequality:

$$|0.01a| \ge 5$$
$$0.01a \le -5 \quad \text{or} \quad 0.01a \ge 5$$
$$a \le -500 \qquad\qquad a \ge 500$$

Graphing the solution set:

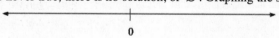

45. Since the inequality $|x| < 0$ is never true, there is no solution, or $\varnothing$. Graphing the solution set:

46. Solving the inequality:

$$|2t + 1| - 3 < 2$$
$$|2t + 1| < 5$$
$$-5 < 2t + 1 < 5$$
$$-6 < 2t < 4$$
$$-3 < t < 2$$

Graphing the solution set:

47.   Solving the equation:

$$2x - 3 = 2(x - 3)$$
$$2x - 3 = 2x - 6$$
$$-3 = -6$$

Since this statement is false, there is no solution, or $\varnothing$.

48.   Solving the equation:

$$3\left(5x - \tfrac{1}{2}\right) = 15x + 2$$
$$15x - \tfrac{3}{2} = 15x + 2$$
$$-\tfrac{3}{2} = 2$$

Since this statement is false, there is no solution, or $\varnothing$.

49.   Since $|4y + 8| = -1$ is never true, there is no solution, or $\varnothing$.

50.   Since $|x| > 0$ except when $x = 0$, the solution set is all real numbers except 0.

51.   Solving the inequality:

$$|5 - 8t| + 4 \le 1$$
$$|5 - 8t| \le -3$$

Since this statement is never true, there is no solution, or $\varnothing$.

52.   Since $|2x + 1| \ge -4$ is always true, the solution set is all real numbers.

Chapter 3
Equations and Inequalities in Two Variables

3.1 Paired Data and the Rectangular Coordinate System

1. Plotting the points:

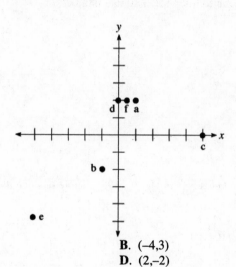

3. A. (4,1)   B. (–4,3)
    C. (–2,–5)   D. (2,–2)
    E. (0,5)   F. (–4,0)
    G. (1,0)

5. The x-intercept is 3 and the y-intercept is –2.

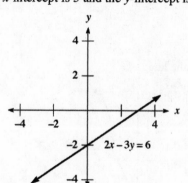

7. The x-intercept is 5 and the y-intercept is –4.

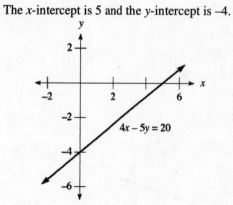

39

9.   The x-intercept is $-\frac{3}{2}$ and the y-intercept is 3.

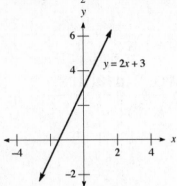

11.   The x-intercept is -4 and the y-intercept is 6.

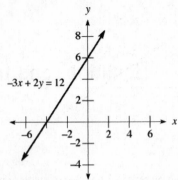

13.   The x-intercept is $\frac{10}{3}$ and the y-intercept is -4.

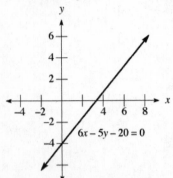

15.   The x-intercept is $\frac{5}{3}$ and the y-intercept is -5.

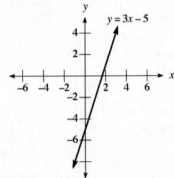

17.   The x-intercept is 2 and the y-intercept is 3.

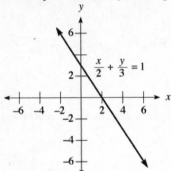

19.   Table b, since its values match the equation.

21.   Graphing the line:

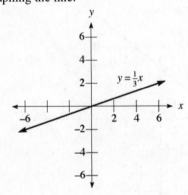

23.   Graphing the line:

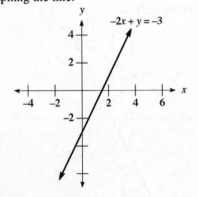

25. Graphing the line:

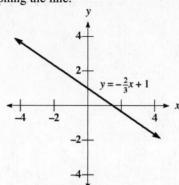

$y = -\frac{2}{3}x + 1$

27. Graphing the line:

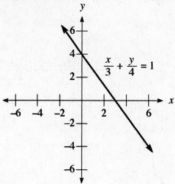

$\frac{x}{3} + \frac{y}{4} = 1$

29. Since the *x*-intercept is 3 and the *y*-intercept is –2, this is the graph of **b**.

31. **a.** Solving the equation:

$$4x + 12 = -16$$
$$4x = -28$$
$$x = -7$$

b. Substituting $y = 0$:

$$4x + 12(0) = -16$$
$$4x = -16$$
$$x = -4$$

c. Substituting $x = 0$:

$$4(0) + 12y = -16$$
$$12y = -16$$
$$y = -\frac{4}{3}$$

d. Graphing the line:

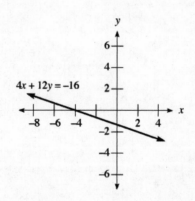

$4x + 12y = -16$

e. Solving for *y*:

$$4x + 12y = -16$$
$$12y = -4x - 16$$
$$y = -\frac{1}{3}x - \frac{4}{3}$$

33. **a.** Graphing the line:

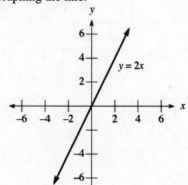

$y = 2x$

b. Graphing the line:

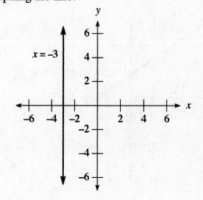

$x = -3$

c.   Graphing the line:

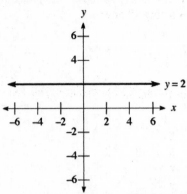

35.   a.   Graphing the line:

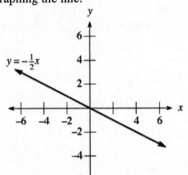

b.   Graphing the line:

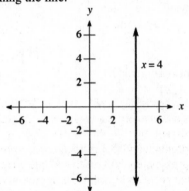

c.   Graphing the line:

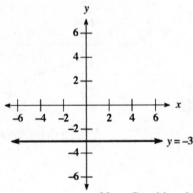

37.   Graphing the line:

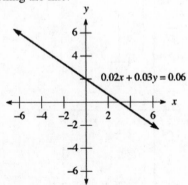

39.   Graphing the line:

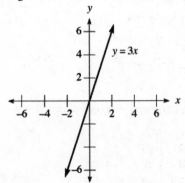

41. **a.** Three ordered pairs are (5,40), (10,80), and (20,160).

  b. She will earn (8)(40) = $320 for working 40 hours.

  c. Solving the equation:
$$8x = 240$$
$$x = 30$$
  She worked 30 hours that week.

  d. She should earn (8)(35) = $280 for working 35 hours, not $260. No, the paycheck was not the correct amount.

43. Constructing a line graph:

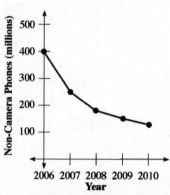

45. Five ordered pairs are (1985,20.2), (1990,34.4), (1995,44.8), (2000,65.4), and (2004,94.4).

47. The coordinates of A are $(6 - 5,2) = (1,2)$ and of B are $(6,2+5) = (6,7)$.

49. The coordinates of A are $(8 - 5,3) = (3,3)$, of B are $(3,3 + 3) = (3,6)$, and of C are $(8,3 + 3) = (8,6)$.

51. **a.** Approximately 35% of the people were online in 1998.

  b. Approximately 70% of the people were online in 2003.

  c. In the years 2000 and 2001, approximately 65% of the people were online.

53. Solving the equation:
$$5x - 4 = -3x + 12$$
$$8x - 4 = 12$$
$$8x = 16$$
$$x = 2$$

55. Solving the equation:
$$\tfrac{1}{2} - \tfrac{1}{8}(3t - 4) = -\tfrac{7}{8}t$$
$$8\left(\tfrac{1}{2} - \tfrac{1}{8}(3t - 4)\right) = 8\left(-\tfrac{7}{8}t\right)$$
$$4 - (3t - 4) = -7t$$
$$4 - 3t + 4 = -7t$$
$$4t + 8 = 0$$
$$4t = -8$$
$$t = -2$$

57. Solving the equation:
$$50 = \frac{K}{24}$$
$$K = 24 \bullet 50 = 1,200$$

59. Solving the equation:
$$2(1) + y = 4$$
$$2 + y = 4$$
$$y = 2$$

61. Solving the equation:
$$4\left(\tfrac{19}{15}\right) - 2y = 4$$
$$\tfrac{76}{15} - 2y = 4$$
$$-2y = -\tfrac{16}{15}$$
$$y = \tfrac{8}{15}$$

63. Writing as a fraction: $-0.06 = -\frac{6}{100}$

65. Substituting $x = 2$:
$$y = 2(2) - 3$$
$$y = 4 - 3$$
$$y = 1$$

67. Simplifying: $\dfrac{1 - (-3)}{-5 - (-2)} = \dfrac{4}{-3} = -\dfrac{4}{3}$

69. Simplifying: $\dfrac{-1 - 4}{3 - 3} = \dfrac{-5}{0}$, which is undefined

71.   **a.**   The number is $\frac{3}{2}$, since $\frac{2}{3} \bullet \frac{3}{2} = 1$.   **b.**   The number is $-\frac{3}{2}$, since $\frac{2}{3} \bullet \left(-\frac{3}{2}\right) = -1$.

73.   Setting $y = 0$, the x-intercept is $\dfrac{c}{a}$. Setting $x = 0$, the y-intercept is $\dfrac{c}{b}$.

75.   Setting $y = 0$, the x-intercept is a. Setting $x = 0$, the y-intercept is b.

3.2 The Slope of a Line

1.   The slope is $\frac{3}{2}$.   **3.**   There is no slope (undefined).

5.   The slope is $\frac{2}{3}$.

7.   Finding the slope: $m = \dfrac{4-1}{4-2} = \dfrac{3}{2}$ **9.**   Finding the slope: $m = \dfrac{2-4}{5-1} = \dfrac{-2}{4} = -\dfrac{1}{2}$

Sketching the graph: Sketching the graph:

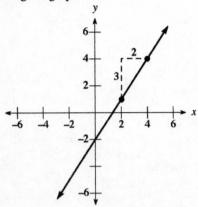 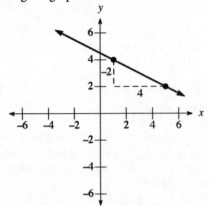

11.   Finding the slope: $m = \dfrac{2-(-3)}{4-1} = \dfrac{2+3}{3} = \dfrac{5}{3}$ **13.**   Finding the slope: $m = \dfrac{3-(-2)}{1-(-3)} = \dfrac{3+2}{1+3} = \dfrac{5}{4}$

Sketching the graph: Sketching the graph:

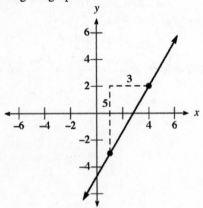 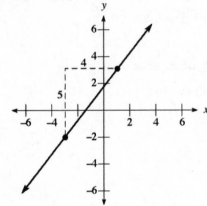

15. Finding the slope: $m = \dfrac{-2-2}{3-(-3)} = \dfrac{-4}{3+3} = -\dfrac{2}{3}$

Sketching the graph:

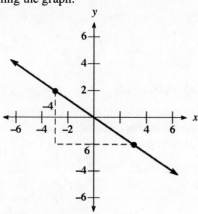

17. Finding the slope: $m = \dfrac{-2-(-5)}{3-2} = \dfrac{-2+5}{1} = 3$

Sketching the graph:

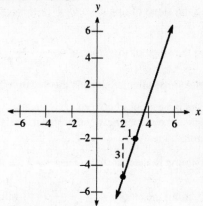

19. Using the slope formula:

$$\dfrac{-2-2}{x-5} = 2$$
$$-4 = 2(x-5)$$
$$-4 = 2x-10$$
$$6 = 2x$$
$$x = 3$$

21. Using the slope formula:

$$\dfrac{3-6}{a-2} = -1$$
$$-3 = -1(a-2)$$
$$-3 = -a+2$$
$$-5 = -a$$
$$a = 5$$

23. Using the slope formula:

$$\dfrac{4b-b}{-1-2} = -2$$
$$3b = -2(-3)$$
$$3b = 6$$
$$b = 2$$

25. Completing the table:

x	y
0	2
3	0

Finding the slope: $m = \dfrac{2-0}{0-3} = -\dfrac{2}{3}$

27. Completing the table:

x	y
0	−5
3	−3

Finding the slope: $m = \dfrac{-5-(-3)}{0-3} = \dfrac{-5+3}{-3} = \dfrac{2}{3}$

29. Graphing the line:

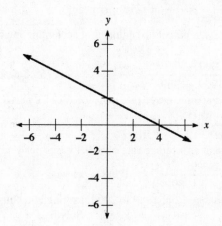

The slope is $m = -\dfrac{1}{2}$.

31. Finding the slope of this line: $m = \dfrac{1-3}{-8-2} = \dfrac{-2}{-10} = \dfrac{1}{5}$. Since the parallel slope is the same, its slope is $\dfrac{1}{5}$.

33. Finding the slope of this line: $m = \dfrac{2-(-6)}{5-5} = \dfrac{8}{0}$, which is undefined

    Since the perpendicular slope is a horizontal line, its slope is 0.

35. Finding the slope of this line: $m = \dfrac{-5-1}{4-(-2)} = \dfrac{-6}{6} = -1$. Since the parallel slope is the same, its slope is -1.

37. Finding the slope of this line: $m = \dfrac{-3-(-5)}{1-(-2)} = \dfrac{2}{3}$

    Since the perpendicular slope is the negative reciprocal, its slope is $-\dfrac{3}{2}$.

39. a. Since the slopes between each successive pairs of points is 2, this could represent ordered pairs from a line.
    b. Since the slopes between each successive pairs of points is not the same, this could not represent ordered pairs from a line.

41. a. Finding the slope: $m = \dfrac{105-0}{6-0} = 17.5$ miles/hour

    b. Finding the slope: $m = \dfrac{80-0}{2-0} = 40$ kilometers/hour

    c. Finding the slope: $m = \dfrac{3600-0}{30-0} = 120$ feet/second

    d. Finding the slope: $m = \dfrac{420-0}{15-0} = 28$ meters/minute

43. For line segment A, the slope is: $m = \dfrac{250-300}{2007-2006} = -50$ million cameras/year

    For line segment B, the slope is: $m = \dfrac{175-250}{2008-2007} = -75$ million cameras/year

    For line segment C, the slope is: $m = \dfrac{125-150}{2010-2009} = -25$ million cameras/year

45. a. Let d represent the distance. Using the slope formula:
    $$\dfrac{0-1106}{d-0} = -\dfrac{7}{100}$$
    $$-7d = -110600$$
    $$d = 15800$$
    The distance to point A is 15,800 feet.

    b. The slope is $-\dfrac{7}{100} = -0.07$.

47. Computing the slope: $m = \dfrac{74\% - 9\%}{2005-1995} = \dfrac{65\%}{10} = 6.5\%$ per year

    Over the 10 years from 1995 to 2005, the percent of adults that access the internet has increased at an average rate of 6.5% per year.

49. a. Computing the slope: $m = \dfrac{10,000-57,686}{1979-1969} = \dfrac{-47,686}{10} = -4,768.6$ cases per year

    New cases of rubella decreased at an average rate of 4,768.6 cases per year from 1969 to 1979.

    b. Computing the slope: $m = \dfrac{9-10,000}{1989-1979} = \dfrac{-9991}{10} = -999.1$ cases per year

    New cases of rubella decreased at an average rate of 999.1 cases per year from 1979 to 1989.

    c. Computing the slope: $m = \dfrac{9-9}{2004-1989} = \dfrac{0}{15} = 0$ cases per year

    New cases of rubella did not change from 1989 to 2004, staying at a constant 9 cases per year.

51. Substituting $x = 4$:
    $$3(4)+2y=12$$
    $$12+2y=12$$
    $$2y=0$$
    $$y=0$$

53. Solving for y:
    $$3x+2y=12$$
    $$2y=-3x+12$$
    $$y=-\tfrac{3}{2}x+6$$

55. Solving for t:
$$A = P + Prt$$
$$A - P = Prt$$
$$t = \frac{A - P}{Pr}$$

57. Simplifying: $2\left(-\frac{1}{2}\right) = -1$

59. Simplifying: $\dfrac{5 - (-3)}{2 - 6} = \dfrac{8}{-4} = -2$

61. Solving for y:
$$\frac{y - b}{x - 0} = m$$
$$y - b = mx$$
$$y = mx + b$$

63. Solving for y:
$$y - 3 = -2(x + 4)$$
$$y - 3 = -2x - 8$$
$$y = -2x - 5$$

65. Solving for y: $y = -\frac{4}{3}(0) + 5 = 0 + 5 = 5$

67. Setting the slope equal to -1:
$$\frac{3x - 2 - 2}{x - 1} = -1$$
$$\frac{3x - 4}{x - 1} = -1$$
$$3x - 4 = -x + 1$$
$$4x = 5$$
$$x = \tfrac{5}{4}$$
$$y = 3\left(\tfrac{5}{4}\right) - 2 = \tfrac{15}{4} - \tfrac{8}{4} = \tfrac{7}{4}$$

The point is $\left(\tfrac{5}{4}, \tfrac{7}{4}\right)$.

69. Graphing the curves:

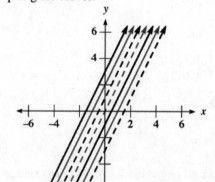

71. Graphing the curves:

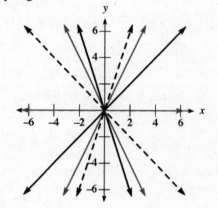

73. Graphing the curves:

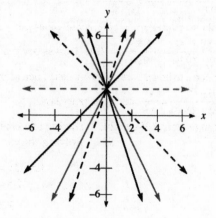

3.3 The Equation of a Line

1. Using the slope-intercept formula: $y = 2x + 3$

3. Using the slope-intercept formula: $y = x - 5$

5. Using the slope-intercept formula: $y = \frac{1}{2}x + \frac{3}{2}$

7. Using the slope-intercept formula: $y = 4$

9. **a.** The parallel slope is 3.

b. The perpendicular slope is $-\frac{1}{3}$.

11. Solving for y:

$$3x + y = -2$$
$$y = -3x - 2$$

a. The parallel slope is –3.

b. The perpendicular slope is $\frac{1}{3}$.

13. Solving for y:

$$2x + 5y = -11$$
$$5y = -2x - 11$$
$$y = -\frac{2}{5}x - \frac{11}{5}$$

a. The parallel slope is $-\frac{2}{5}$.

b. The perpendicular slope is $\frac{5}{2}$.

15. The slope is 3, the y-intercept is –2, and the perpendicular slope is $-\frac{1}{3}$.

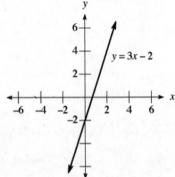

17. The slope is $\frac{2}{3}$, the y-intercept is –4, and the perpendicular slope is $-\frac{3}{2}$.

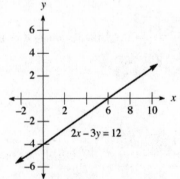

19. The slope is $-\frac{4}{5}$, the y-intercept is 4, and the perpendicular slope is $\frac{5}{4}$.

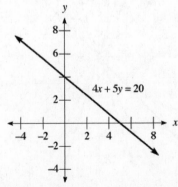

21.   The slope is $\frac{1}{2}$ and the y-intercept is -4. Using the slope-intercept form, the equation is $y = \frac{1}{2}x - 4$.

23.   The slope is $-\frac{2}{3}$ and the y-intercept is 3. Using the slope-intercept form, the equation is $y = -\frac{2}{3}x + 3$.

25.   Using the point-slope formula:

$$\begin{aligned} y - (-5) &= 2(x - (-2)) \\ y + 5 &= 2(x + 2) \\ y + 5 &= 2x + 4 \\ y &= 2x - 1 \end{aligned}$$

27.   Using the point-slope formula:

$$\begin{aligned} y - 1 &= -\frac{1}{2}(x - (-4)) \\ y - 1 &= -\frac{1}{2}(x + 4) \\ y - 1 &= -\frac{1}{2}x - 2 \\ y &= -\frac{1}{2}x - 1 \end{aligned}$$

29.   Using the point-slope formula:

$$\begin{aligned} y - 2 &= -3\left(x - \left(-\frac{1}{3}\right)\right) \\ y - 2 &= -3\left(x + \frac{1}{3}\right) \\ y - 2 &= -3x - 1 \\ y &= -3x + 1 \end{aligned}$$

31.   Using the point-slope formula:

$$\begin{aligned} y - (-2) &= \frac{2}{3}(x - (-4)) \\ y + 2 &= \frac{2}{3}(x + 4) \\ y + 2 &= \frac{2}{3}x + \frac{8}{3} \\ y &= \frac{2}{3}x + \frac{2}{3} \end{aligned}$$

33.   Using the point-slope formula:

$$\begin{aligned} y - 2 &= -\frac{1}{4}(x - (-5)) \\ y - 2 &= -\frac{1}{4}(x + 5) \\ y - 2 &= -\frac{1}{4}x - \frac{5}{4} \\ y &= -\frac{1}{4}x + \frac{3}{4} \end{aligned}$$

35.   First find the slope: $m = \dfrac{-1 - (-4)}{1 - (-2)} = \dfrac{-1 + 4}{1 + 2} = \dfrac{3}{3} = 1$. Using the point-slope formula:

$$\begin{aligned} y - (-1) &= 1(x - 1) \\ y + 1 &= x - 1 \\ -x + y &= -2 \\ x - y &= 2 \end{aligned}$$

37.   First find the slope: $m = \dfrac{1 - (-5)}{2 - (-1)} = \dfrac{1 + 5}{2 + 1} = \dfrac{6}{3} = 2$. Using the point-slope formula:

$$\begin{aligned} y - 1 &= 2(x - 2) \\ y - 1 &= 2x - 4 \\ -2x + y &= -3 \\ 2x - y &= 3 \end{aligned}$$

39.   First find the slope: $m = \dfrac{-1 - \left(-\frac{1}{5}\right)}{-\frac{1}{3} - \frac{1}{3}} = \dfrac{-1 + \frac{1}{5}}{-\frac{2}{3}} = \dfrac{-\frac{4}{5}}{-\frac{2}{3}} = \frac{4}{5} \cdot \frac{3}{2} = \frac{6}{5}$. Using the point-slope formula:

$$\begin{aligned} y - (-1) &= \frac{6}{5}\left(x - \left(-\frac{1}{3}\right)\right) \\ y + 1 &= \frac{6}{5}\left(x + \frac{1}{3}\right) \\ 5(y + 1) &= 6\left(x + \frac{1}{3}\right) \\ 5y + 5 &= 6x + 2 \\ 6x - 5y &= 3 \end{aligned}$$

41. **a.** For the x-intercept, substitute $y = 0$:
$$3x - 2(0) = 10$$
$$3x = 10$$
$$x = \frac{10}{3}$$

For the y-intercept, substitute $x = 0$:
$$3(0) - 2y = 10$$
$$-2y = 10$$
$$y = -5$$

b. Substituting $y = 1$:
$$3x - 2(1) = 10$$
$$3x - 2 = 10$$
$$3x = 12$$
$$x = 4$$
Another solution is (4,1). Other answers are possible.

c. Solving for y:
$$3x - 2y = 10$$
$$-2y = -3x + 10$$
$$y = \frac{3}{2}x - 5$$

d. Substituting $x = 2$: $y = \frac{3}{2}(2) - 5 = 3 - 5 = -2$. No, the point (2,2) is not a solution to the equation.

43. **a.** For the x-intercept, substitute $y = 0$:
$$\frac{3x}{4} - \frac{0}{2} = 1$$
$$\frac{3x}{4} = 1$$
$$3x = 4$$
$$x = \frac{4}{3}$$

For the y-intercept, substitute $x = 0$:
$$\frac{3(0)}{4} - \frac{y}{2} = 1$$
$$-\frac{y}{2} = 1$$
$$-y = 2$$
$$y = -2$$

b. Substituting $x = 2$:
$$\frac{3(2)}{4} - \frac{y}{2} = 1$$
$$\frac{3}{2} - \frac{y}{2} = 1$$
$$-\frac{y}{2} = -\frac{1}{2}$$
$$y = 1$$

Another solution is (2,1). Other answers are possible.

c. Solving for y:
$$\frac{3x}{4} - \frac{y}{2} = 1$$
$$4\left(\frac{3x}{4} - \frac{y}{2}\right) = 4(1)$$
$$3x - 2y = 4$$
$$-2y = -3x + 4$$
$$y = \frac{3}{2}x - 2$$

d. Substituting $x = 1$: $y = \frac{3}{2}(1) - 2 = \frac{3}{2} - 2 = -\frac{1}{2}$. No, the point (1,2) is not a solution to the equation.

45.   **a.**   Solving for x:
$$-2x + 1 = -3$$
$$-2x = -4$$
$$x = 2$$

b.   Solving for y:
$$-2x + y = -3$$
$$y = 2x - 3$$

c.   The slope-intercept form is $y = 2x - 3$, so the y-intercept is –3.

d.   The slope-intercept form is $y = 2x - 3$, so the slope is 2.

e.   Sketching the graph:

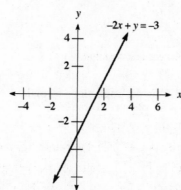

47.   Two points on the line are (0,–4) and (2,0). Finding the slope: $m = \dfrac{0 - (-4)}{2 - 0} = \dfrac{4}{2} = 2$

Using the slope-intercept form, the equation is $y = 2x - 4$.

49.   Two points on the line are (0,4) and (–2,0). Finding the slope: $m = \dfrac{0 - 4}{-2 - 0} = \dfrac{-4}{-2} = 2$

Using the slope-intercept form, the equation is $y = 2x + 4$.

51.   The slope is 0 and the y-intercept is –2.

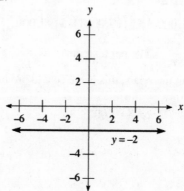

53.   First find the slope:
$$3x - y = 5$$
$$-y = -3x + 5$$
$$y = 3x - 5$$

So the slope is 3. Using (–1,4) in the point-slope formula:
$$y - 4 = 3(x - (-1))$$
$$y - 4 = 3(x + 1)$$
$$y - 4 = 3x + 3$$
$$y = 3x + 7$$

55.   First find the slope:
$$2x - 5y = 10$$
$$-5y = -2x + 10$$
$$y = \tfrac{2}{5}x - 2$$

So the perpendicular slope is $-\tfrac{5}{2}$. Using $(-4,-3)$ in the point-slope formula:
$$y - (-3) = -\tfrac{5}{2}\left(x - (-4)\right)$$
$$y + 3 = -\tfrac{5}{2}(x + 4)$$
$$y + 3 = -\tfrac{5}{2}x - 10$$
$$y = -\tfrac{5}{2}x - 13$$

57.   The perpendicular slope is $\tfrac{1}{4}$. Using $(-1,0)$ in the point-slope formula:
$$y - 0 = \tfrac{1}{4}\left(x - (-1)\right)$$
$$y = \tfrac{1}{4}(x + 1)$$
$$y = \tfrac{1}{4}x + \tfrac{1}{4}$$

59.   Using the points $(3,0)$ and $(0,2)$, first find the slope: $m = \dfrac{2-0}{0-3} = -\tfrac{2}{3}$

Using the slope-intercept formula, the equation is: $y = -\tfrac{2}{3}x + 2$

61.   **a.**   Using the points $(0,32)$ and $(25,77)$, first find the slope: $m = \dfrac{77-32}{25-0} = \dfrac{45}{25} = \tfrac{9}{5}$

Using the slope-intercept formula, the equation is: $F = \tfrac{9}{5}C + 32$

   b.   Substituting $C = 30$: $F = \tfrac{9}{5}(30) + 32 = 54 + 32 = 86°$

63.   **a.**   Substituting $n = 10{,}000$: $C = 125{,}000 + 6.5(10{,}000) = \$190{,}000$

   b.   Finding the average cost: $\dfrac{\$190{,}000}{10{,}000} = \$19$ per textbook

   c.   Since each textbook costs \$6.50 in materials, this is the cost to produce the next textbook.

65.   **a.**   Using $(2002,75)$ and $(2004,144)$: $m = \dfrac{144 - 75}{2004 - 2002} = \tfrac{69}{2}$

Using the point-slope formula:
$$y - 75 = \tfrac{69}{2}(x - 2002)$$
$$y - 75 = \tfrac{69}{2}x - 69{,}069$$
$$y = \tfrac{69}{2}x - 68{,}994$$

   b.   Substituting $x = 2006$: $y = \tfrac{69}{2}(2006) - 68{,}994 = 213$ books

67.   **a.**   Using $(1995,9\%)$ and $(2005,74\%)$, find the slope: $m = \dfrac{74\% - 9\%}{2005 - 1995} = \dfrac{65}{10} = 6.5$

Using the point-slope formula:
$$y - 9 = 6.5(x - 1995)$$
$$y - 9 = 6.5x - 12{,}967.5$$
$$y = 6.5x - 12{,}958.5$$

   b.   Substituting $x = 2006$: $y = 6.5(2006) - 12{,}958.5 = 80.5\%$

   In 2006, 80.5% of adults will be online.

   c.   Substituting $x = 2010$: $y = 6.5(2010) - 12{,}958.5 = 106.5\%$

   This answer is larger than 100%, which is impossible.

69. Let w represent the width and $4w + 3$ represent the length. Using the perimeter formula:
$$2w + 2(4w + 3) = 56$$
$$2w + 8w + 6 = 56$$
$$10w = 50$$
$$w = 5$$
The width is 5 inches and the length is 23 inches.

71. The total amount collected is: $732.50 – $66 = $666.50
Let x represent the sales. Since this amount includes the sales tax:
$$x + 0.075x = 666.50$$
$$1.075x = 666.50$$
$$x = 620$$
The amount which is sales tax is therefore: $666.50 – $620 = $46.50

73. Since $0 + 0 \le 4$ and $4 + 0 \le 4$, but $2 + 3 > 4$, the points (0,0) and (4,0) are solutions.

75. Since $0 \le \frac{1}{2}(0)$ and $0 \le \frac{1}{2}(2)$, but $0 > \frac{1}{2}(-2)$, the points (0,0) and (2,0) are solutions.

77. First find the midpoint: $\left(\dfrac{1+7}{2}, \dfrac{4+8}{2} \right) = \left(\frac{8}{2}, \frac{12}{2} \right) = (4,6)$. Now find the slope: $m = \dfrac{8-4}{7-1} = \frac{4}{6} = \frac{2}{3}$

So the perpendicular slope is $-\frac{3}{2}$. Using (4,6) in the point-slope formula:
$$y - 6 = -\tfrac{3}{2}(x - 4)$$
$$y - 6 = -\tfrac{3}{2}x + 6$$
$$y = -\tfrac{3}{2}x + 12$$

79. First find the midpoint: $\left(\dfrac{-5-1}{2}, \dfrac{1+4}{2} \right) = \left(\dfrac{-6}{2}, \dfrac{5}{2} \right) = \left(-3, \tfrac{5}{2} \right)$. Now find the slope: $m = \dfrac{4-1}{-1+5} = \frac{3}{4}$

So the perpendicular slope is $-\frac{4}{3}$. Using $\left(-3, \tfrac{5}{2} \right)$ in the point-slope formula:
$$y - \tfrac{5}{2} = -\tfrac{4}{3}(x + 3)$$
$$y - \tfrac{5}{2} = -\tfrac{4}{3}x - 4$$
$$y = -\tfrac{4}{3}x - \tfrac{3}{2}$$

3.4 Linear Inequalities in Two Variables

1. Graphing the solution set:

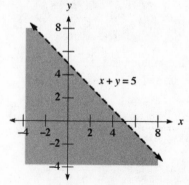

3. Graphing the solution set:

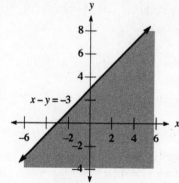

5. Graphing the solution set:

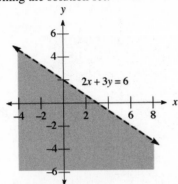

7. Graphing the solution set:

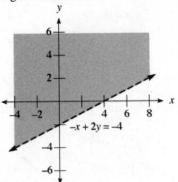

9. Graphing the solution set:

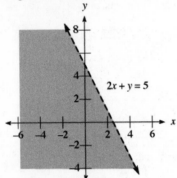

11. Graphing the solution set:

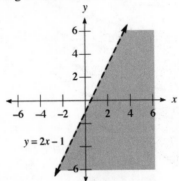

13. Graphing the solution set:

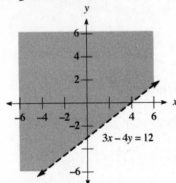

15. Graphing the solution set:

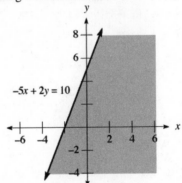

17. Graphing the solution set:

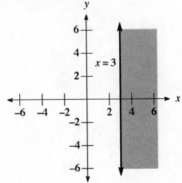

19. Graphing the solution set:

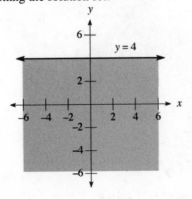

21.   Graphing the solution set:

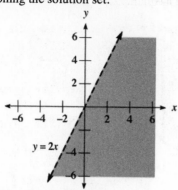

23.   Graphing the solution set:

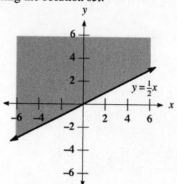

25.   Graphing the solution set:

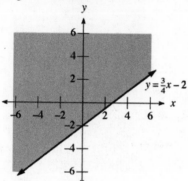

27.   Graphing the solution set:

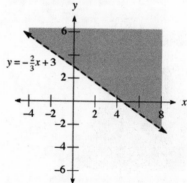

29.   Graphing the solution set:

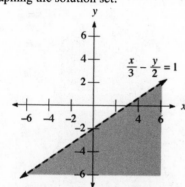

31.   Graphing the solution set:

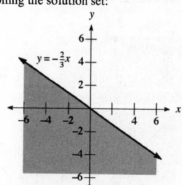

33.   Graphing the solution set:

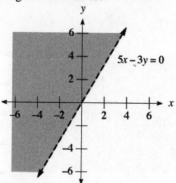

35.   Graphing the solution set:

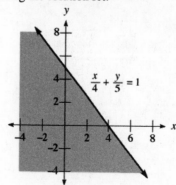

37.   The inequality is $x + y > 4$.

39.   The inequality is $-x + 2y \leq 4$ or $y \leq \frac{1}{2}x + 2$.

41. **a.** Solving the inequality:
$$\frac{1}{3} + \frac{y}{2} < 1$$
$$\frac{y}{2} < \frac{2}{3}$$
$$y < \frac{4}{3}$$

b. Solving the inequality:
$$\frac{1}{3} - \frac{y}{2} < 1$$
$$-\frac{y}{2} < \frac{2}{3}$$
$$y > -\frac{4}{3}$$

c. Solving for y:
$$\frac{x}{3} + \frac{y}{2} = 1$$
$$6\left(\frac{x}{3} + \frac{y}{2}\right) = 6(1)$$
$$2x + 3y = 6$$
$$3y = -2x + 6$$
$$y = -\frac{2}{3}x + 2$$

d. Graphing the inequality:

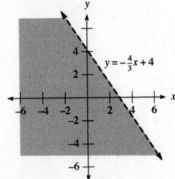

43. Graphing the region:

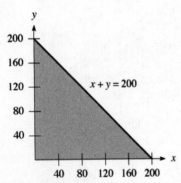

45. Graphing the region:

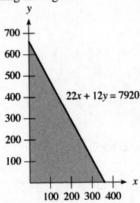

47. Solving the inequality:
$$\frac{1}{3} + \frac{y}{5} \leq \frac{26}{15}$$
$$15\left(\frac{1}{3} + \frac{y}{5}\right) \leq 15\left(\frac{26}{15}\right)$$
$$5 + 3y \leq 26$$
$$3y \leq 21$$
$$y \leq 7$$

49. Solving the inequality:
$$5t - 4 > 3t - 8$$
$$2t - 4 > -8$$
$$2t > -4$$
$$t > -2$$

51. Solving the inequality:
$$-9 < -4 + 5t < 6$$
$$-5 < 5t < 10$$
$$-1 < t < 2$$

53.   Completing the table:

x	y
0	0
10	75
20	150

55.   Completing the table:

x	y
0	0
1	1
1	−1

57.   Graphing the inequality:

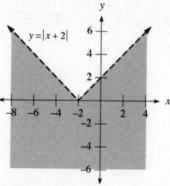

59.   Graphing the inequality:

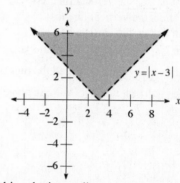

61.   Graphing the inequality:

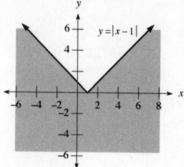

63.   Graphing the inequality:

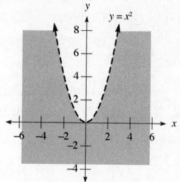

65.   Shading the region:

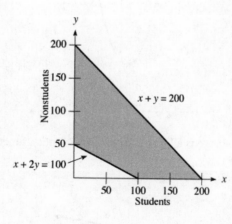

3.5 Introduction to Functions

1. The domain is {1,2,4} and the range is {1,3,5}. This is a function.
3. The domain is {−1,1,2} and the range is {−5,3}. This is a function.
5. The domain is {3,7} and the range is {−1,4}. This is not a function.
7. The domain is {a,b,c,d} and the range is {3,4,5}. This is a function.
9. The domain is {a} and the range is {1,2,3,4}. This is not a function.
11. Yes, since it passes the vertical line test.   **13.**   No, since it fails the vertical line test.
15. No, since it fails the vertical line test.   **17.**   Yes, since it passes the vertical line test.
19. Yes, since it passes the vertical line test.
21. The domain is $\{x \mid -5 \le x \le 5\}$ and the range is $\{y \mid 0 \le y \le 5\}$.
23. The domain is $\{x \mid -5 \le x \le 3\}$ and the range is $\{y \mid y = 3\}$.
25. The domain is all real numbers and the range is $\{y \mid y \ge -1\}$. This is a function.

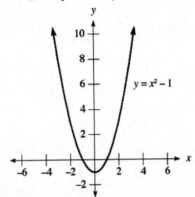

27. The domain is all real numbers and the range is $\{y \mid y \ge 4\}$. This is a function.

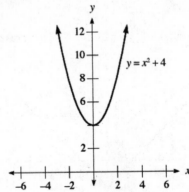

29. The domain is $\{x \mid x \ge -1\}$ and the range is all real numbers. This is not a function.

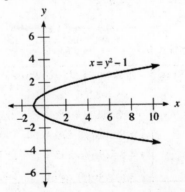

31. The domain is $\{x \mid x \geq 4\}$ and the range is all real numbers. This is not a function.

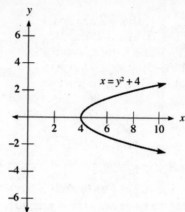

33. The domain is all real numbers and the range is $\{y \mid y \geq 0\}$. This is a function.

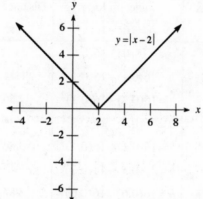

35. The domain is all real numbers and the range is $\{y \mid y \geq -2\}$. This is a function.

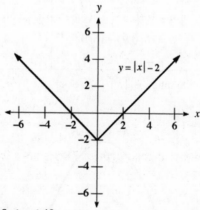

37. **a.** The equation is $y = 8.5x$ for $10 \leq x \leq 40$.

b. Completing the table:

Hours Worked	Function Rule	Gross Pay ($)
x	$y = 8.5x$	y
10	$y = 8.5(10) = 85$	85
20	$y = 8.5(20) = 170$	170
30	$y = 8.5(30) = 255$	255
40	$y = 8.5(40) = 340$	340

c.   Constructing a line graph:

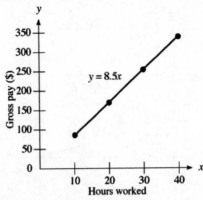

d.   The domain is $\{x \mid 10 \le x \le 40\}$ and the range is $\{y \mid 85 \le y \le 340\}$.

e.   The minimum is $85 and the maximum is $340.

Time (sec)	Function Rule	Distance (ft)
t	$h = 16t - 16t^2$	h
0	$h = 16(0) - 16(0)^2$	0
0.1	$h = 16(0.1) - 16(0.1)^2$	1.44
0.2	$h = 16(0.2) - 16(0.2)^2$	2.56
0.3	$h = 16(0.3) - 16(0.3)^2$	3.36
0.4	$h = 16(0.4) - 16(0.4)^2$	3.84
0.5	$h = 16(0.5) - 16(0.5)^2$	4
0.6	$h = 16(0.6) - 16(0.6)^2$	3.84
0.7	$h = 16(0.7) - 16(0.7)^2$	3.36
0.8	$h = 16(0.8) - 16(0.8)^2$	2.56
0.9	$h = 16(0.9) - 16(0.9)^2$	1.44
1	$h = 16(1) - 16(1)^2$	0

39.   a.   Completing the table: (rows shown at 0.4)

b.   The domain is $\{t \mid 0 \le t \le 1\}$ and the range is $\{h \mid 0 \le h \le 4\}$.

c.   Graphing the function:

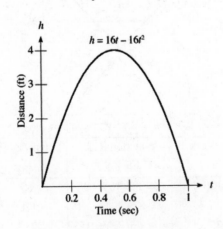

41.   **a.**   Graphing the function:

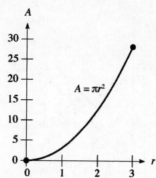

b.   The domain is $\{r \mid 0 \le r \le 3\}$ and the range is $\{A \mid 0 \le A \le 9\pi\}$.

43.   **a.**   Yes, since it passes the vertical line test.

b.   The domain is $\{t \mid 0 \le t \le 6\}$ and the range is $\{h \mid 0 \le h \le 60\}$.

c.   At time $t = 3$ the ball reaches its maximum height.

d.   The maximum height is $h = 60$.   **e.**   At time $t = 6$ the ball hits the ground.

45.   **a.**   Figure III   **b.**   Figure I

c.   Figure II   **d.**   Figure IV

47.   Substituting $x = 4$: $y = 3(4) - 2 = 12 - 2 = 10$ **49.**   Substituting $x = -4$: $y = 3(-4) - 2 = -12 - 2 = -14$

51.   Substituting $x = 2$: $y = (2)^2 - 3 = 4 - 3 = 1$ **53.**   Substituting $x = 0$: $y = (0)^2 - 3 = 0 - 3 = -3$

55.   Solving for y: **57.**   Substituting $x = 0$ and $y = 0$:

$$\frac{8}{5} - 2y = 4$$
$$-2y = \frac{12}{5}$$
$$y = -\frac{6}{5}$$

$$0 = a(0 - 80)^2 + 70$$
$$0 = 6400a + 70$$
$$6400a = -70$$
$$a = -\frac{7}{640}$$

59.   Simplifying: $7.5(20) = 150$ **61.**   Simplifying: $4(3.14)(9) \approx 113$

63.   Simplifying: $4(-2) - 1 = -8 - 1 = -9$

65.   **a.**   Substituting $t = 10$: $s = \dfrac{60}{10} = 6$ **b.**   Substituting $t = 8$: $s = \dfrac{60}{8} = 7.5$

67.   **a.**   Substituting $x = 5$: $(5)^2 + 2 = 25 + 2 = 27$ **b.**   Substituting $x = -2$: $(-2)^2 + 2 = 4 + 2 = 6$

69.   The domain is all real numbers and the range is $\{y \mid y \le 5\}$. This is a function.

71.   The domain is $\{x \mid x \geq 3\}$ and the range is all real numbers. This is not a function.

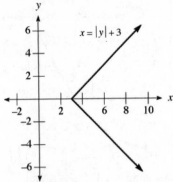

73.   The domain is $\{x \mid -4 \leq x \leq 4\}$ and the range is $\{y \mid -4 \leq y \leq 4\}$. This is not a function.

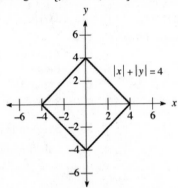

3.6 Function Notation

1.   Evaluating the function: $f(2) = 2(2) - 5 = 4 - 5 = -1$

3.   Evaluating the function: $f(-3) = 2(-3) - 5 = -6 - 5 = -11$

5.   Evaluating the function: $g(-1) = (-1)^2 + 3(-1) + 4 = 1 - 3 + 4 = 2$

7.   Evaluating the function: $g(-3) = (-3)^2 + 3(-3) + 4 = 9 - 9 + 4 = 4$

9.   First evaluate each function:
$$g(4) = (4)^2 + 3(4) + 4 = 16 + 12 + 4 = 32 \qquad f(4) = 2(4) - 5 = 8 - 5 = 3$$
Now evaluating: $g(4) + f(4) = 32 + 3 = 35$

11.   First evaluate each function:
$$f(3) = 2(3) - 5 = 6 - 5 = 1 \qquad\qquad g(2) = (2)^2 + 3(2) + 4 = 4 + 6 + 4 = 14$$
Now evaluating: $f(3) - g(2) = 1 - 14 = -13$

13.   Evaluating the function: $f(0) = 3(0)^2 - 4(0) + 1 = 0 - 0 + 1 = 1$

15.   Evaluating the function: $g(-4) = 2(-4) - 1 = -8 - 1 = -9$

17.   Evaluating the function: $f(-1) = 3(-1)^2 - 4(-1) + 1 = 3 + 4 + 1 = 8$

19.   Evaluating the function: $g(10) = 2(10) - 1 = 20 - 1 = 19$

21.   Evaluating the function: $f(3) = 3(3)^2 - 4(3) + 1 = 27 - 12 + 1 = 16$

23.   Evaluating the function: $g\left(\frac{1}{2}\right) = 2\left(\frac{1}{2}\right) - 1 = 1 - 1 = 0$ **25.**   Evaluating the function: $f(a) = 3a^2 - 4a + 1$

27. $f(1) = 4$

29. $g\left(\frac{1}{2}\right) = 0$

31. $g(-2) = 2$

33. Evaluating the function: $f(0) = 2(0)^2 - 8 = 0 - 8 = -8$

35. Evaluating the function: $g(-4) = \frac{1}{2}(-4) + 1 = -2 + 1 = -1$

37. Evaluating the function: $f(a) = 2a^2 - 8$

39. Evaluating the function: $f(b) = 2b^2 - 8$

41. Evaluating the function: $f\big[g(2)\big] = f\left[\frac{1}{2}(2) + 1\right] = f(2) = 2(2)^2 - 8 = 8 - 8 = 0$

43. Evaluating the function: $g\big[f(-1)\big] = g\left[2(-1)^2 - 8\right] = g(-6) = \frac{1}{2}(-6) + 1 = -3 + 1 = -2$

45. Evaluating the function: $g\big[f(0)\big] = g\left[2(0)^2 - 8\right] = g(-8) = \frac{1}{2}(-8) + 1 = -4 + 1 = -3$

47. Graphing the function:

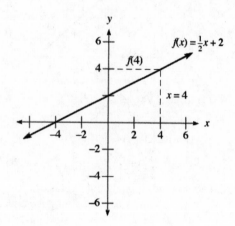

49. Finding where $f(x) = x$:

$$\frac{1}{2}x + 2 = x$$
$$2 = \frac{1}{2}x$$
$$x = 4$$

51. Graphing the function:

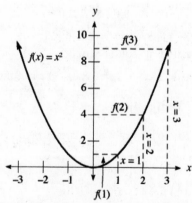

53. Evaluating: $V(3) = 150 \cdot 2^{3/3} = 150 \cdot 2 = 300$; The painting is worth \$300 in 3 years.

Evaluating: $V(6) = 150 \cdot 2^{6/3} = 150 \cdot 4 = 600$; The painting is worth \$600 in 6 years.

55. **a.** True   **b.** True

   c. True   **d.** False

   e. True

57. **a.** Evaluating: $V(3.75) = -3300(3.75) + 18000 = \$5,625$

b. Evaluating: $V(5) = -3300(5) + 18000 = \$1,500$

c. The domain of this function is $\{t \mid 0 \le t \le 5\}$.

d. Sketching the graph:

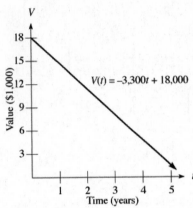

$V(t) = -3,300t + 18,000$

e. The range of this function is $\{V(t) \mid 1,500 \le V(t) \le 18,000\}$.

f. Solving $V(t) = 10000$:

$$-3300t + 18000 = 10000$$
$$-3300t = -8000$$
$$t \approx 2.42$$

The copier will be worth \$10,000 after approximately 2.42 years.

59. Solving the equation:

$$|3x - 5| = 7$$
$$3x - 5 = -7, 7$$
$$3x = -2, 12$$
$$x = -\tfrac{2}{3}, 4$$

61. Solving the equation:

$$|4y + 2| - 8 = -2$$
$$|4y + 2| = 6$$
$$4y + 2 = -6, 6$$
$$4y = -8, 4$$
$$y = -2, 1$$

63. Solving the equation:

$$5 + |6t + 2| = 3$$
$$|6t + 2| = -2$$

Since this last equation is impossible, there is no solution, or $\varnothing$.

65. Simplifying: $(35x - 0.1x^2) - (8x + 500) = 35x - 0.1x^2 - 8x - 500 = -0.1x^2 + 27x - 500$

67. Simplifying: $(4x^2 + 3x + 2) + (2x^2 - 5x - 6) = 4x^2 + 3x + 2 + 2x^2 - 5x - 6 = 6x^2 - 2x - 4$

69. Simplifying: $(4x^2 + 3x + 2) - (2x^2 - 5x - 6) = 4x^2 + 3x + 2 - 2x^2 + 5x + 6 = 2x^2 + 8x + 8$

71. Simplifying: $0.6(m - 70) = 0.6m - 42$

73. Simplifying: $(4x - 3)(x - 1) = 4x^2 - 4x - 3x + 3 = 4x^2 - 7x + 3$

75. **a.** $f(2) = 2$ **b.** $f(-4) = 0$

77. **a.** Completing the table:

Weight (ounces)	0.6	1.0	1.1	2.5	3.0	4.8	5.0	5.3
Cost (cents)	39	39	63	87	87	135	135	159

b. The letter weighs over 2 ounces, but not over 3 ounces. As an inequality, this can be written as $2 < x \le 3$.

c. The domain is $\{x \mid 0 < x \le 6\}$.   **d.** The range is $\{39, 63, 87, 111, 135, 159\}$.

3.7 Algebra and Composition with Functions

1.   Writing the formula: $f + g = f(x) + g(x) = (4x - 3) + (2x + 5) = 6x + 2$

3.   Writing the formula: $g - f = g(x) - f(x) = (2x + 5) - (4x - 3) = -2x + 8$.

5.   Writing the formula: $g + f = g(x) + f(x) = (x - 2) + (3x - 5) = 4x - 7$

7.   Writing the formula: $g + h = g(x) + h(x) = (x - 2) + (3x^2) = 3x^2 + x - 2$

9.   Writing the formula: $g - f = g(x) - f(x) = (x - 2) - (3x - 5) = -2x + 3$

11. Writing the formula: $fh = f(x) \cdot h(x) = (3x - 5)(3x^2) = 9x^3 - 15x^2$

13. Writing the formula: $h / f = \dfrac{h(x)}{f(x)} = \dfrac{3x^2}{3x - 5}$ 15.   Writing the formula: $f / h = \dfrac{f(x)}{h(x)} = \dfrac{3x - 5}{3x^2}$

17. Writing the formula: $f + g + h = f(x) + g(x) + h(x) = (3x - 5) + (x - 2) + (3x^2) = 3x^2 + 4x - 7$

19. Evaluating: $(f + g)(2) = f(2) + g(2) = (2 \cdot 2 + 1) + (4 \cdot 2 + 2) = 5 + 10 = 15$

21. Evaluating: $(fg)(3) = f(3) \cdot g(3) = (2 \cdot 3 + 1)(4 \cdot 3 + 2) = 7 \cdot 14 = 98$

23. Evaluating: $(h / g)(1) = \dfrac{h(1)}{g(1)} = \dfrac{4(1)^2 + 4(1) + 1}{4(1) + 2} = \dfrac{9}{6} = \dfrac{3}{2}$

25. Evaluating: $(fh)(0) = f(0) \cdot h(0) = (2(0) + 1)(4(0)^2 + 4(0) + 1) = (1)(1) = 1$

27. Evaluating: $(f + g + h)(2) = f(2) + g(2) + h(2) = (2(2) + 1) + (4(2) + 2) + (4(2)^2 + 4(2) + 1) = 5 + 10 + 25 = 40$

29. Evaluating:

$$(h + fg)(3) = h(3) + f(3) \cdot g(3) = (4(3)^2 + 4(3) + 1) + (2(3) + 1) \cdot (4(3) + 2) = 49 + 7 \cdot 14 = 49 + 98 = 147$$

31. **a.**   Evaluating: $(f \circ g)(5) = f(g(5)) = f(5 + 4) = f(9) = 9^2 = 81$

     b.   Evaluating: $(g \circ f)(5) = g(f(5)) = g(5^2) = g(25) = 25 + 4 = 29$

     c.   Evaluating: $(f \circ g)(x) = f(g(x)) = f(x + 4) = (x + 4)^2$

     d.   Evaluating: $(g \circ f)(x) = g(f(x)) = g(x^2) = x^2 + 4$

33. **a.**   Evaluating: $(f \circ g)(0) = f(g(0)) = f(4 \cdot 0 - 1) = f(-1) = (-1)^2 + 3(-1) = 1 - 3 = -2$

     b.   Evaluating: $(g \circ f)(0) = g(f(0)) = g(0^2 + 3 \cdot 0) = g(0) = 4(0) - 1 = -1$

     c.   Evaluating: $(g \circ f)(x) = g(f(x)) = g(x^2 + 3x) = 4(x^2 + 3x) - 1 = 4x^2 + 12x - 1$

35. Evaluating each composition:

$$(f \circ g)(x) = f(g(x)) = f\left(\frac{x + 4}{5}\right) = 5\left(\frac{x + 4}{5}\right) - 4 = x + 4 - 4 = x$$

$$(g \circ f)(x) = g(f(x)) = g(5x - 4) = \frac{5x - 4 + 4}{5} = \frac{5x}{5} = x$$

Thus $(f \circ g)(x) = (g \circ f)(x) = x$.

37. Using the graph: $f(2) + 5 = 1 + 5 = 6$ 39.   Using the graph: $f(-3) + g(-3) = 3 + (-1) = 2$

41. Using the graph: $(f \circ g)(0) = f(g(0)) = f(-3) = 3$ 43.   Using the graph: $f(x) = -3$ when $x = -8$

45. Using the graph: $f(-3) + 2 = 4 + 2 = 6$ 47.   Using the graph: $f(2) + g(2) = 3 + 2 = 5$

49. Using the graph: $(f \circ g)(0) = f(g(0)) = f(2) = 3$ 51.   Using the graph: $f(x) = 1$ when $x = -6$

53.   **a.**   Finding the revenue: $R(x) = x(11.5 - 0.05x) = 11.5x - 0.05x^2$

   b.   Finding the cost: $C(x) = 2x + 200$

   c.   Finding the profit: $P(x) = R(x) - C(x) = (11.5x - 0.05x^2) - (2x + 200) = -0.05x^2 + 9.5x - 200$

   d.   Finding the average cost: $\bar{C}(x) = \dfrac{C(x)}{x} = \dfrac{2x + 200}{x} = 2 + \dfrac{200}{x}$

55.   **a.**   The function is $M(x) = 220 - x$.   **b.**   Evaluating: $M(24) = 220 - 24 = 196$ beats per minute

   c.   The training heart rate function is: $T(M) = 62 + 0.6(M - 62) = 0.6M + 24.8$

   Finding the composition: $T(M(x)) = T(220 - x) = 0.6(220 - x) + 24.8 = 156.8 - 0.6x$

   Evaluating: $T(M(24)) = 156.8 - 0.6(24) \approx 142$ beats per minute

   d.   Evaluating: $T(M(36)) = 156.8 - 0.6(36) \approx 135$ beats per minute

   e.   Evaluating: $T(M(48)) = 156.8 - 0.6(48) \approx 128$ beats per minute

57.   Solving the inequality:

$$|x - 3| < 1$$
$$-1 < x - 3 < 1$$
$$2 < x < 4$$

59.   Solving the inequality:

$$|6 - x| > 2$$

$6 - x < -2$	or	$6 - x > 2$
$-x < -8$		$-x > -4$
$x > 8$		$x < 4$

61.   Solving the inequality:

$$|7x - 1| \le 6$$
$$-6 \le 7x - 1 \le 6$$
$$-5 \le 7x \le 7$$
$$-\frac{5}{7} \le x \le 1$$

63.   Simplifying: $16(3.5)^2 = 16(12.25) = 196$

65.   Simplifying: $\dfrac{180}{45} = 4$

67.   Simplifying: $\dfrac{0.0005(200)}{(0.25)^2} = \dfrac{0.1}{0.0625} = 1.6$

69.   Solving for K:

$$15 = K(5)$$
$$K = 3$$

71.   Solving for K:

$$50 = \dfrac{K}{48}$$
$$K = 50 \bullet 48 = 2,400$$

3.8 Variation

1.   Direct variation

3.   Direct variation

5.   Direct variation

7.   Direct variation

9.   Direct variation

11.   Inverse variation

13.   The variation equation is $y = Kx$. Substituting $x = 2$ and $y = 10$:

$$10 = K \bullet 2$$
$$K = 5$$

So $y = 5x$. Substituting $x = 6$: $y = 5 \bullet 6 = 30$

15.   The variation equation is $y = Kx$. Substituting $x = 4$ and $y = -32$:

$$-32 = K \bullet 4$$
$$K = -8$$

So $y = -8x$. Substituting $y = -40$:

$$-40 = -8x$$
$$x = 5$$

17. The variation equation is $r = \dfrac{K}{s}$. Substituting $s = 4$ and $r = -3$:

$$-3 = \dfrac{K}{4}$$
$$K = -12$$

So $r = \dfrac{-12}{s}$. Substituting $s = 2$: $r = \dfrac{-12}{2} = -6$

19. The variation equation is $r = \dfrac{K}{s}$. Substituting $s = 3$ and $r = 8$:

$$8 = \dfrac{K}{3}$$
$$K = 24$$

So $r = \dfrac{24}{s}$. Substituting $r = 48$:

$$48 = \dfrac{24}{s}$$
$$48s = 24$$
$$s = \dfrac{1}{2}$$

21. The variation equation is $d = Kr^2$. Substituting $r = 5$ and $d = 10$:

$$10 = K \bullet 5^2$$
$$10 = 25K$$
$$K = \dfrac{2}{5}$$

So $d = \dfrac{2}{5}r^2$. Substituting $r = 10$: $d = \dfrac{2}{5}(10)^2 = \dfrac{2}{5} \bullet 100 = 40$

23. The variation equation is $d = Kr^2$. Substituting $r = 2$ and $d = 100$:

$$100 = K \bullet 2^2$$
$$100 = 4K$$
$$K = 25$$

So $d = 25r^2$. Substituting $r = 3$: $d = 25(3)^2 = 25 \bullet 9 = 225$

25. The variation equation is $y = \dfrac{K}{|x|}$. Substituting $x = 3$ and $y = 6$:

$$6 = \dfrac{K}{|3|}$$
$$6 = \dfrac{K}{3}$$
$$K = 18$$

So $y = \dfrac{18}{|x|}$. Substituting $x = 9$: $y = \dfrac{18}{|9|} = \dfrac{18}{9} = 2$

27. The variation equation is $y = \dfrac{K}{|x|}$. Substituting $x = -5$ and $y = 20$:

$$20 = \dfrac{K}{|-5|}$$
$$20 = \dfrac{K}{5}$$
$$K = 100$$

So $y = \dfrac{100}{|x|}$. Substituting $x = 10$: $y = \dfrac{100}{|10|} = \dfrac{100}{10} = 10$

29. The variation equation is $y = \dfrac{K}{x^2}$. Substituting $x = 3$ and $y = 45$:

$$45 = \frac{K}{3^2}$$
$$45 = \frac{K}{9}$$
$$K = 405$$

So $y = \dfrac{405}{x^2}$. Substituting $x = 5$: $y = \dfrac{405}{5^2} = \dfrac{405}{25} = \dfrac{81}{5}$

31. The variation equation is $y = \dfrac{K}{x^2}$. Substituting $x = 3$ and $y = 18$:

$$18 = \frac{K}{3^2}$$
$$18 = \frac{K}{9}$$
$$K = 162$$

So $y = \dfrac{162}{x^2}$. Substituting $x = 2$: $y = \dfrac{162}{2^2} = \dfrac{162}{4} = 40.5$

33. The variation equation is $z = Kxy^2$. Substituting $x = 3$, $y = 3$, and $z = 54$:

$$54 = K(3)(3)^2$$
$$54 = 27K$$
$$K = 2$$

So $z = 2xy^2$. Substituting $x = 2$ and $y = 4$: $z = 2(2)(4)^2 = 64$

35. The variation equation is $z = Kxy^2$. Substituting $x = 1$, $y = 4$, and $z = 64$:

$$64 = K(1)(4)^2$$
$$64 = 16K$$
$$K = 4$$

So $z = 4xy^2$. Substituting $z = 32$ and $y = 1$:

$$32 = 4x(1)^2$$
$$32 = 4x$$
$$x = 8$$

37. Let l represent the length and f represent the force. The variation equation is $l = Kf$. Substituting $f = 5$ and $l = 3$:

$$3 = K \bullet 5$$
$$K = \frac{3}{5}$$

So $l = \frac{3}{5}f$. Substituting $l = 10$:

$$10 = \frac{3}{5}f$$
$$50 = 3f$$
$$f = \frac{50}{3}$$

The force required is $\frac{50}{3}$ pounds.

39.   **a.**   The variation equation is $T = 4P$.

     b.   Graphing the equation:

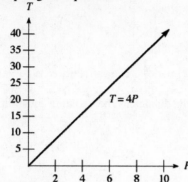

     c.   Substituting $T = 280$:

$$280 = 4P$$
$$P = 70$$

The pressure is 70 pounds per square inch.

41.   Let v represent the volume and p represent the pressure. The variation equation is $v = \dfrac{K}{p}$.

Substituting $p = 36$ and $v = 25$:

$$25 = \frac{K}{36}$$
$$K = 900$$

The equation is $v = \dfrac{900}{p}$. Substituting $v = 75$:

$$75 = \frac{900}{p}$$
$$75p = 900$$
$$p = 12$$

The pressure is 12 pounds per square inch.

43.   **a.**   The variation equation is $f = \dfrac{80}{d}$.

     b.   Graphing the equation:

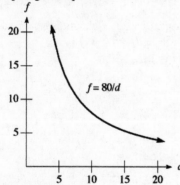

     c.   Substituting $d = 10$:

$$f = \frac{80}{10}$$
$$f = 8$$

The f-stop is 8.

45. Let A represent the surface area, h represent the height, and r represent the radius. The variation equation is $A = Khr$.
Substituting $A = 94$, $r = 3$, and $h = 5$:

$$94 = K(3)(5)$$
$$94 = 15K$$
$$K = \frac{94}{15}$$

The equation is $A = \frac{94}{15}hr$. Substituting $r = 2$ and $h = 8$: $A = \frac{94}{15}(8)(2) = \frac{1504}{15}$. The surface area is $\frac{1504}{15}$ square inches

47. Let R represent the resistance, l represent the length, and d represent the diameter. The variation equation is $R = \frac{Kl}{d^2}$.

Substituting $R = 10$, $l = 100$, and $d = 0.01$:

$$10 = \frac{K(100)}{(0.01)^2}$$
$$0.001 = 100K$$
$$K = 0.00001$$

The equation is $R = \frac{0.00001l}{d^2}$. Substituting $l = 60$ and $d = 0.02$: $R = \frac{0.00001(60)}{(0.02)^2} = 1.5$. The resistance is 1.5 ohms.

49. Let p represent the pitch and w represent the wavelength. The variation equation is $p = \frac{K}{w}$.

Substituting $p = 420$ and $w = 2.2$:

$$420 = \frac{K}{2.2}$$
$$K = 924$$

The equation is $p = \frac{924}{w}$. Substituting $p = 720$:

$$720 = \frac{924}{w}$$
$$720w = 924$$
$$w \approx 1.28$$

The wavelength is approximately 1.28 meters.

51. The variation equation is $F = \frac{Gm_1 m_2}{d^2}$.

53. Solving the equation:

$$x - 5 = 7$$
$$x = 12$$

55. Solving the equation:

$$5 - \tfrac{4}{7}a = -11$$
$$7\left(5 - \tfrac{4}{7}a\right) = 7(-11)$$
$$35 - 4a = -77$$
$$-4a = -112$$
$$a = 28$$

57. Solving the equation:

$$5(x - 1) - 2(2x + 3) = 5x - 4$$
$$5x - 5 - 4x - 6 = 5x - 4$$
$$x - 11 = 5x - 4$$
$$-4x = 7$$
$$x = -\tfrac{7}{4}$$

59. Solving for w:

$$P = 2l + 2w$$
$$P - 2l = 2w$$
$$w = \frac{P - 2l}{2}$$

61.   Solving the inequality:

$$-5t \leq 30$$
$$t \geq -6$$

The solution set is $[-6, \infty)$. Graphing the solution set:

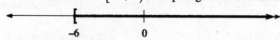

63.   Solving the inequality:
$$1.6x - 2 < 0.8x + 2.8$$
$$0.8x - 2 < 2.8$$
$$0.8x < 4.8$$
$$x < 6$$

The solution set is $(-\infty, 6)$. Graphing the solution set:

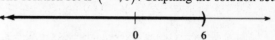

65.   Solving the equation:
$$\left| \tfrac{1}{4}x - 1 \right| = \tfrac{1}{2}$$
$$\tfrac{1}{4}x - 1 = -\tfrac{1}{2}, \tfrac{1}{2}$$
$$\tfrac{1}{4}x = \tfrac{1}{2}, \tfrac{3}{2}$$
$$x = 2, 6$$

67.   Solving the equation:

$$\left| 3 - 2x \right| + 5 = 2$$
$$\left| 3 - 2x \right| = -3$$

Since this statement is false, there is no solution, or $\varnothing$.

69.   Solving the inequality:

$$\left| \tfrac{x}{5} + 1 \right| \geq \tfrac{4}{5}$$

$$\tfrac{x}{5} + 1 \leq -\tfrac{4}{5} \qquad \text{or} \qquad \tfrac{x}{5} + 1 \geq \tfrac{4}{5}$$
$$x + 5 \leq -4 \qquad\qquad\qquad x + 5 \geq 4$$
$$x \leq -9 \qquad\qquad\qquad\quad x \geq -1$$

Graphing the solution set:

71.   Since $\left| 3 - 4t \right| > -5$ is always true, the solution set is all real numbers. Graphing the solution set:

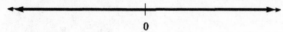

73.   **a.**   The distance appears to vary directly with the square of the speed.

      b.   Let $d = Ks^2$. Substituting $d = 108$ and $s = 40$:

$$108 = K(40)^2$$
$$1600K = 108$$
$$K = 0.0675$$

So the equation is $d = 0.0675 s^2$.

      c.   Substituting $s = 55$: $d = 0.0675(55)^2 \approx 204.2$ feet away

      d.   Substituting $s = 56$: $d = 0.0675(56)^2 \approx 211.7$ feet. This is approximately 7.5 feet further away.

Chapter 3 Review/Test

1. Graphing the line:

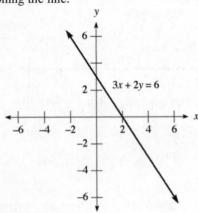

2. Graphing the line:

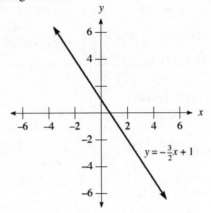

3. Graphing the line:

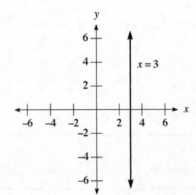

4. Finding the slope: $m = \dfrac{6-2}{3-5} = \dfrac{4}{-2} = -2$

5. Finding the slope: $m = \dfrac{2-2}{3-(-4)} = \dfrac{0}{7} = 0$

6. Solving for x:

$$\frac{-3-x}{1-4} = 2$$
$$\frac{-3-x}{-3} = 2$$
$$-3-x = -6$$
$$-x = -3$$
$$x = 3$$

7. Solving for x:

$$\frac{x-7}{2+4} = -\frac{1}{3}$$
$$\frac{x-7}{6} = -\frac{1}{3}$$
$$x-7 = -2$$
$$x = 5$$

8. Finding the slope: $m = \dfrac{-2-8}{5-3} = \dfrac{-10}{2} = -5$

9. Solving for y:

$$\frac{y-3y}{2-5} = 4$$
$$\frac{-2y}{-3} = 4$$
$$-2y = -12$$
$$y = 6$$

10. Using the slope-intercept formula, the slope is $y = 3x + 5$.

11. Using the slope-intercept formula, the slope is $y = -2x$.

12. Solving for y:
$$3x - y = 6$$
$$-y = -3x + 6$$
$$y = 3x - 6$$

The slope is $m = 3$ and the y-intercept is $b = -6$.

13. Solving for y:
$$2x - 3y = 9$$
$$-3y = -2x + 9$$
$$y = \frac{2}{3}x - 3$$

The slope is $m = \frac{2}{3}$ and the y-intercept is $b = -3$.

14. Using the point-slope formula:
$$y - 4 = 2(x - 2)$$
$$y - 4 = 2x - 4$$
$$y = 2x$$

15. Using the point-slope formula:
$$y - 1 = -\frac{1}{3}(x + 3)$$
$$y - 1 = -\frac{1}{3}x - 1$$
$$y = -\frac{1}{3}x$$

16. First find the slope: $m = \frac{-5 - 5}{-3 - 2} = \frac{-10}{-5} = 2$. Now using the point-slope formula:
$$y - 5 = 2(x - 2)$$
$$y - 5 = 2x - 4$$
$$y = 2x + 1$$

17. First find the slope: $m = \frac{7 - 7}{4 - (-3)} = \frac{0}{7} = 0$. Since the line is horizontal, its equation is $y = 7$.

18. First find the slope: $m = \frac{-4 - (-1)}{-3 - (-5)} = \frac{-4 + 1}{-3 + 5} = -\frac{3}{2}$. Now using the point-slope formula:
$$y + 1 = -\frac{3}{2}(x + 5)$$
$$y + 1 = -\frac{3}{2}x - \frac{15}{2}$$
$$y = -\frac{3}{2}x - \frac{17}{2}$$

19. First find the slope by solving for y:
$$2x - y = 4$$
$$-y = -2x + 4$$
$$y = 2x - 4$$

The parallel slope is also $m = 2$. Now using the point-slope formula:
$$y + 3 = 2(x - 2)$$
$$y + 3 = 2x - 4$$
$$y = 2x - 7$$

20. The perpendicular slope is $m = \frac{1}{3}$. Using the point-slope formula:
$$y - 0 = \frac{1}{3}(x - 2)$$
$$y = \frac{1}{3}x - \frac{2}{3}$$

21. Graphing the inequality:

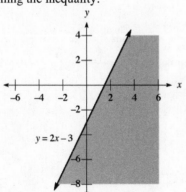

22. Graphing the inequality:

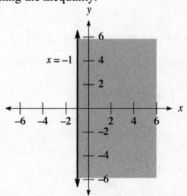

23. The domain is $\{2, 3, 4\}$ and the range is $\{2, 3, 4\}$. This is a function.

24. The domain is $\{-4,-2,6\}$ and the range is $\{0,3\}$. This is a function.

25. $f(-3)=0$

26. $f(2)+g(2)=-1+2=1$

27. Evaluating the function: $f(0)=2(0)^2-4(0)+1=0-0+1=1$

28. Evaluating the function: $g(a)=3a+2$

29. Evaluating the function: $f\big[g(0)\big]=f\big[3(0)+2\big]=f(2)=2(2)^2-4(2)+1=8-8+1=1$

30. Evaluating the function: $f\big[g(1)\big]=f\big[3(1)+2\big]=f(5)=2(5)^2-4(5)+1=50-20+1=31$

31. The variation equation is $y=Kx$. Substituting $x=2$ and $y=6$:
$$6=K\bullet 2$$
$$K=3$$
The equation is $y=3x$. Substituting $x=8$: $y=3\bullet 8=24$

32. The variation equation is $y=Kx$. Substituting $x=5$ and $y=-3$:
$$-3=K\bullet 5$$
$$K=-\tfrac{3}{5}$$
The equation is $y=-\tfrac{3}{5}x$. Substituting $x=-10$: $y=-\tfrac{3}{5}(-10)=6$

33. The variation equation is $y=\dfrac{K}{x^2}$. Substituting $x=2$ and $y=9$:
$$9=\frac{K}{2^2}$$
$$9=\frac{K}{4}$$
$$K=36$$
The equation is $y=\dfrac{36}{x^2}$. Substituting $x=3$: $y=\dfrac{36}{3^2}=\dfrac{36}{9}=4$

34. The variation equation is $y=\dfrac{K}{x^2}$. Substituting $x=5$ and $y=4$:
$$4=\frac{K}{5^2}$$
$$4=\frac{K}{25}$$
$$K=100$$
The equation is $y=\dfrac{100}{x^2}$. Substituting $x=2$: $y=\dfrac{100}{2^2}=\dfrac{100}{4}=25$

35. The variation equation is $t=Kd$. Substituting $t=42$ and $d=2$:
$$42=K\bullet 2$$
$$K=21$$
The equation is $t=21d$. Substituting $d=4$: $t=21\bullet 4=84$. The tension is 84 pounds.

36. Let I represent the intensity and d represent the distance. The variation equation is $I=\dfrac{K}{d^2}$.

Substituting $I=9$ and $d=4$:
$$9=\frac{K}{4^2}$$
$$9=\frac{K}{16}$$
$$K=144$$
The equation is $I=\dfrac{144}{d^2}$. Substituting $d=3$: $I=\dfrac{144}{3^2}=\dfrac{144}{9}=16$. The intensity is 16 foot-candles.

Chapter 4
Systems of Linear Equations and Inequalities

4.1 Systems of Linear Equations in Two Variables

1. The intersection point is (4,3).

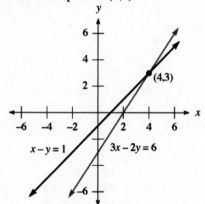

3. The intersection point is (–5,–6).

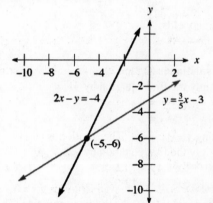

5. The intersection point is (4,2).

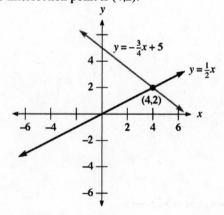

7. The lines are parallel. There is no solution to the system.

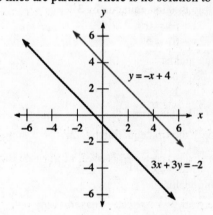

75

9. The lines coincide. Any solution to one of the equations is a solution to the other.

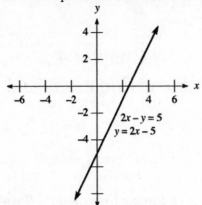

11. Solving the two equations:
$$x + y = 5$$
$$3x - y = 3$$
Adding yields:
$$4x = 8$$
$$x = 2$$
The solution is $(2,3)$.

13. Multiply the first equation by -1:
$$-3x - y = 4$$
$$4x + y = 5$$
Adding yields: $x = 1$. The solution is $(1,1)$.

15. Multiply the first equation by -2:
$$-6x + 4y = -12$$
$$6x - 4y = 12$$
Adding yields $0 = 0$, so the lines coincide. The solution is $\{(x,y) \mid 3x - 2y = 6\}$.

17. Multiply the first equation by 3:
$$3x + 6y = 0$$
$$2x - 6y = 5$$
Adding yields:
$$5x = 5$$
$$x = 1$$
The solution is $\left(1, -\frac{1}{2}\right)$.

19. Multiply the first equation by -2:
$$-4x + 10y = -32$$
$$4x - 3y = 11$$
Adding yields:
$$7y = -21$$
$$y = -3$$
The solution is $\left(\frac{1}{2}, -3\right)$.

21. Multiply the first equation by 3 and the second equation by -2:
$$18x + 9y = -3$$
$$-18x - 10y = -2$$
Adding yields:
$$-y = -5$$
$$y = 5$$
The solution is $\left(-\frac{8}{3}, 5\right)$.

23. Multiply the first equation by 2 and the second equation by 3:
$$8x + 6y = 28$$
$$27x - 6y = 42$$
Adding yields:
$$35x = 70$$
$$x = 2$$
The solution is (2,2).

25. Multiply the first equation by 2:
$$4x - 10y = 6$$
$$-4x + 10y = 3$$
Adding yields $0 = 9$, which is false (parallel lines). There is no solution ($\varnothing$).

27. To clear each equation of fractions, multiply the first equation by 12 and the second equation by 30:
$$3x - 2y = -24$$
$$-5x + 6y = 120$$
Multiply the first equation by 3:
$$9x - 6y = -72$$
$$-5x + 6y = 120$$
Adding yields $4x = 48$, so $x = 12$. Substituting into the first equation:
$$36 - 2y = -24$$
$$-2y = -60$$
$$y = 30$$
The solution is (12,30).

29. To clear each equation of fractions, multiply the first equation by 6 and the second equation by 20:
$$3x + 2y = 78$$
$$8x + 5y = 200$$
Multiply the first equation by 5 and the second equation by –2:
$$15x + 10y = 390$$
$$-16x - 10y = -400$$
Adding yields:
$$-x = -10$$
$$x = 10$$
The solution is (10,24).

31. Substituting into the first equation:

$$7(2y + 9) - y = 24$$
$$14y + 63 - y = 24$$
$$13y = -39$$
$$y = -3$$

The solution is (3,–3).

33. Substituting into the first equation:
$$6x - \left(-\tfrac{3}{4}x - 1\right) = 10$$
$$6x + \tfrac{3}{4}x + 1 = 10$$
$$\tfrac{27}{4}x = 9$$
$$27x = 36$$
$$x = \tfrac{4}{3}$$
The solution is $\left(\tfrac{4}{3}, -2\right)$.

35. Substituting $z = -3y + 17$ into the second equation:
$$5y + 20(-3y + 17) = 65$$
$$5y - 60y + 340 = 65$$
$$-55y + 340 = 65$$
$$-55y = -275$$
$$y = 5$$
The solution is (5,2).

37. Substituting into the first equation:

$$4x - 4 = 3x - 2$$
$$x - 4 = -2$$
$$x = 2$$

The solution is (2,4).

39. Solving the first equation for y yields $y = 2x - 5$. Substituting into the second equation:
$$4x - 2(2x - 5) = 10$$
$$4x - 4x + 10 = 10$$
$$10 = 10$$
Since this statement is true, the two lines coincide. The solution is $\{(x, y) \mid 2x - y = 5\}$.

41. Substituting into the first equation:
$$\tfrac{1}{3}\left(\tfrac{3}{2}y\right) - \tfrac{1}{2}y = 0$$
$$\tfrac{1}{2}y - \tfrac{1}{2}y = 0$$
$$0 = 0$$
Since this statement is true, the two lines coincide. The solution is $\left\{(x, y) \mid x = \tfrac{3}{2}y\right\}$.

43. Multiply the first equation by 2 and the second equation by 7:
$$8x - 14y = 6$$
$$35x + 14y = -21$$
Adding yields:
$$43x = -15$$
$$x = -\tfrac{15}{43}$$
Substituting into the original second equation:
$$5\left(-\tfrac{15}{43}\right) + 2y = -3$$
$$-\tfrac{75}{43} + 2y = -3$$
$$2y = -\tfrac{54}{43}$$
$$y = -\tfrac{27}{43}$$
The solution is $\left(-\tfrac{15}{43}, -\tfrac{27}{43}\right)$.

45. Multiply the first equation by 3 and the second equation by 8:
$$27x - 24y = 12$$
$$16x + 24y = 48$$
Adding yields:
$$43x = 60$$
$$x = \tfrac{60}{43}$$
Substituting into the original second equation:
$$2\left(\tfrac{60}{43}\right) + 3y = 6$$
$$\tfrac{120}{43} + 3y = 6$$
$$3y = \tfrac{138}{43}$$
$$y = \tfrac{46}{43}$$
The solution is $\left(\tfrac{60}{43}, \tfrac{46}{43}\right)$.

47. Multiply the first equation by 2 and the second equation by 5:
$$6x - 10y = 4$$
$$35x + 10y = 5$$
Adding yields:
$$41x = 9$$
$$x = \frac{9}{41}$$
Substituting into the original second equation:
$$7\left(\frac{9}{41}\right) + 2y = 1$$
$$\frac{63}{41} + 2y = 1$$
$$2y = -\frac{22}{41}$$
$$y = -\frac{11}{41}$$
The solution is $\left(\frac{9}{41}, -\frac{11}{41}\right)$.

49. Multiply the second equation by 3:
$$x - 3y = 7$$
$$6x + 3y = -18$$
Adding yields:
$$7x = -11$$
$$x = -\frac{11}{7}$$
Substituting into the original second equation:
$$2\left(-\frac{11}{7}\right) + y = -6$$
$$-\frac{22}{7} + y = -6$$
$$y = -\frac{20}{7}$$
The solution is $\left(-\frac{11}{7}, -\frac{20}{7}\right)$.

51. Substituting into the first equation:
$$-\frac{1}{3}x + 2 = \frac{1}{2}x + \frac{1}{3}$$
$$6\left(-\frac{1}{3}x + 2\right) = 6\left(\frac{1}{2}x + \frac{1}{3}\right)$$
$$-2x + 12 = 3x + 2$$
$$-5x = -10$$
$$x = 2$$
Substituting into the first equation: $y = \frac{1}{2}(2) + \frac{1}{3} = 1 + \frac{1}{3} = \frac{4}{3}$. The solution is $\left(2, \frac{4}{3}\right)$.

53. Substituting into the first equation:
$$3\left(\frac{2}{3}y - 4\right) - 4y = 12$$
$$2y - 12 - 4y = 12$$
$$-2y - 12 = 12$$
$$-2y = 24$$
$$y = -12$$
Substituting into the second equation: $x = \frac{2}{3}(-12) - 4 = -8 - 4 = -12$. The solution is $(-12, -12)$.

55. Multiply the first equation by 2:
$$8x - 6y = -14$$
$$-8x + 6y = -11$$
$$0 = -25$$
Since this statement is false, there is no solution ($\varnothing$).

57.   Multiply the first equation by -20:
$$-60y - 20z = -340$$
$$5y + 20z = 65$$

Adding yields:
$$-55y = -275$$
$$y = 5$$

Substituting into the first equation:
$$3(5) + z = 17$$
$$15 + z = 17$$
$$z = 2$$

The solution is $y = 5$, $z = 2$.

59.   Substitute into the first equation:
$$\tfrac{3}{4}x - \tfrac{1}{3}\left(\tfrac{1}{4}x\right) = 1$$
$$\tfrac{3}{4}x - \tfrac{1}{12}x = 1$$
$$\tfrac{2}{3}x = 1$$
$$x = \tfrac{3}{2}$$

Substituting into the second equation: $y = \tfrac{1}{4}\left(\tfrac{3}{2}\right) = \tfrac{3}{8}$. The solution is $\left(\tfrac{3}{2}, \tfrac{3}{8}\right)$.

61.   To clear each equation of fractions, multiply the first equation by 12 and the second equation by 12:
$$3x - 6y = 4$$
$$4x - 3y = -8$$

Multiply the second equation by -2:
$$3x - 6y = 4$$
$$-8x + 6y = 16$$

Adding yields:
$$-5x = 20$$
$$x = -4$$

Substituting into the first equation:
$$3(-4) - 6y = 4$$
$$-12 - 6y = 4$$
$$-6y = 16$$
$$y = -\tfrac{8}{3}$$

The solution is $\left(-4, -\tfrac{8}{3}\right)$.

63. **a.** Simplifying: $(3x - 4y) - 3(x - y) = 3x - 4y - 3x + 3y = -y$

   b. Substituting $x = 0$:
$$3(0) - 4y = 8$$
$$-4y = 8$$
$$y = -2$$

   c. From part **b**, the y-intercept is $(0, -2)$.

   d. Graphing the line:

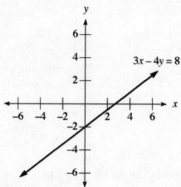

   e. Multiply the second equation by -3:
$$3x - 4y = 8$$
$$-3x + 3y = -6$$
   Adding yields:
$$-y = 2$$
$$y = -2$$
   Substituting into the first equation:
$$3x - 4(-2) = 8$$
$$3x + 8 = 8$$
$$3x = 0$$
$$x = 0$$
   The lines intersect at the point $(0, -2)$.

65. Multiply the second equation by 100:
$$x + y = 10000$$
$$6x + 5y = 56000$$
   Multiply the first equation by -5:
$$-5x - 5y = -50000$$
$$6x + 5y = 56000$$
   Adding yields $x = 6000$. The solution is $(6000, 4000)$.

67. Multiplying the first equation by $\frac{2}{3}$ yields the equation $4x - 6y = 2$. For the lines to coincide, the value is $c = 2$.

69. Substituting $x = 4 - y$ into the second equation:
$$(4 - y) - 2y = 4$$
$$4 - 3y = 4$$
$$-3y = 0$$
$$y = 0$$
   The solution is $(4, 0)$.

71. Finding the slope: $m = \dfrac{5 - (-1)}{-2 - (-4)} = \dfrac{5 + 1}{-2 + 4} = \dfrac{6}{2} = 3$

73.   Solving for y:

$$2x - 3y = 6$$
$$-3y = -2x + 6$$
$$y = \frac{2}{3}x - 2$$

The slope is $m = \frac{2}{3}$ and the y-intercept is $b = -2$.

75.   Using the point-slope formula:

$$y - 2 = \frac{2}{3}(x + 6)$$
$$y - 2 = \frac{2}{3}x + 4$$
$$y = \frac{2}{3}x + 6$$

77.   First find the slope: $m = \dfrac{-2 - 0}{0 - 3} = \dfrac{2}{3}$. The slope-intercept form is $y = \frac{2}{3}x - 2$.

79.   Simplifying: $2 - 2(6) = 2 - 12 = -10$

81.   Simplifying: $(x + 3y) - 1(x - 2z) = x + 3y - x + 2z = 3y + 2z$

83.   Solving the equation:

$$-9y = -9$$
$$y = 1$$

85.   Solving the equation:

$$3(1) + 2z = 9$$
$$3 + 2z = 9$$
$$2z = 6$$
$$z = 3$$

87.   Applying the distributive property: $2(5x - z) = 10x - 2z$

89.   Applying the distributive property: $3(3x + y - 2z) = 9x + 3y - 6z$

91.   Substituting the points $(1, -2)$ and $(3, 1)$ results in the two equations:

$$a - 2b = 7$$
$$3a + b = 7$$

Multiply the second equation by 2:

$$a - 2b = 7$$
$$6a + 2b = 14$$

Adding the two equations:

$$7a = 21$$
$$a = 3$$

Substituting into the second equation:

$$9 + b = 7$$
$$b = -2$$

So $a = 3$ and $b = -2$.

93.   Substituting the points $(3, 24)$ and $(6, 0)$:

$$24 = a(3)^2 + b(3) \qquad\qquad 0 = a(6)^2 + b(6)$$
$$24 = 9a + 3b \qquad\qquad\qquad 0 = 36a + 6b$$
$$3a + b = 8 \qquad\qquad\qquad\quad b = -6a$$

Substituting into the first equation:

$$3a + (-6a) = 8$$
$$-3a = 8$$
$$a = -\frac{8}{3}$$
$$b = -6\left(-\frac{8}{3}\right) = 16$$

Thus $a = -\frac{8}{3}, b = 16$.

4.2 Systems of Linear Equations in Three Variables

1.   Adding the first two equations and the first and third equations results in the system:

$$2x + 3z = 5$$
$$2x - 2z = 0$$

Solving the second equation yields $x = z$, now substituting:

$$2z + 3z = 5$$
$$5z = 5$$
$$z = 1$$

So $x = 1$, now substituting into the original first equation:

$$1 + y + 1 = 4$$
$$y + 2 = 4$$
$$y = 2$$

The solution is (1,2,1).

3.   Adding the first two equations and the first and third equations results in the system:

$$2x + 3z = 13$$
$$3x - 3z = -3$$

Adding yields:

$$5x = 10$$
$$x = 2$$

Substituting to find z:

$$2(2) + 3z = 13$$
$$4 + 3z = 13$$
$$3z = 9$$
$$z = 3$$

Substituting into the original first equation:

$$2 + y + 3 = 6$$
$$y + 5 = 6$$
$$y = 1$$

The solution is (2,1,3).

5.   Adding the second and third equations:

$$5x + z = 11$$

Multiplying the second equation by 2:

$$x + 2y + z = 3$$
$$4x - 2y + 4z = 12$$

Adding yields:

$$5x + 5z = 15$$
$$x + z = 3$$

So the system becomes:

$$5x + z = 11$$
$$x + z = 3$$

Multiply the second equation by -1:

$$5x + z = 11$$
$$-x - z = -3$$

Adding yields:

$$4x = 8$$
$$x = 2$$

Substituting to find z:

$$5(2) + z = 11$$
$$z + 10 = 11$$
$$z = 1$$

Substituting into the original first equation:
$$2+2y+1=3$$
$$2y+3=3$$
$$2y=0$$
$$y=0$$
The solution is $(2,0,1)$.

7. Multiply the second equation by -1 and add it to the first equation:
$$2x+3y-2z=4$$
$$-x-3y+3z=-4$$
Adding results in the equation $x+z=0$. Multiply the second equation by 2 and add it to the third equation:
$$2x+6y-6z=8$$
$$3x-6y+z=-3$$
Adding results in the equation:
$$5x-5z=5$$
$$x-z=1$$
So the system becomes:
$$x-z=1$$
$$x+z=0$$
Adding yields:
$$2x=1$$
$$x=\tfrac{1}{2}$$
Substituting to find z:
$$\tfrac{1}{2}+z=0$$
$$z=-\tfrac{1}{2}$$
Substituting into the original first equation:
$$2\left(\tfrac{1}{2}\right)+3y-2\left(-\tfrac{1}{2}\right)=4$$
$$1+3y+1=4$$
$$3y+2=4$$
$$3y=2$$
$$y=\tfrac{2}{3}$$
The solution is $\left(\tfrac{1}{2},\tfrac{2}{3},-\tfrac{1}{2}\right)$.

9. Multiply the first equation by 2 and add it to the second equation:
$$-2x+8y-6z=4$$
$$2x-8y+6z=1$$
Adding yields $0=5$, which is false. There is no solution (inconsistent system).

11. To clear the system of fractions, multiply the first equation by 2 and the second equation by 3:
$$x-2y+2z=0$$
$$6x+y+3z=6$$
$$x+y+z=-4$$
Multiply the third equation by 2 and add it to the first equation:
$$x-2y+2z=0$$
$$2x+2y+2z=-8$$
Adding yields the equation $3x+4z=-8$. Multiply the third equation by -1 and add it to the second equation:
$$6x+y+3z=6$$
$$-x-y-z=4$$
Adding yields the equation $5x+2z=10$. So the system becomes:
$$3x+4z=-8$$
$$5x+2z=10$$

Multiply the second equation by –2:
$$3x + 4z = -8$$
$$-10x - 4z = -20$$

Adding yields:
$$-7x = -28$$
$$x = 4$$

Substituting to find z:
$$3(4) + 4z = -8$$
$$12 + 4z = -8$$
$$4z = -20$$
$$z = -5$$

Substituting into the original third equation:
$$4 + y - 5 = -4$$
$$y - 1 = -4$$
$$y = -3$$

The solution is $(4, -3, -5)$.

13.   Multiply the first equation by –2 and add it to the third equation:
$$-4x + 2y + 6z = -2$$
$$4x - 2y - 6z = 2$$

Adding yields $0 = 0$, which is true. Since there are now less equations than unknowns, there is no unique solution (dependent system).

15.   Multiply the second equation by 3 and add it to the first equation:
$$2x - y + 3z = 4$$
$$3x + 6y - 3z = -9$$

Adding yields the equation $5x + 5y = -5$, or $x + y = -1$.

Multiply the second equation by 2 and add it to the third equation:
$$2x + 4y - 2z = -6$$
$$4x + 3y + 2z = -5$$

Adding yields the equation $6x + 7y = -11$. So the system becomes:
$$6x + 7y = -11$$
$$x + y = -1$$

Multiply the second equation by –6:
$$6x + 7y = -11$$
$$-6x - 6y = 6$$

Adding yields $y = -5$. Substituting to find x:
$$6x + 7(-5) = -11$$
$$6x - 35 = -11$$
$$6x = 24$$
$$x = 4$$

Substituting into the original first equation:
$$2(4) - (-5) + 3z = 4$$
$$13 + 3z = 4$$
$$3z = -9$$
$$z = -3$$

The solution is $(4, -5, -3)$.

17.   Adding the second and third equations results in the equation $x + y = 9$. Since this is the same as the first equation, there are less equations than unknowns. There is no unique solution (dependent system).

19. Adding the second and third equations results in the equation $4x + y = 3$. So the system becomes:

$$4x + y = 3$$
$$2x + y = 2$$

Multiplying the second equation by -1:

$$4x + y = 3$$
$$-2x - y = -2$$

Adding yields:

$$2x = 1$$
$$x = \frac{1}{2}$$

Substituting to find y:

$$2\left(\frac{1}{2}\right) + y = 2$$
$$1 + y = 2$$
$$y = 1$$

Substituting into the original second equation:

$$1 + z = 3$$
$$z = 2$$

The solution is $\left(\frac{1}{2}, 1, 2\right)$.

21. Multiply the third equation by 2 and adding it to the second equation:

$$6y - 4z = 1$$
$$2x + 4z = 2$$

Adding yields the equation $2x + 6y = 3$. So the system becomes:

$$2x - 3y = 0$$
$$2x + 6y = 3$$

Multiply the first equation by 2:

$$4x - 6y = 0$$
$$2x + 6y = 3$$

Adding yields:

$$6x = 3$$
$$x = \frac{1}{2}$$

Substituting to find y:

$$2\left(\frac{1}{2}\right) + 6y = 3$$
$$1 + 6y = 3$$
$$6y = 2$$
$$y = \frac{1}{3}$$

Substituting into the original third equation to find z:

$$\frac{1}{2} + 2z = 1$$
$$2z = \frac{1}{2}$$
$$z = \frac{1}{4}$$

The solution is $\left(\frac{1}{2}, \frac{1}{3}, \frac{1}{4}\right)$.

23. Multiply the first equation by –2 and add it to the second equation:

$$-2x - 2y + 2z = -4$$
$$2x + y + 3z = 4$$

Adding yields $-y + 5z = 0$. Multiply the first equation by –1 and add it to the third equation:

$$-x - y + z = -2$$
$$x - 2y + 2z = 6$$

Adding yields $-3y + 3z = 4$. So the system becomes:

$$-y + 5z = 0$$
$$-3y + 3z = 4$$

Multiply the first equation by –3:

$$3y - 15z = 0$$
$$-3y + 3z = 4$$

Adding yields:

$$-12z = 4$$
$$z = -\tfrac{1}{3}$$

Substituting to find y:

$$-3y + 3\left(-\tfrac{1}{3}\right) = 4$$
$$-3y - 1 = 4$$
$$-3y = 5$$
$$y = -\tfrac{5}{3}$$

Substituting into the original first equation:

$$x - \tfrac{5}{3} + \tfrac{1}{3} = 2$$
$$x - \tfrac{4}{3} = 2$$
$$x = \tfrac{10}{3}$$

The solution is $\left(\tfrac{10}{3}, -\tfrac{5}{3}, -\tfrac{1}{3}\right)$.

25. Multiply the third equation by –1 and add it to the first equation:

$$2x + 3y = -\tfrac{1}{2}$$
$$-3y - 2z = \tfrac{3}{4}$$

Adding yields the equation $2x - 2z = \tfrac{1}{4}$. So the system becomes:

$$2x - 2z = \tfrac{1}{4}$$
$$4x + 8z = 2$$

Multiply the first equation by 4:

$$8x - 8z = 1$$
$$4x + 8z = 2$$

Adding yields:

$$12x = 3$$
$$x = \tfrac{1}{4}$$

Substituting to find z:

$$4\left(\tfrac{1}{4}\right) + 8z = 2$$
$$1 + 8z = 2$$
$$8z = 1$$
$$z = \tfrac{1}{8}$$

Substituting to find y:

$$2\left(\tfrac{1}{4}\right) + 3y = -\tfrac{1}{2}$$
$$\tfrac{1}{2} + 3y = -\tfrac{1}{2}$$
$$3y = -1$$
$$y = -\tfrac{1}{3}$$

The solution is $\left(\tfrac{1}{4}, -\tfrac{1}{3}, \tfrac{1}{8}\right)$.

27. To clear each equation of fractions, multiply the first equation by 6, the second equation by 4, and the third equation by 12:

$$2x + 3y - z = 24$$
$$x - 3y + 2z = 6$$
$$6x - 8y - 3z = -64$$

Multiply the first equation by 2 and add it to the second equation:

$$4x + 6y - 2z = 48$$
$$x - 3y + 2z = 6$$

Adding yields the equation $5x + 3y = 54$. Multiply the first equation by -3 and add it to the third equation:

$$-6x - 9y + 3z = -72$$
$$6x - 8y - 3z = -64$$

Adding yields:

$$-17y = -136$$
$$y = 8$$

Substituting to find x:

$$5x + 3(8) = 54$$
$$5x + 24 = 54$$
$$5x = 30$$
$$x = 6$$

Substituting to find z:

$$6 - 3(8) + 2z = 6$$
$$-18 + 2z = 6$$
$$2z = 24$$
$$z = 12$$

The solution is $(6, 8, 12)$.

29. To clear each equation of fractions, multiply the first equation by 6, the second equation by 6, and the third equation by 12:

$$6x - 3y - 2z = -8$$
$$2x + 6y - 3z = 30$$
$$-3x + 8y - 12z = -9$$

Multiply the first equation by 2 and add it to the second equation:

$$12x - 6y - 4z = -16$$
$$2x + 6y - 3z = 30$$

Adding yields the equation:

$$14x - 7z = 14$$
$$2x - z = 2$$

Multiply the first equation by 8 and the third equation by 3:

$$48x - 24y - 16z = -64$$
$$-9x + 24y - 36z = -27$$

Adding yields the equation $39x - 52z = -91$. So the system becomes:
$$2x - z = 2$$
$$39x - 52z = -91$$
Multiply the first equation by -52:
$$-104x + 52z = -104$$
$$39x - 52z = -91$$
Adding yields:
$$-65x = -195$$
$$x = 3$$
Substituting to find z:
$$6 - z = 2$$
$$z = 4$$
Substituting to find y:
$$6 + 6y - 12 = 30$$
$$6y - 6 = 30$$
$$6y = 36$$
$$y = 6$$
The solution is $(3,6,4)$.

31.   To clear each equation of fractions, multiply the first equation by 6, the second equation by 10, and the third equation by 12:
$$3x + 4y = 15$$
$$2x - 5z = -3$$
$$4y - 3z = 9$$

Multiply the third equation by -1 and add it to the first equation:
$$3x + 4y = 15$$
$$-4y + 3z = -9$$
Adding yields:
$$3x + 3z = 6$$
$$x + z = 2$$
So the system becomes:
$$x + z = 2$$
$$2x - 5z = -3$$
Multiply the first equation by 5:
$$5x + 5z = 10$$
$$2x - 5z = -3$$
Adding yields:
$$7x = 7$$
$$x = 1$$
Substituting yields $z = 1$. Substituting to find y:
$$3 + 4y = 15$$
$$4y = 12$$
$$y = 3$$
The solution is $(1,3,1)$.

33. To clear each equation of fractions, multiply the first equation by 4, the second equation by 12, and the third equation by 6:

$$2x - y + 2z = -8$$
$$3x - y - 4z = 3$$
$$x + 2y - 3z = 9$$

Multiply the first equation by -1 and add it to the first equation:

$$-2x + y - 2z = 8$$
$$3x - y - 4z = 3$$

Adding yields the equation $x - 6z = 11$. Multiply the first equation by 2 and add it to the third equation:

$$4x - 2y + 4z = -16$$
$$x + 2y - 3z = 9$$

Adding yields the equation $5x + z = -7$. So the system becomes:

$$5x + z = -7$$
$$x - 6z = 11$$

Multiply the first equation by 6:

$$30x + 6z = -42$$
$$x - 6z = 11$$

Adding yields:

$$31x = -31$$
$$x = -1$$

Substituting to find z:

$$-5 + z = -7$$
$$z = -2$$

Substituting to find y:

$$-1 + 2y + 6 = 9$$
$$2y + 5 = 9$$
$$2y = 4$$
$$y = 2$$

The solution is $(-1, 2, -2)$.

35. Divide the second equation by 5 and the third equation by 10 to produce the system:

$$x - y - z = 0$$
$$x + 4y = 16$$
$$2y - z = 5$$

Multiply the third equation by -1 and add it to the first equation:

$$x - y - z = 0$$
$$-2y + z = -5$$

Adding yields the equation $x - 3y = -5$. So the system becomes:

$$x + 4y = 16$$
$$x - 3y = -5$$

Multiply the second equation by -1:

$$x + 4y = 16$$
$$-x + 3y = 5$$

Adding yields:

$$7y = 21$$
$$y = 3$$

Substituting to find x:

$$x + 12 = 16$$
$$x = 4$$

Substituting to find z:

$$6 - z = 5$$
$$z = 1$$

The currents are 4 amps, 3 amps, and 1 amp.

37. The variation equation is $y = Kx^2$. Substituting $x = 5$ and $y = 75$:
$$75 = K \cdot 5^2$$
$$75 = 25K$$
$$K = 3$$
So $y = 3x^2$. Substituting $x = 7$: $y = 3 \cdot 7^2 = 3 \cdot 49 = 147$

39. The variation equation is $y = \dfrac{K}{x}$. Substituting $x = 25$ and $y = 10$:
$$10 = \frac{K}{25}$$
$$K = 250$$
So $y = \dfrac{250}{x}$. Substituting $y = 5$:
$$5 = \frac{250}{x}$$
$$5x = 250$$
$$x = 50$$

41. The variation equation is $z = Kxy^2$. Substituting $z = 40$, $x = 5$, and $y = 2$:
$$40 = K \cdot 5 \cdot 2^2$$
$$40 = 20K$$
$$K = 2$$
So $z = 2xy^2$. Substituting $x = 2$ and $y = 5$: $z = 2 \cdot 2 \cdot 5^2 = 100$

43. Simplifying: $1(4) - 3(2) = 4 - 6 = -2$

45. Simplifying: $1(1) - 3(-2) + (-2)(-2) = 1 + 6 + 4 = 11$

47. Simplifying: $-3(-1-1) + 4(-2+2) - 5\left[2-(-2)\right] = -3(-2) + 4(0) - 5(4) = 6 + 0 - 20 = -14$

49. Solving the equation:
$$-5x = 20$$
$$x = -4$$

51. Adding the first and second, the first and third, and third and fourth equations results in the system:
$$2x + 3y + 2w = 16$$
$$2x + 3w = 14$$
$$2x - 3y - w = -8$$
Adding the first and third equations results in the system:
$$4x + w = 8$$
$$2x + 3w = 14$$
Multiply the second equation by -2:
$$4x + w = 8$$
$$-4x - 6w = -28$$
Adding yields:
$$-5w = -20$$
$$w = 4$$
Substituting to find x:
$$4x + 4 = 8$$
$$4x = 4$$
$$x = 1$$
Substituting to find y:
$$2 + 3y + 8 = 16$$
$$3y + 10 = 16$$
$$3y = 6$$
$$y = 2$$

Substituting to find z:
$$1 + 2 + z + 4 = 10$$
$$z + 7 = 10$$
$$z = 3$$
The solution is $(1,2,3,4)$.

4.3 Introduction to Determinants

1. Evaluating the determinant: $\begin{vmatrix} 1 & 0 \\ 2 & 3 \end{vmatrix} = 1 \bullet 3 - 0 \bullet 2 = 3 - 0 = 3$

3. Evaluating the determinant: $\begin{vmatrix} 2 & 1 \\ 3 & 4 \end{vmatrix} = 2 \bullet 4 - 1 \bullet 3 = 8 - 3 = 5$

5. Evaluating the determinant: $\begin{vmatrix} 0 & 1 \\ 1 & 0 \end{vmatrix} = 0 \bullet 0 - 1 \bullet 1 = 0 - 1 = -1$

7. Evaluating the determinant: $\begin{vmatrix} -3 & 2 \\ 6 & -4 \end{vmatrix} = (-3) \bullet (-4) - 6 \bullet 2 = 12 - 12 = 0$

9. Solving the equation:
$$\begin{vmatrix} 2x & 1 \\ x & 3 \end{vmatrix} = 10$$
$$6x - x = 10$$
$$5x = 10$$
$$x = 2$$

11. Solving the equation:
$$\begin{vmatrix} 1 & 2x \\ 2 & -3x \end{vmatrix} = 21$$
$$-3x - 4x = 21$$
$$-7x = 21$$
$$x = -3$$

13. Solving the equation:
$$\begin{vmatrix} 2x & -4 \\ x & 2 \end{vmatrix} = -16$$
$$2x(2) + 4x = -16$$
$$4x + 4x = -16$$
$$8x = -16$$
$$x = -2$$

15. Solving the equation:
$$\begin{vmatrix} 11x & -7x \\ 3 & -2 \end{vmatrix} = 3$$
$$11x(-2) + 7x(3) = 3$$
$$-22x + 21x = 3$$
$$-x = 3$$
$$x = -3$$

17. Duplicating the first two columns:
$$\begin{vmatrix} 1 & 2 & 0 \\ 0 & 2 & 1 \\ 1 & 1 & 1 \end{vmatrix}\begin{matrix} 1 & 2 \\ 0 & 2 \\ 1 & 1 \end{matrix} = 1 \bullet 2 \bullet 1 + 2 \bullet 1 \bullet 1 + 0 \bullet 0 \bullet 1 - 1 \bullet 2 \bullet 0 - 1 \bullet 1 \bullet 1 - 1 \bullet 0 \bullet 2 = 2 + 2 + 0 - 0 - 1 - 0 = 3$$

19. Duplicating the first two columns:
$$\begin{vmatrix} 1 & 2 & 3 \\ 3 & 2 & 1 \\ 1 & 1 & 1 \end{vmatrix}\begin{matrix} 1 & 2 \\ 3 & 2 \\ 1 & 1 \end{matrix} = 1 \bullet 2 \bullet 1 + 2 \bullet 1 \bullet 1 + 3 \bullet 3 \bullet 1 - 1 \bullet 2 \bullet 3 - 1 \bullet 1 \bullet 1 - 1 \bullet 3 \bullet 2 = 2 + 2 + 9 - 6 - 1 - 6 = 0$$

21. Expanding across the first row:
$$\begin{vmatrix} 0 & 1 & 2 \\ 1 & 0 & 1 \\ -1 & 2 & 0 \end{vmatrix} = 0\begin{vmatrix} 0 & 1 \\ 2 & 0 \end{vmatrix} - 1\begin{vmatrix} 1 & 1 \\ -1 & 0 \end{vmatrix} + 2\begin{vmatrix} 1 & 0 \\ -1 & 2 \end{vmatrix} = 0(0-2) - 1(0+1) + 2(2-0) = 0 - 1 + 4 = 3$$

23. Expanding across the first row:
$$\begin{vmatrix} 3 & 0 & 2 \\ 0 & -1 & -1 \\ 4 & 0 & 0 \end{vmatrix} = 3\begin{vmatrix} -1 & -1 \\ 0 & 0 \end{vmatrix} - 0\begin{vmatrix} 0 & -1 \\ 4 & 0 \end{vmatrix} + 2\begin{vmatrix} 0 & -1 \\ 4 & 0 \end{vmatrix} = 3(0-0) - 0(0+4) + 2(0+4) = 0 - 0 + 8 = 8$$

25. Expanding across the first row: $\begin{vmatrix} 2 & -1 & 0 \\ 1 & 0 & -2 \\ 0 & 1 & 2 \end{vmatrix} = 2\begin{vmatrix} 0 & -2 \\ 1 & 2 \end{vmatrix} + 1\begin{vmatrix} 1 & -2 \\ 0 & 2 \end{vmatrix} + 0\begin{vmatrix} 1 & 0 \\ 0 & 1 \end{vmatrix} = 2(0+2)+1(2-0)+0 = 4+2 = 6$

27. Expanding across the first row:

$\begin{vmatrix} 1 & 3 & 7 \\ -2 & 6 & 4 \\ 3 & 7 & -1 \end{vmatrix} = 1\begin{vmatrix} 6 & 4 \\ 7 & -1 \end{vmatrix} - 3\begin{vmatrix} -2 & 4 \\ 3 & -1 \end{vmatrix} + 7\begin{vmatrix} -2 & 6 \\ 3 & 7 \end{vmatrix} = 1(-6-28)-3(2-12)+7(-14-18) = -34+30-224 = -228$

29. The determinant equation is:

$\begin{vmatrix} y & x \\ m & 1 \end{vmatrix} = b$

$y - mx = b$

$\quad y = mx + b$

31. **a.** Writing the determinant equation:

$\begin{vmatrix} x & -1.7 \\ 2 & 0.3 \end{vmatrix} = y$

$0.3x + 3.4 = y$

$\quad y = 0.3x + 3.4$

b. Substituting $x = 2$: $y = 0.3(2) + 3.4 = 0.6 + 3.4 = 4$ billion dollars

33. The domain is $\{1, 3, 4\}$ and the range is $\{2, 4\}$. This is a function.

35. The domain is $\{1, 2, 3\}$ and the range is $\{1, 2, 3\}$. This is a function.

37. Since this passes the vertical line test, it is a function. **39.** Since this fails the vertical line test, it is not a function.

41. Simplifying: $2(3) - 4(4) = 6 - 16 = -10$ **43.** Simplifying: $1(-1)(-3) = 3$

45. Simplifying: $-\dfrac{10}{22} = -\dfrac{2 \bullet 5}{2 \bullet 11} = -\dfrac{5}{11}$ **47.** Simplifying: $6(1) - 1(-5) + 1(2) = 6 + 5 + 2 = 13$

49. Evaluating the determinant: $\begin{vmatrix} 3 & -5 \\ 2 & 4 \end{vmatrix} = 3 \bullet 4 - (-5) \bullet 2 = 12 + 10 = 22$

51. Expanding across the first row:

$\begin{vmatrix} 6 & 1 & 1 \\ 3 & -1 & 1 \\ -4 & 2 & -3 \end{vmatrix} = 6\begin{vmatrix} -1 & 1 \\ 2 & -3 \end{vmatrix} - 1\begin{vmatrix} 3 & 1 \\ -4 & -3 \end{vmatrix} + 1\begin{vmatrix} 3 & -1 \\ -4 & 2 \end{vmatrix} = 6(3-2) - 1(-9+4) + 1(6-4) = 6 + 5 + 2 = 13$

53. Expanding across row 1:

$\begin{vmatrix} 2 & 0 & 1 & -3 \\ -1 & 2 & 0 & 1 \\ -3 & 0 & 1 & 0 \\ 1 & 1 & 0 & 0 \end{vmatrix} = 2\begin{vmatrix} 2 & 0 & 1 \\ 0 & 1 & 0 \\ 1 & 0 & 0 \end{vmatrix} - 0 + 1\begin{vmatrix} -1 & 2 & 1 \\ -3 & 0 & 0 \\ 1 & 1 & 0 \end{vmatrix} + 3\begin{vmatrix} -1 & 2 & 0 \\ -3 & 0 & 1 \\ 1 & 1 & 0 \end{vmatrix}$

$= 2 \bullet 1\begin{vmatrix} 2 & 1 \\ 1 & 0 \end{vmatrix} + 1 \bullet 3\begin{vmatrix} 2 & 1 \\ 1 & 0 \end{vmatrix} + 3\left(-1\begin{vmatrix} 0 & 1 \\ 1 & 0 \end{vmatrix} - 2\begin{vmatrix} -3 & 1 \\ 1 & 0 \end{vmatrix} \right)$

$= 2(-1) + 3(-1) + 3(1+2)$

$= -2 - 3 + 9$

$= 4$

55.   Expanding down column 3:

$$\begin{vmatrix} 2 & 0 & 1 & -3 \\ -1 & 2 & 0 & 1 \\ -3 & 0 & 1 & 0 \\ 1 & 1 & 0 & 0 \end{vmatrix} = 1\begin{vmatrix} -1 & 2 & 1 \\ -3 & 0 & 0 \\ 1 & 1 & 0 \end{vmatrix} + 1\begin{vmatrix} 2 & 0 & -3 \\ -1 & 2 & 1 \\ 1 & 1 & 0 \end{vmatrix}$$

$$= 1 \cdot 3\begin{vmatrix} 2 & 1 \\ 1 & 0 \end{vmatrix} + 1 \cdot \left(2\begin{vmatrix} 2 & 1 \\ 1 & 0 \end{vmatrix} - 3\begin{vmatrix} -1 & 2 \\ 1 & 1 \end{vmatrix} \right)$$

$$= 3(-1) + 1(-2 + 9)$$

$$= -3 + 7$$

$$= 4$$

4.4 Cramer's Rule

1.   First find the determinants:

$$D = \begin{vmatrix} 2 & -3 \\ 4 & -2 \end{vmatrix} = -4 + 12 = 8$$

$$D_x = \begin{vmatrix} 3 & -3 \\ 10 & -2 \end{vmatrix} = -6 + 30 = 24$$

$$D_y = \begin{vmatrix} 2 & 3 \\ 4 & 10 \end{vmatrix} = 20 - 12 = 8$$

Now use Cramer's rule:

$$x = \frac{D_x}{D} = \frac{24}{8} = 3 \qquad y = \frac{D_y}{D} = \frac{8}{8} = 1$$

The solution is (3,1).

3.   First find the determinants:

$$D = \begin{vmatrix} 5 & -2 \\ -10 & 4 \end{vmatrix} = 20 - 20 = 0$$

$$D_x = \begin{vmatrix} 4 & -2 \\ 1 & 4 \end{vmatrix} = 16 + 2 = 18$$

$$D_y = \begin{vmatrix} 5 & 4 \\ -10 & 1 \end{vmatrix} = 5 + 40 = 45$$

Since $D = 0$ and other determinants are nonzero, there is no solution, or $\varnothing$.

5.   First find the determinants:

$$D = \begin{vmatrix} 4 & -7 \\ 5 & 2 \end{vmatrix} = 8 + 35 = 43$$

$$D_x = \begin{vmatrix} 3 & -7 \\ -3 & 2 \end{vmatrix} = 6 - 21 = -15$$

$$D_y = \begin{vmatrix} 4 & 3 \\ 5 & -3 \end{vmatrix} = -12 - 15 = -27$$

Now use Cramer's rule:

$$x = \frac{D_x}{D} = -\frac{15}{43} \qquad y = \frac{D_y}{D} = -\frac{27}{43}$$

The solution is $\left(-\frac{15}{43}, -\frac{27}{43} \right)$.

7.   First find the determinants:

$$D = \begin{vmatrix} 9 & -8 \\ 2 & 3 \end{vmatrix} = 27 + 16 = 43$$

$$D_x = \begin{vmatrix} 4 & -8 \\ 6 & 3 \end{vmatrix} = 12 + 48 = 60$$

$$D_y = \begin{vmatrix} 9 & 4 \\ 2 & 6 \end{vmatrix} = 54 - 8 = 46$$

Now use Cramer's rule:

$$x = \frac{D_x}{D} = \frac{60}{43} \qquad y = \frac{D_y}{D} = \frac{46}{43}$$

The solution is $\left(\frac{60}{43}, \frac{46}{43} \right)$.

9.   First find the determinants:

$$D = \begin{vmatrix} 1 & 1 & 1 \\ 1 & -1 & -1 \\ 2 & 2 & -1 \end{vmatrix} = 1\begin{vmatrix} -1 & -1 \\ 2 & -1 \end{vmatrix} - 1\begin{vmatrix} 1 & -1 \\ 2 & -1 \end{vmatrix} + 1\begin{vmatrix} 1 & -1 \\ 2 & 2 \end{vmatrix} = 3 - 1 + 4 = 6$$

$$D_x = \begin{vmatrix} 4 & 1 & 1 \\ 2 & -1 & -1 \\ 2 & 2 & -1 \end{vmatrix} = 4\begin{vmatrix} -1 & -1 \\ 2 & -1 \end{vmatrix} - 1\begin{vmatrix} 2 & -1 \\ 2 & -1 \end{vmatrix} + 1\begin{vmatrix} 2 & -1 \\ 2 & 2 \end{vmatrix} = 12 - 0 + 6 = 18$$

$$D_y = \begin{vmatrix} 1 & 4 & 1 \\ 1 & 2 & -1 \\ 2 & 2 & -1 \end{vmatrix} = 1\begin{vmatrix} 2 & -1 \\ 2 & -1 \end{vmatrix} - 4\begin{vmatrix} 1 & -1 \\ 2 & -1 \end{vmatrix} + 1\begin{vmatrix} 1 & 2 \\ 2 & 2 \end{vmatrix} = 0 - 4 - 2 = -6$$

$$D_z = \begin{vmatrix} 1 & 1 & 4 \\ 1 & -1 & 2 \\ 2 & 2 & 2 \end{vmatrix} = 1\begin{vmatrix} -1 & 2 \\ 2 & 2 \end{vmatrix} - 1\begin{vmatrix} 1 & 2 \\ 2 & 2 \end{vmatrix} + 4\begin{vmatrix} 1 & -1 \\ 2 & 2 \end{vmatrix} = -6 + 2 + 16 = 12$$

Now use Cramer's rule:

$$x = \frac{D_x}{D} = \frac{18}{6} = 3 \qquad y = \frac{D_y}{D} = \frac{-6}{6} = -1 \qquad z = \frac{D_z}{D} = \frac{12}{6} = 2$$

The solution is $(3, -1, 2)$.

11.   First find the determinants:

$$D = \begin{vmatrix} 1 & 1 & -1 \\ -1 & 1 & 1 \\ 1 & 1 & 1 \end{vmatrix} = 1\begin{vmatrix} 1 & 1 \\ 1 & 1 \end{vmatrix} - 1\begin{vmatrix} -1 & 1 \\ 1 & 1 \end{vmatrix} - 1\begin{vmatrix} -1 & 1 \\ 1 & 1 \end{vmatrix} = 0 + 2 + 2 = 4$$

$$D_x = \begin{vmatrix} 2 & 1 & -1 \\ 3 & 1 & 1 \\ 4 & 1 & 1 \end{vmatrix} = 2\begin{vmatrix} 1 & 1 \\ 1 & 1 \end{vmatrix} - 1\begin{vmatrix} 3 & 1 \\ 4 & 1 \end{vmatrix} - 1\begin{vmatrix} 3 & 1 \\ 4 & 1 \end{vmatrix} = 0 + 1 + 1 = 2$$

$$D_y = \begin{vmatrix} 1 & 2 & -1 \\ -1 & 3 & 1 \\ 1 & 4 & 1 \end{vmatrix} = 1\begin{vmatrix} 3 & 1 \\ 4 & 1 \end{vmatrix} - 2\begin{vmatrix} -1 & 1 \\ 1 & 1 \end{vmatrix} - 1\begin{vmatrix} -1 & 3 \\ 1 & 4 \end{vmatrix} = -1 + 4 + 7 = 10$$

$$D_z = \begin{vmatrix} 1 & 1 & 2 \\ -1 & 1 & 3 \\ 1 & 1 & 4 \end{vmatrix} = 1\begin{vmatrix} 1 & 3 \\ 1 & 4 \end{vmatrix} - 1\begin{vmatrix} -1 & 3 \\ 1 & 4 \end{vmatrix} + 2\begin{vmatrix} -1 & 1 \\ 1 & 1 \end{vmatrix} = 1 + 7 - 4 = 4$$

Now use Cramer's rule:

$$x = \frac{D_x}{D} = \frac{2}{4} = \tfrac{1}{2} \qquad y = \frac{D_y}{D} = \frac{10}{4} = \tfrac{5}{2} \qquad z = \frac{D_z}{D} = \frac{4}{4} = 1$$

The solution is $\left(\tfrac{1}{2}, \tfrac{5}{2}, 1\right)$.

13.   First find the determinants:

$$D = \begin{vmatrix} 3 & -1 & 2 \\ 6 & -2 & 4 \\ 1 & -5 & 2 \end{vmatrix} = 3\begin{vmatrix} -2 & 4 \\ -5 & 2 \end{vmatrix} + 1\begin{vmatrix} 6 & 4 \\ 1 & 2 \end{vmatrix} + 2\begin{vmatrix} 6 & -2 \\ 1 & -5 \end{vmatrix} = 48 + 8 - 56 = 0$$

$$D_x = \begin{vmatrix} 4 & -1 & 2 \\ 8 & -2 & 4 \\ 1 & -5 & 2 \end{vmatrix} = 4\begin{vmatrix} -2 & 4 \\ -5 & 2 \end{vmatrix} + 1\begin{vmatrix} 8 & 4 \\ 1 & 2 \end{vmatrix} + 2\begin{vmatrix} 8 & -2 \\ 1 & -5 \end{vmatrix} = 64 + 12 - 76 = 0$$

$$D_y = \begin{vmatrix} 3 & 4 & 2 \\ 6 & 8 & 4 \\ 1 & 1 & 2 \end{vmatrix} = 3\begin{vmatrix} 8 & 4 \\ 1 & 2 \end{vmatrix} - 4\begin{vmatrix} 6 & 4 \\ 1 & 2 \end{vmatrix} + 2\begin{vmatrix} 6 & 8 \\ 1 & 1 \end{vmatrix} = 36 - 32 - 4 = 0$$

$$D_z = \begin{vmatrix} 3 & -1 & 4 \\ 6 & -2 & 8 \\ 1 & -5 & 1 \end{vmatrix} = 3\begin{vmatrix} -2 & 8 \\ -5 & 1 \end{vmatrix} + 1\begin{vmatrix} 6 & 8 \\ 1 & 1 \end{vmatrix} + 4\begin{vmatrix} 6 & -2 \\ 1 & -5 \end{vmatrix} = 114 - 2 - 112 = 0$$

Since $D = 0$ and the other determinants are also 0, there is no unique solution (dependent).

15.   First find the determinants:

$$D = \begin{vmatrix} 2 & -1 & 3 \\ 1 & -5 & -2 \\ -4 & -2 & 1 \end{vmatrix} = 2\begin{vmatrix} -5 & -2 \\ -2 & 1 \end{vmatrix} + 1\begin{vmatrix} 1 & -2 \\ -4 & 1 \end{vmatrix} + 3\begin{vmatrix} 1 & -5 \\ -4 & -2 \end{vmatrix} = -18 - 7 - 66 = -91$$

$$D_x = \begin{vmatrix} 4 & -1 & 3 \\ 1 & -5 & -2 \\ 3 & -2 & 1 \end{vmatrix} = 4\begin{vmatrix} -5 & -2 \\ -2 & 1 \end{vmatrix} + 1\begin{vmatrix} 1 & -2 \\ 3 & 1 \end{vmatrix} + 3\begin{vmatrix} 1 & -5 \\ 3 & -2 \end{vmatrix} = -36 + 7 + 39 = 10$$

$$D_y = \begin{vmatrix} 2 & 4 & 3 \\ 1 & 1 & -2 \\ -4 & 3 & 1 \end{vmatrix} = 2\begin{vmatrix} 1 & -2 \\ 3 & 1 \end{vmatrix} - 4\begin{vmatrix} 1 & -2 \\ -4 & 1 \end{vmatrix} + 3\begin{vmatrix} 1 & 1 \\ -4 & 3 \end{vmatrix} = 14 + 28 + 21 = 63$$

$$D_z = \begin{vmatrix} 2 & -1 & 4 \\ 1 & -5 & 1 \\ -4 & -2 & 3 \end{vmatrix} = 2\begin{vmatrix} -5 & 1 \\ -2 & 3 \end{vmatrix} + 1\begin{vmatrix} 1 & 1 \\ -4 & 3 \end{vmatrix} + 4\begin{vmatrix} 1 & -5 \\ -4 & -2 \end{vmatrix} = -26 + 7 - 88 = -107$$

Now use Cramer's rule:

$$x = \frac{D_x}{D} = -\frac{10}{91} \qquad y = \frac{D_y}{D} = -\frac{63}{91} = -\frac{9}{13} \qquad z = \frac{D_z}{D} = \frac{-107}{-91} = \frac{107}{91}$$

The solution is $\left(-\frac{10}{91}, -\frac{9}{13}, \frac{107}{91}\right)$.

17. First find the determinants:

$$D = \begin{vmatrix} -1 & -7 & 0 \\ 1 & 0 & 3 \\ 0 & 2 & 1 \end{vmatrix} = -1\begin{vmatrix} 0 & 3 \\ 2 & 1 \end{vmatrix} + 7\begin{vmatrix} 1 & 3 \\ 0 & 1 \end{vmatrix} + 0\begin{vmatrix} 1 & 0 \\ 0 & 2 \end{vmatrix} = 6 + 7 + 0 = 13$$

$$D_x = \begin{vmatrix} 1 & -7 & 0 \\ 11 & 0 & 3 \\ 0 & 2 & 1 \end{vmatrix} = 1\begin{vmatrix} 0 & 3 \\ 2 & 1 \end{vmatrix} + 7\begin{vmatrix} 11 & 3 \\ 0 & 1 \end{vmatrix} + 0\begin{vmatrix} 11 & 0 \\ 0 & 2 \end{vmatrix} = -6 + 77 + 0 = 71$$

$$D_y = \begin{vmatrix} -1 & 1 & 0 \\ 1 & 11 & 3 \\ 0 & 0 & 1 \end{vmatrix} = -1\begin{vmatrix} 11 & 3 \\ 0 & 1 \end{vmatrix} - 1\begin{vmatrix} 1 & 3 \\ 0 & 1 \end{vmatrix} + 0\begin{vmatrix} 1 & 11 \\ 0 & 0 \end{vmatrix} = -11 - 1 + 0 = -12$$

$$D_z = \begin{vmatrix} -1 & -7 & 1 \\ 1 & 0 & 11 \\ 0 & 2 & 0 \end{vmatrix} = -1\begin{vmatrix} 0 & 11 \\ 2 & 0 \end{vmatrix} + 7\begin{vmatrix} 1 & 11 \\ 0 & 0 \end{vmatrix} + 1\begin{vmatrix} 1 & 0 \\ 0 & 2 \end{vmatrix} = 22 + 0 + 2 = 24$$

Now use Cramer's rule:

$$x = \frac{D_x}{D} = \frac{71}{13} \qquad y = \frac{D_y}{D} = -\frac{12}{13} \qquad z = \frac{D_z}{D} = \frac{24}{13}$$

The solution is $\left(\frac{71}{13}, -\frac{12}{13}, \frac{24}{13}\right)$.

19. First find the determinants:

$$D = \begin{vmatrix} 1 & -1 & 0 \\ 3 & 0 & 1 \\ 0 & 1 & -2 \end{vmatrix} = 1\begin{vmatrix} 0 & 1 \\ 1 & -2 \end{vmatrix} + 1\begin{vmatrix} 3 & 1 \\ 0 & -2 \end{vmatrix} + 0\begin{vmatrix} 3 & 0 \\ 0 & 1 \end{vmatrix} = -1 - 6 + 0 = -7$$

$$D_x = \begin{vmatrix} 2 & -1 & 0 \\ 11 & 0 & 1 \\ -3 & 1 & -2 \end{vmatrix} = 2\begin{vmatrix} 0 & 1 \\ 1 & -2 \end{vmatrix} + 1\begin{vmatrix} 11 & 1 \\ -3 & -2 \end{vmatrix} + 0\begin{vmatrix} 11 & 0 \\ -3 & 1 \end{vmatrix} = -2 - 19 + 0 = -21$$

$$D_y = \begin{vmatrix} 1 & 2 & 0 \\ 3 & 11 & 1 \\ 0 & -3 & -2 \end{vmatrix} = 1\begin{vmatrix} 11 & 1 \\ -3 & -2 \end{vmatrix} - 2\begin{vmatrix} 3 & 1 \\ 0 & -2 \end{vmatrix} + 0\begin{vmatrix} 3 & 11 \\ 0 & -3 \end{vmatrix} = -19 + 12 + 0 = -7$$

$$D_z = \begin{vmatrix} 1 & -1 & 2 \\ 3 & 0 & 11 \\ 0 & 1 & -3 \end{vmatrix} = 1\begin{vmatrix} 0 & 11 \\ 1 & -3 \end{vmatrix} + 1\begin{vmatrix} 3 & 11 \\ 0 & -3 \end{vmatrix} + 2\begin{vmatrix} 3 & 0 \\ 0 & 1 \end{vmatrix} = -11 - 9 + 6 = -14$$

Now use Cramer's rule:

$$x = \frac{D_x}{D} = \frac{-21}{-7} = 3 \qquad y = \frac{D_y}{D} = \frac{-7}{-7} = 1 \qquad z = \frac{D_z}{D} = \frac{-14}{-7} = 2$$

The solution is $(3, 1, 2)$.

21. First rewrite the system as:
$$-10x + y = 100$$
$$-12x + y = 0$$

Now find the determinants:

$$D = \begin{vmatrix} -10 & 1 \\ -12 & 1 \end{vmatrix} = -10 + 12 = 2$$

$$D_x = \begin{vmatrix} 100 & 1 \\ 0 & 1 \end{vmatrix} = 100 - 0 = 100$$

$$D_y = \begin{vmatrix} -10 & 100 \\ -12 & 0 \end{vmatrix} = 0 + 1200 = 1200$$

Now using Cramer's rule:

$$x = \frac{D_x}{D} = \frac{100}{2} = 50 \qquad\qquad y = \frac{D_y}{D} = \frac{1200}{2} = 600$$

The company must sell 50 items per week to break even.

23. Evaluating the function: $f(0) = \frac{1}{2}(0) + 3 = 3$ **25.** Evaluating the function: $g(2) = 2^2 - 4 = 0$

27. Evaluating the function: $f(-4) = \frac{1}{2}(-4) + 3 = 1$

29. Evaluating the function: $f\big[g(2)\big] = f(0) = \frac{1}{2}(0) + 3 = 3$

31. Translating into symbols: $3x + 2$ **33.** Simplifying: $25 - \dfrac{385}{9} = \dfrac{225}{9} - \dfrac{385}{9} = -\dfrac{160}{9}$

35. Simplifying: $0.08(4,000) = 320$

37. Applying the distributive property: $10(0.2x + 0.5y) = 2x + 5y$

39. Solving the equation:

$$x + (3x + 2) = 26$$
$$4x + 2 = 26$$
$$4x = 24$$
$$x = 6$$

41. Substituting $z = -3y + 17$ into the second equation:

$$5y + 20(-3y + 17) = 65$$
$$5y - 60y + 340 = 65$$
$$-55y + 340 = 65$$
$$-55y = -275$$
$$y = 5$$

The solution is $y = 5$, $z = 2$.

43. First find the determinants:

$$D = \begin{vmatrix} a & b \\ b & a \end{vmatrix} = a^2 - b^2 = (a+b)(a-b)$$

$$D_x = \begin{vmatrix} -1 & b \\ 1 & a \end{vmatrix} = -a + b = -(a-b)$$

$$D_y = \begin{vmatrix} a & -1 \\ b & 1 \end{vmatrix} = a + b$$

Now using Cramer's rule:

$$x = \frac{D_x}{D} = \frac{-(a+b)}{(a+b)(a-b)} = \frac{1}{b-a} \qquad\qquad y = \frac{D_y}{D} = \frac{a+b}{(a+b)(a-b)} = \frac{1}{a-b}$$

The solution is $\left(\dfrac{1}{b-a}, \dfrac{1}{a-b} \right)$.

45.   First find the determinants:

$$D = \begin{vmatrix} a^2 & b \\ b^2 & a \end{vmatrix} = a^3 - b^3 = (a-b)(a^2 + ab + b^2)$$

$$D_x = \begin{vmatrix} 1 & b \\ 1 & a \end{vmatrix} = a - b$$

$$D_y = \begin{vmatrix} a^2 & 1 \\ b^2 & 1 \end{vmatrix} = a^2 - b^2 = (a-b)(a+b)$$

Now using Cramer's rule:

$$x = \frac{D_x}{D} = \frac{a-b}{(a-b)(a^2 + ab + b^2)} = \frac{1}{a^2 + ab + b^2} \qquad\qquad y = \frac{D_y}{D} = \frac{(a-b)(a+b)}{(a-b)(a^2 + ab + b^2)} = \frac{a+b}{a^2 + ab + b^2}$$

The solution is $\left(\dfrac{1}{a^2 + ab + b^2}, \dfrac{a+b}{a^2 + ab + b^2} \right)$.

47.   The system is:
$$x + 2y = 1$$
$$3x + 4y = 0$$

4.5 Applications

1.   Let x and y represent the two numbers. The system of equations is:
$$y = 2x + 3$$
$$x + y = 18$$

Substituting into the second equation:
$$x + 2x + 3 = 18$$
$$3x = 15$$
$$x = 5$$
$$y = 2(5) + 3 = 13$$

The two numbers are 5 and 13.

3.   Let x and y represent the two numbers. The system of equations is:
$$y - x = 6$$
$$2x = 4 + y$$

The second equation is $y = 2x - 4$. Substituting into the first equation:
$$2x - 4 - x = 6$$
$$x = 10$$
$$y = 2(10) - 4 = 16$$

The two numbers are 10 and 16.

5.   Let x, y, and z represent the three numbers. The system of equations is:
$$x + y + z = 8$$
$$2x = z - 2$$
$$x + z = 5$$

The third equation is $z = 5 - x$. Substituting into the second equation:
$$2x = 5 - x - 2$$
$$3x = 3$$
$$x = 1$$
$$z = 5 - 1 = 4$$

Substituting into the first equation:
$$1 + y + 4 = 8$$
$$y = 3$$

The three numbers are 1, 3, and 4.

7.   Let a represent the number of adult tickets and c represent the number of children's tickets. The system of equations is:
$$a + c = 925$$
$$2a + c = 1150$$
Multiply the first equation by -1:
$$-a - c = -925$$
$$2a + c = 1150$$
Adding yields:
$$a = 225$$
$$c = 700$$
There were 225 adult tickets and 700 children's tickets sold.

9.   Let x represent the amount invested at 6% and y represent the amount invested at 7%. The system of equations is:
$$x + y = 20000$$
$$0.06x + 0.07y = 1280$$
Multiplying the first equation by -0.06:
$$-0.06x - 0.06y = -1200$$
$$0.06x + 0.07y = 1280$$
Adding yields:
$$0.01y = 80$$
$$y = 8000$$
$$x = 12000$$
Mr. Jones invested $12,000 at 6% and $8,000 at 7%.

11.   Let x represent the amount invested at 6% and $2x$ represent the amount invested at 7.5%. The equation is:
$$0.075(2x) + 0.06(x) = 840$$
$$0.21x = 840$$
$$x = 4000$$
$$2x = 8000$$
Susan invested $4,000 at 6% and $8,000 at 7.5%.

13.   Let x, y and z represent the amounts invested in the three accounts. The system of equations is:
$$x + y + z = 2200$$
$$z = 3x$$
$$0.06x + 0.08y + 0.09z = 178$$
Substituting into the first equation:
$$x + y + 3x = 2200$$
$$4x + y = 2200$$
Substituting into the third equation:
$$0.06x + 0.08y + 0.09(3x) = 178$$
$$0.33x + 0.08y = 178$$
The system of equations becomes:
$$4x + y = 2200$$
$$0.33x + 0.08y = 178$$
Multiply the first equation by -0.08:
$$-0.32x - 0.08y = -176$$
$$0.33x + 0.08y = 178$$
Adding yields:
$$0.01x = 2$$
$$x = 200$$
$$z = 3(200) = 600$$
$$y = 2200 - 4(200) = 1400$$
He invested $200 at 6%, $1,400 at 8%, and $600 at 9%.

15.   Let x represent the amount of 20% alcohol and y represent the amount of 50% alcohol. The system of equations is:
$$x + y = 9$$
$$0.20x + 0.50y = 0.30(9)$$
Multiplying the first equation by –0.2:
$$-0.20x - 0.20y = -1.8$$
$$0.20x + 0.50y = 2.7$$
Adding yields:
$$0.30y = 0.9$$
$$y = 3$$
$$x = 6$$
The mixture contains 3 gallons of 50% alcohol and 6 gallons of 20% alcohol.

17.   Let x represent the amount of 20% disinfectant and y represent the amount of 14% disinfectant.
The system of equations is:
$$x + y = 15$$
$$0.20x + 0.14y = 0.16(15)$$
Multiplying the first equation by –0.14:
$$-0.14x - 0.14y = -2.1$$
$$0.20x + 0.14y = 2.4$$
Adding yields:
$$0.06x = 0.3$$
$$x = 5$$
$$y = 10$$
The mixture contains 5 gallons of 20% disinfectant and 10 gallons of 14% disinfectant.

19.   Let x represent the amount 40% copper alloy and y represent the amount of 60% copper alloy. The system of equations is:
$$x + y = 50$$
$$0.4x + 0.6y = 0.55(50)$$
Multiplying the first equation by –0.4:
$$-0.4x - 0.4y = -20$$
$$0.4x + 0.6y = 27.5$$
Adding yields:
$$0.2y = 7.5$$
$$y = 37.5$$
$$x = 12.5$$
The mixture contains 12.5 pounds of 40% copper alloy and 37.5 pounds of 60% copper alloy.

21.   Let b represent the rate of the boat and c represent the rate of the current. The system of equations is:
$$2(b + c) = 24$$
$$3(b - c) = 18$$
The system of equations simplifies to:
$$b + c = 12$$
$$b - c = 6$$
Adding yields:
$$2b = 18$$
$$b = 9$$
$$c = 3$$
The rate of the boat is 9 mph and the rate of the current is 3 mph.

23.   Let a represent the rate of the airplane and w represent the rate of the wind. The system of equations is:

$$2(a+w) = 600$$
$$\frac{5}{2}(a-w) = 600$$

The system of equations simplifies to:

$$a + w = 300$$
$$a - w = 240$$

Adding yields:

$$2a = 540$$
$$a = 270$$
$$w = 30$$

The rate of the airplane is 270 mph and the rate of the wind is 30 mph.

25.   Let n represent the number of nickels and d represent the number of dimes. The system of equations is:

$$n + d = 20$$
$$0.05n + 0.10d = 1.40$$

Multiplying the first equation by –0.05:

$$-0.05n - 0.05d = -1$$
$$0.05n + 0.10d = 1.40$$

Adding yields:

$$0.05d = 0.40$$
$$d = 8$$
$$n = 12$$

Bob has 12 nickels and 8 dimes.

27.   Let n, d, and q represent the number of nickels, dimes, and quarters. The system of equations is:

$$n + d + q = 9$$
$$0.05n + 0.10d + 0.25q = 1.20$$
$$d = n$$

Substituting into the first equation:

$$n + n + q = 9$$
$$2n + q = 9$$

Substituting into the second equation:

$$0.05n + 0.10n + 0.25q = 1.20$$
$$0.15n + 0.25q = 1.20$$

The system of equations becomes:

$$2n + q = 9$$
$$0.15n + 0.25q = 1.20$$

Multiplying the first equation by –0.25:

$$-0.50n - 0.25q = -2.25$$
$$0.15n + 0.25q = 1.20$$

Adding yields:

$$-0.35n = -1.05$$
$$n = 3$$
$$d = 3$$
$$q = 9 - 2(3) = 3$$

The collection contains 3 nickels, 3 dimes, and 3 quarters.

29. Let n, d, and q represent the number of nickels, dimes, and quarters. The system of equations is:
$$n + d + q = 140$$
$$0.05n + 0.10d + 0.25q = 10.00$$
$$d = 2q$$

Substituting into the first equation:
$$n + 2q + q = 140$$
$$n + 3q = 140$$

Substituting into the second equation:
$$0.05n + 0.10(2q) + 0.25q = 10.00$$
$$0.05n + 0.45q = 10.00$$

The system of equations becomes:
$$n + 3q = 140$$
$$0.05n + 0.45q = 10.00$$

Multiplying the first equation by –0.05:
$$-0.05n - 0.15q = -7$$
$$0.05n + 0.45q = 10$$

Adding yields:
$$0.30q = 3$$
$$q = 10$$
$$d = 2(10) = 20$$
$$n = 140 - 3(10) = 110$$

There are 110 nickels in the collection.

31. Let $x = mp + b$ represent the relationship. Using the points (2,300) and (1.5,400) results in the system:
$$300 = 2m + b$$
$$400 = 1.5m + b$$

Multiplying the second equation by –1:
$$300 = 2m + b$$
$$-400 = -1.5m - b$$

Adding yields:
$$-100 = 0.5m$$
$$m = -200$$
$$b = 300 - 2(-200) = 700$$

The equation is $x = -200p + 700$. Substituting $p = 3$: $x = -200(3) + 700 = 100$ items

33. The system of equations is:
$$a + b + c = 128$$
$$9a + 3b + c = 128$$
$$25a + 5b + c = 0$$

Multiply the first equation by –1 and add it to the second equation:
$$-a - b - c = -128$$
$$9a + 3b + c = 128$$

Adding yields:
$$8a + 2b = 0$$
$$4a + b = 0$$

Multiply the first equation by –1 and add it to the third equation:
$$-a - b - c = -128$$
$$25a + 5b + c = 0$$

Adding yields:
$$24a + 4b = -128$$
$$6a + b = -32$$

The system simplifies to:
$$4a + b = 0$$
$$6a + b = -32$$

Multiplying the first equation by -1:
$$-4a - b = 0$$
$$6a + b = -32$$
Adding yields:
$$2a = -32$$
$$a = -16$$
Substituting to find b:
$$4(-16) + b = 0$$
$$b = 64$$
Substituting to find c:
$$-16 + 64 + c = 128$$
$$c = 80$$
The equation for the height is $h = -16t^2 + 64t + 80$.

35. Graphing the inequality:

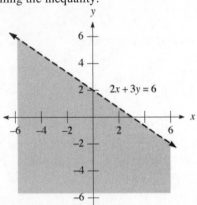

37. Graphing the inequality:

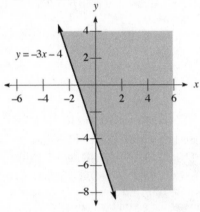

39. Graphing the inequality:

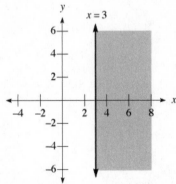

41. No, the graph does not include the boundary line.

43. Substituting $x = 4 - y$ into the second equation:
$$(4 - y) - 2y = 4$$
$$4 - 3y = 4$$
$$-3y = 0$$
$$y = 0$$
The solution is $(4,0)$.

45. Solving the inequality:
$$20x + 9,300 > 18,000$$
$$20x > 8,700$$
$$x > 435$$

47.   Graphing the data:

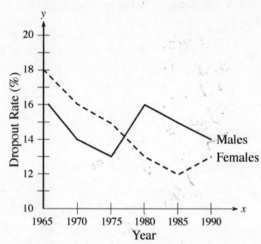

49.   **a.**   Using the points (1975,13) and (1980,16) in the model $M = mx + b$ results in the system:

$1975m + b = 13$
$1980m + b = 16$

Multiply the first equation by -1:

$-1975m - b = -13$
$1980m + b = 16$

Adding yields:

$5m = 3$
$m = 0.6$
$b = 16 - 1980(0.6) = -1172$

The equation is $M = 0.6x - 1172$.

b.   Using the points (1975,15) and (1980,13) in the model $F = mx + b$ results in the system:

$1975m + b = 15$
$1980m + b = 13$

Multiply the first equation by -1:

$-1975m - b = -15$
$1980m + b = 13$

Adding yields:

$5m = -2$
$m = -0.4$
$b = 13 - 1980(-0.4) = 805$

The equation is $F = -0.4x + 805$.

c.   Finding where the two rates are equal:

$0.6x - 1172 = -0.4x + 805$
$x = 1977$

The dropout rates are equal in the year 1977.

4.6 Systems of Linear Inequalities

1. Graphing the solution set:

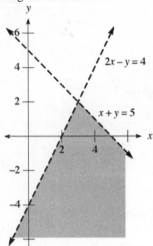

3. Graphing the solution set:

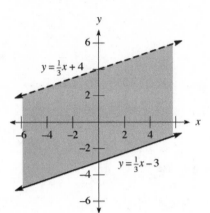

5. Graphing the solution set:

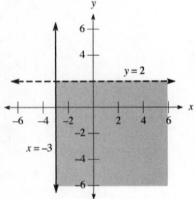

7. Graphing the solution set:

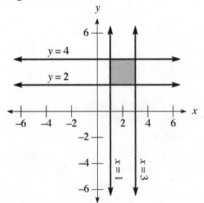

9. Graphing the solution set:

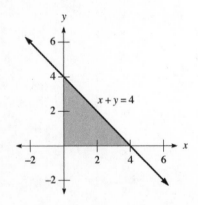

11. Graphing the solution set:

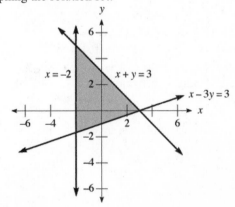

13.   Graphing the solution set:

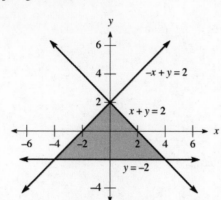

15.   Graphing the solution set:

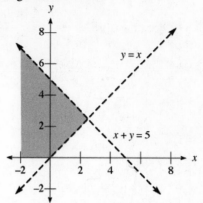

17.   Graphing the solution set:

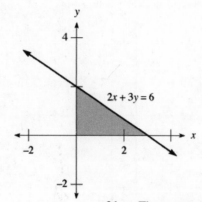

19.   The system of inequalities is:
$$x + y \leq 4$$
$$-x + y < 4$$

21.   The system of inequalities is:
$$x + y \geq 4$$
$$-x + y < 4$$

23.   a.   The system of inequalities is:
$$0.55x + 0.65y \leq 40$$
$$x \geq 2y$$
$$x > 15$$
$$y \geq 0$$

Graphing the solution set:

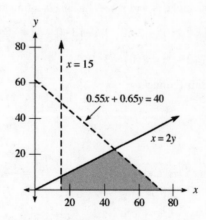

b.   Substitute $x = 20$:
$$2y \leq 20$$
$$y \leq 10$$
The most he can purchase is 10 65-cent stamps.

25.   The *x*-intercept is 3, the *y*-intercept is 6, and the slope is –2.

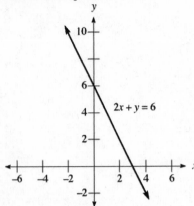

27.   The *x*-intercept is –2, there is no *y*-intercept, and there is no slope.

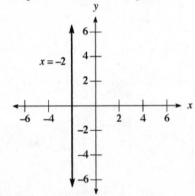

29.   First find the slope: $m = \dfrac{-1-2}{4-(-3)} = \dfrac{-3}{4+3} = -\dfrac{3}{7}$. Using the point-slope formula:

$$y - 2 = -\tfrac{3}{7}(x+3)$$
$$y - 2 = -\tfrac{3}{7}x - \tfrac{9}{7}$$
$$y = -\tfrac{3}{7}x + \tfrac{5}{7}$$

31.   Since the line is vertical, its equation is $x = 4$.

33.   The domain is all real numbers and the range is $\{y \mid y \geq -9\}$. This is a function.

35.   Evaluating the function: $h(0) + g(0) = \left[3 \cdot 0^2 - 2 \cdot 0 - 8\right] + \left[3 \cdot 0 + 4\right] = -8 + 4 = -4$

37.   Evaluating the function: $g\left[f(2)\right] = g(2-2) = g(0) = 3 \cdot 0 + 4 = 4$

39.   The variation equation is $z = Kxy^3$. Substituting $x = 5$, $y = 2$, and $z = 15$:

$$15 = K(5)(2)^3$$
$$15 = 40K$$
$$K = \tfrac{3}{8}$$

The equation is $z = \tfrac{3}{8}xy^3$. Substituting $x = 2$ and $y = 3$: $z = \tfrac{3}{8}(2)(3)^3 = \tfrac{3}{8} \cdot 54 = \tfrac{81}{4}$

Chapter 4 Review/Test

1. Adding the two equations yields:
    $$3x = 18$$
    $$x = 6$$
    Substituting to find y:
    $$6 + y = 4$$
    $$y = -2$$
    The solution is $(6, -2)$.

2. Multiply the second equation by -1:
    $$3x + y = 2$$
    $$-2x - y = 0$$
    Adding yields $x = 2$. Substituting to find y:
    $$6 + y = 2$$
    $$y = -4$$
    The solution is $(2, -4)$.

3. Multiply the second equation by 2:
    $$2x - 4y = 5$$
    $$-2x + 4y = 6$$
    Adding yields $0 = 11$, which is false. There is no solution (parallel lines).

4. Multiply the second equation by 2:
    $$5x - 2y = 7$$
    $$6x + 2y = 4$$
    Adding yields:
    $$11x = 11$$
    $$x = 1$$
    Substituting to find y:
    $$3 + y = 2$$
    $$y = -1$$
    The solution is $(1, -1)$.

5. Multiply the second equation by -2:
    $$6x - 5y = -5$$
    $$-6x - 2y = -2$$
    Adding yields:
    $$-7y = -7$$
    $$y = 1$$
    Substituting to find x:
    $$3x + 1 = 1$$
    $$3x = 0$$
    $$x = 0$$
    The solution is $(0, 1)$.

6. Divide the first equation by 2 and the second equation by 3:
    $$3x + 2y = 4$$
    $$3x + 2y = 4$$
    Since these two lines are identical, there is no unique solution (lines coincide).

7. Multiply the first equation by 4 and the second equation by 3:
$$12x - 28y = 8$$
$$-12x + 18y = -18$$
Adding yields:
$$-10y = -10$$
$$y = 1$$
Substitute to find x:
$$3x - 7 = 2$$
$$3x = 9$$
$$x = 3$$
The solution is $(3,1)$.

8. Multiply the first equation by 2 and the second equation by -3:
$$12x + 10y = 18$$
$$-12x - 9y = -18$$
Adding yields $y = 0$. Substitute to find x:
$$6x = 9$$
$$x = \frac{3}{2}$$
The solution is $\left(\frac{3}{2}, 0\right)$.

9. Multiply the first equation by 3 and the second equation by 4:
$$-21x + 12y = -3$$
$$20x - 12y = 0$$
Adding yields $x = 3$. Substitute to find y:
$$-21 + 4y = -1$$
$$4y = 20$$
$$y = 5$$
The solution is $(3,5)$.

10. To clear each equation of fractions, multiply the first equation by 4 and the second equation by 4:
$$2x - 3y = -16$$
$$x + 6y = 52$$
Multiply the first equation by 2:
$$4x - 6y = -32$$
$$x + 6y = 52$$
Adding yields:
$$5x = 20$$
$$x = 4$$
Substitute to find y:
$$4 + 6y = 52$$
$$6y = 48$$
$$y = 8$$
The solution is $(4,8)$.

11. To clear each equation of fractions, multiply the first equation by 6 and the second equation by 6:
$$4x - y = 0$$
$$8x + 5y = 84$$
Multiply the first equation by –2:
$$-8x + 2y = 0$$
$$8x + 5y = 84$$
Adding yields:
$$7y = 84$$
$$y = 12$$
Substitute to find x:
$$4x - 12 = 0$$
$$4x = 12$$
$$x = 3$$
The solution is (3,12).

12. To clear each equation of fractions, multiply the first equation by 6 and the second equation by 20:
$$-3x + 2y = -13$$
$$16x + 15y = 18$$
Multiply the first equation by –15 and the second equation by 2:
$$45x - 30y = 195$$
$$32x + 30y = 36$$
Adding yields:
$$77x = 231$$
$$x = 3$$
Substitute to find y:
$$-9 + 2y = -13$$
$$2y = -4$$
$$y = -2$$
The solution is (3,–2).

13. Substitute into the first equation:
$$x + x - 1 = 2$$
$$2x - 1 = 2$$
$$2x = 3$$
$$x = \frac{3}{2}$$
$$y = \frac{3}{2} - 1 = \frac{1}{2}$$
The solution is $\left(\frac{3}{2}, \frac{1}{2}\right)$.

14. Substitute into the first equation:
$$2x - 3(2x - 7) = 5$$
$$2x - 6x + 21 = 5$$
$$-4x = -16$$
$$x = 4$$
$$y = 2(4) - 7 = 1$$
The solution is (4,1).

15. Write the first equation as $y = 4 - x$. Substitute into the second equation:
$$2x + 5(4 - x) = 2$$
$$2x + 20 - 5x = 2$$
$$-3x = -18$$
$$x = 6$$
$$y = 4 - 6 = -2$$
The solution is (6,–2).

16. Write the first equation as $y = 3 - x$. Substitute into the second equation:

$$2x + 5(3 - x) = -6$$
$$2x + 15 - 5x = -6$$
$$-3x = -21$$
$$x = 7$$
$$y = 3 - 7 = -4$$

The solution is (7,–4).

17. Substitute into the first equation:

$$3(-3y + 4) + 7y = 6$$
$$-9y + 12 + 7y = 6$$
$$-2y = -6$$
$$y = 3$$
$$x = -9 + 4 = -5$$

The solution is (–5,3).

18. Substitute into the first equation:

$$5x - (5x - 3) = 4$$
$$5x - 5x + 3 = 4$$
$$3 = 4$$

Since this statement is false, there is no solution (parallel lines).

19. Adding the first and second equations yields:

$$2x - 2z = -2$$
$$x - z = -1$$

Adding the second and third equations yields $2x - 5z = -14$. So the system becomes:

$$x - z = -1$$
$$2x - 5z = -14$$

Multiply the first equation by –2:

$$-2x + 2z = 2$$
$$2x - 5z = -14$$

Adding yields:

$$-3z = -12$$
$$z = 4$$

Substitute to find x:

$$x - 4 = -1$$
$$x = 3$$

Substitute to find y:

$$3 + y + 4 = 6$$
$$y = -1$$

The solution is (3,–1,4).

20. Multiply the first equation by –1 and add it to the second equation:

$$-3x - 2y - z = -4$$
$$2x - 4y + z = -1$$

Adding yields:

$$-x - 6y = -5$$
$$x + 6y = 5$$

Multiply the first equation by –3 and add it to the third equation:

$$-9x - 6y - 3z = -12$$
$$x + 6y + 3z = -4$$

Adding yields:

$$-8x = -16$$
$$x = 2$$

Substituting to find y:
$$2 + 6y = 5$$
$$6y = 3$$
$$y = \frac{1}{2}$$

Substituting to find z:
$$2 + 6\left(\frac{1}{2}\right) + 3z = -4$$
$$3z + 5 = -4$$
$$3z = -9$$
$$z = -3$$

The solution is $\left(2, \frac{1}{2}, -3\right)$.

21.   Multiply the second equation by 2 and add it to the first equation:
$$5x + 8y - 4z = -7$$
$$14x + 8y + 4z = -4$$

Adding yields the equation $19x + 16y = -11$. Multiply the first equation by 2 and add it to the third equation:
$$10x + 16y - 8z = -14$$
$$3x - 2y + 8z = 8$$

Adding yields the equation $13x + 14y = -6$. So the system of equations becomes:
$$19x + 16y = -11$$
$$13x + 14y = -6$$

Multiply the first equation by 7 and the second equation by -8:
$$133x + 112y = -77$$
$$-104x - 112y = 48$$

Adding yields:
$$29x = -29$$
$$x = -1$$

Substituting to find y:
$$-13 + 14y = -6$$
$$14y = 7$$
$$y = \frac{1}{2}$$

Substituting to find z:
$$7(-1) + 4\left(\frac{1}{2}\right) + 2z = -2$$
$$-5 + 2z = -2$$
$$2z = 3$$
$$z = \frac{3}{2}$$

The solution is $\left(-1, \frac{1}{2}, \frac{3}{2}\right)$.

22.   Multiply the first equation by -2 and add it to the the second equation:
$$-10x + 6y + 12z = -10$$
$$4x - 6y - 3z = 4$$

Adding yields:
$$-6x + 9z = -6$$
$$2x - 3z = 2$$

Multiply the first equation by 3 and add it to the third equation:
$$15x - 9y - 18z = 15$$
$$-x + 9y + 9z = 7$$

Adding yields the equation $14x - 9z = 22$. So the system becomes:
$$2x - 3z = 2$$
$$14x - 9z = 22$$

Multiply the first equation by –3:
$$-6x + 9z = -6$$
$$14x - 9z = 22$$
Adding yields:
$$8x = 16$$
$$x = 2$$
Substituting to find z:
$$4 - 3z = 2$$
$$-3z = -2$$
$$z = \tfrac{2}{3}$$
Substituting to find y:
$$5(2) - 3y - 6\left(\tfrac{2}{3}\right) = 5$$
$$6 - 3y = 5$$
$$-3y = -1$$
$$y = \tfrac{1}{3}$$
The solution is $\left(2, \tfrac{1}{3}, \tfrac{2}{3}\right)$.

23. Multiply the second equation by 2 and add it to the third equation:
$$-6x + 8y - 2z = 4$$
$$6x - 8y + 2z = -4$$

Adding yields $0 = 0$. Since this is a true statement, there is no unique solution (dependent system).

24. Multiply the second equation by 4 and add it to the first equation:
$$4x - 6y + 8z = 4$$
$$20x + 4y - 8z = 16$$

Adding yields:
$$24x - 2y = 20$$
$$12x - y = 10$$

Multiply the second equation by 6 and add it to the third equation:
$$30x + 6y - 12z = 24$$
$$6x - 9y + 12z = 6$$

Adding yields:
$$36x - 3y = 30$$
$$12x - y = 10$$

Since these two equations are identical, there is no unique solution (dependent system).

25. Multiply the third equation by 2 and add it to the second equation:
$$3x - 2z = -2$$
$$10y + 2z = -2$$

Adding yields the equation $3x + 10y = -4$. So the system becomes:
$$2x - y = 5$$
$$3x + 10y = -4$$

Multiplying the first equation by 10:
$$20x - 10y = 50$$
$$3x + 10y = -4$$

Adding yields:
$$23x = 46$$
$$x = 2$$

Substituting to find y:
$$4 - y = 5$$
$$y = -1$$

Substituting to find z:
$$-5 + z = -1$$
$$z = 4$$
The solution is $(2, -1, 4)$.

26. Adding the first and second equations yields $x - z = -1$. Since this is the same as the third equation, there is no unique solution (dependent system).

27. Evaluating the determinant: $\begin{vmatrix} 2 & 3 \\ -5 & 4 \end{vmatrix} = 2 \bullet 4 - 3(-5) = 8 + 15 = 23$

28. Evaluating the determinant: $\begin{vmatrix} 3 & 0 \\ 5 & -1 \end{vmatrix} = 3(-1) - 0 \bullet 5 = -3 - 0 = -3$

29. Evaluating the determinant: $\begin{vmatrix} 1 & 0 \\ -7 & -3 \end{vmatrix} = 1(-3) - 0(-7) = -3 - 0 = -3$

30. Evaluating the determinant: $\begin{vmatrix} 3 & 0 & 2 \\ -1 & 4 & 0 \\ 2 & 0 & 0 \end{vmatrix} = 2 \begin{vmatrix} 0 & 2 \\ 4 & 0 \end{vmatrix} = 2(0 - 8) = -16$

31. Evaluating the determinant: $\begin{vmatrix} 3 & -1 & 0 \\ 0 & 2 & -4 \\ 6 & 0 & 2 \end{vmatrix} = 3 \begin{vmatrix} 2 & -4 \\ 0 & 2 \end{vmatrix} + 1 \begin{vmatrix} 0 & -4 \\ 6 & 2 \end{vmatrix} = 3(4 - 0) + 1(0 + 24) = 12 + 24 = 36$

32. Evaluating the determinant: $\begin{vmatrix} -3 & -2 & 0 \\ 0 & -4 & 2 \\ 5 & 1 & 1 \end{vmatrix} = -3 \begin{vmatrix} -4 & 2 \\ 1 & 1 \end{vmatrix} + 2 \begin{vmatrix} 0 & 2 \\ 5 & 1 \end{vmatrix} = -3(-4 - 2) + 2(0 - 10) = 18 - 20 = -2$

33. Solving for x:
$$\begin{vmatrix} 2 & 3x \\ -1 & 2x \end{vmatrix} = 4$$
$$4x + 3x = 4$$
$$7x = 4$$
$$x = \frac{4}{7}$$

34. Solving for x:
$$\begin{vmatrix} 4x & x \\ 3 & 1 \end{vmatrix} = -4$$
$$4x(1) - x(3) = -4$$
$$4x - 3x = -4$$
$$x = -4$$

35. First find the determinants:
$$D = \begin{vmatrix} 3 & -5 \\ 7 & -2 \end{vmatrix} = -6 + 35 = 29$$
$$D_x = \begin{vmatrix} 4 & -5 \\ 3 & -2 \end{vmatrix} = -8 + 15 = 7$$
$$D_y = \begin{vmatrix} 3 & 4 \\ 7 & 3 \end{vmatrix} = 9 - 28 = -19$$

Now using Cramer's rule:
$$x = \frac{D_x}{D} = \frac{7}{29} \qquad y = \frac{D_y}{D} = -\frac{19}{29}$$

The solution is $\left(\frac{7}{29}, -\frac{19}{29} \right)$.

36. First find the determinants:
$$D = \begin{vmatrix} 7 & -5 \\ 4 & 3 \end{vmatrix} = 21 + 20 = 41$$
$$D_x = \begin{vmatrix} 8 & -5 \\ 2 & 3 \end{vmatrix} = 24 + 10 = 34$$
$$D_y = \begin{vmatrix} 7 & 8 \\ 4 & 2 \end{vmatrix} = 14 - 32 = -18$$

Now using Cramer's rule:
$$x = \frac{D_x}{D} = \frac{34}{41} \qquad y = \frac{D_y}{D} = -\frac{18}{41}$$

The solution is $\left(\frac{34}{41}, -\frac{18}{41} \right)$.

37. First find the determinants:

$$D = \begin{vmatrix} 3 & -6 \\ 2 & -4 \end{vmatrix} = -12 + 12 = 0$$

$$D_x = \begin{vmatrix} 9 & -6 \\ 6 & -4 \end{vmatrix} = -36 + 36 = 0$$

$$D_y = \begin{vmatrix} 3 & 9 \\ 2 & 6 \end{vmatrix} = 18 - 18 = 0$$

Since all of the determinants are zero, there is no unique solution (lines coincide).

38. First find the determinants:

$$D = \begin{vmatrix} 6 & -9 \\ 7 & 3 \end{vmatrix} = 18 + 63 = 81$$

$$D_x = \begin{vmatrix} 5 & -9 \\ 4 & 3 \end{vmatrix} = 15 + 36 = 51$$

$$D_y = \begin{vmatrix} 6 & 5 \\ 7 & 4 \end{vmatrix} = 24 - 35 = -11$$

Now using Cramer's rule:

$$x = \frac{D_x}{D} = \frac{51}{81} = \frac{17}{27} \qquad y = \frac{D_y}{D} = -\frac{11}{81}$$

The solution is $\left(\frac{17}{27}, -\frac{11}{81} \right)$.

39. First find the determinants:

$$D = \begin{vmatrix} -6 & 3 \\ 5 & -8 \end{vmatrix} = 48 - 15 = 33$$

$$D_x = \begin{vmatrix} 7 & 3 \\ -2 & -8 \end{vmatrix} = -56 + 6 = -50$$

$$D_y = \begin{vmatrix} -6 & 7 \\ 5 & -2 \end{vmatrix} = 12 - 35 = -23$$

Now using Cramer's rule:

$$x = \frac{D_x}{D} = -\frac{50}{33} \qquad y = \frac{D_y}{D} = -\frac{23}{33}$$

The solution is $\left(-\frac{50}{33}, -\frac{23}{33} \right)$.

40. First find the determinants:

$$D = \begin{vmatrix} 2 & -1 & 3 \\ 5 & 2 & -1 \\ -1 & -3 & 2 \end{vmatrix} = 2\begin{vmatrix} 2 & -1 \\ -3 & 2 \end{vmatrix} + 1\begin{vmatrix} 5 & -1 \\ -1 & 2 \end{vmatrix} + 3\begin{vmatrix} 5 & 2 \\ -1 & -3 \end{vmatrix} = 2(4-3) + 1(10-1) + 3(-15+2) = 2 + 9 - 39 = -28$$

$$D_x = \begin{vmatrix} 4 & -1 & 3 \\ 3 & 2 & -1 \\ 1 & -3 & 2 \end{vmatrix} = 4\begin{vmatrix} 2 & -1 \\ -3 & 2 \end{vmatrix} + 1\begin{vmatrix} 3 & -1 \\ 1 & 2 \end{vmatrix} + 3\begin{vmatrix} 3 & 2 \\ 1 & -3 \end{vmatrix} = 4(4-3) + 1(6+1) + 3(-9-2) = 4 + 7 - 33 = -22$$

$$D_y = \begin{vmatrix} 2 & 4 & 3 \\ 5 & 3 & -1 \\ -1 & 1 & 2 \end{vmatrix} = 2\begin{vmatrix} 3 & -1 \\ 1 & 2 \end{vmatrix} - 4\begin{vmatrix} 5 & -1 \\ -1 & 2 \end{vmatrix} + 3\begin{vmatrix} 5 & 3 \\ -1 & 1 \end{vmatrix} = 2(6+1) - 4(10-1) + 3(5+3) = 14 - 36 + 24 = 2$$

$$D_z = \begin{vmatrix} 2 & -1 & 4 \\ 5 & 2 & 3 \\ -1 & -3 & 1 \end{vmatrix} = 2\begin{vmatrix} 2 & 3 \\ -3 & 1 \end{vmatrix} + 1\begin{vmatrix} 5 & 3 \\ -1 & 1 \end{vmatrix} + 4\begin{vmatrix} 5 & 2 \\ -1 & -3 \end{vmatrix} = 2(2+9) + 1(5+3) + 4(-15+2) = 22 + 8 - 52 = -22$$

Now using Cramer's rule:

$$x = \frac{D_x}{D} = \frac{-22}{-28} = \frac{11}{14} \qquad y = \frac{D_y}{D} = \frac{2}{-28} = -\frac{1}{14} \qquad z = \frac{D_z}{D} = \frac{-22}{-28} = \frac{11}{14}$$

The solution is $\left(\frac{11}{14}, -\frac{1}{14}, \frac{11}{14} \right)$.

41. First find the determinants:

$$D = \begin{vmatrix} 4 & -5 & 0 \\ 2 & 0 & 3 \\ 0 & 3 & -1 \end{vmatrix} = 4\begin{vmatrix} 0 & 3 \\ 3 & -1 \end{vmatrix} + 5\begin{vmatrix} 2 & 3 \\ 0 & -1 \end{vmatrix} + 0\begin{vmatrix} 2 & 0 \\ 0 & 3 \end{vmatrix} = 4(0-9) + 5(-2-0) + 0 = -36 - 10 = -46$$

$$D_x = \begin{vmatrix} -3 & -5 & 0 \\ 4 & 0 & 3 \\ 8 & 3 & -1 \end{vmatrix} = -3\begin{vmatrix} 0 & 3 \\ 3 & -1 \end{vmatrix} + 5\begin{vmatrix} 4 & 3 \\ 8 & -1 \end{vmatrix} + 0\begin{vmatrix} 4 & 0 \\ 8 & 3 \end{vmatrix} = -3(0-9) + 5(-4-24) + 0 = 27 - 140 = -113$$

$$D_y = \begin{vmatrix} 4 & -3 & 0 \\ 2 & 4 & 3 \\ 0 & 8 & -1 \end{vmatrix} = 4\begin{vmatrix} 4 & 3 \\ 8 & -1 \end{vmatrix} + 3\begin{vmatrix} 2 & 3 \\ 0 & -1 \end{vmatrix} + 0\begin{vmatrix} 2 & 4 \\ 0 & 8 \end{vmatrix} = 4(-4-24) + 3(-2-0) + 0 = -112 - 6 = -118$$

$$D_z = \begin{vmatrix} 4 & -5 & -3 \\ 2 & 0 & 4 \\ 0 & 3 & 8 \end{vmatrix} = 4\begin{vmatrix} 0 & 4 \\ 3 & 8 \end{vmatrix} + 5\begin{vmatrix} 2 & 4 \\ 0 & 8 \end{vmatrix} - 3\begin{vmatrix} 2 & 0 \\ 0 & 3 \end{vmatrix} = 4(0-12) + 5(16-0) - 3(6-0) = -48 + 80 - 18 = 14$$

Now using Cramer's rule:

$$x = \frac{D_x}{D} = \frac{-113}{-46} = \frac{113}{46} \qquad y = \frac{D_y}{D} = \frac{-118}{-46} = \frac{59}{23} \qquad z = \frac{D_z}{D} = \frac{14}{-46} = -\frac{7}{23}$$

The solution is $\left(\frac{113}{46}, \frac{59}{23}, -\frac{7}{23}\right)$.

42. First find the determinants:

$$D = \begin{vmatrix} 2 & -4 & 0 \\ 4 & 0 & -2 \\ 0 & 4 & -1 \end{vmatrix} = 2\begin{vmatrix} 0 & -2 \\ 4 & -1 \end{vmatrix} + 4\begin{vmatrix} 4 & -2 \\ 0 & -1 \end{vmatrix} + 0\begin{vmatrix} 4 & 0 \\ 0 & 4 \end{vmatrix} = 2(0+8) + 4(-4-0) + 0 = 16 - 16 = 0$$

$$D_x = \begin{vmatrix} 2 & -4 & 0 \\ 3 & 0 & -2 \\ 2 & 4 & -1 \end{vmatrix} = 2\begin{vmatrix} 0 & -2 \\ 4 & -1 \end{vmatrix} + 4\begin{vmatrix} 3 & -2 \\ 2 & -1 \end{vmatrix} + 0\begin{vmatrix} 3 & 0 \\ 2 & 4 \end{vmatrix} = 2(0+8) + 4(-3+4) + 0 = 16 + 4 = 20$$

$$D_y = \begin{vmatrix} 2 & 2 & 0 \\ 4 & 3 & -2 \\ 0 & 2 & -1 \end{vmatrix} = 2\begin{vmatrix} 3 & -2 \\ 2 & -1 \end{vmatrix} - 2\begin{vmatrix} 4 & -2 \\ 0 & -1 \end{vmatrix} + 0\begin{vmatrix} 4 & 3 \\ 0 & 2 \end{vmatrix} = 2(-3+4) - 2(-4-0) + 0 = 2 + 8 = 10$$

$$D_z = \begin{vmatrix} 2 & -4 & 2 \\ 4 & 0 & 3 \\ 0 & 4 & 2 \end{vmatrix} = 2\begin{vmatrix} 0 & 3 \\ 4 & 2 \end{vmatrix} + 4\begin{vmatrix} 4 & 3 \\ 0 & 2 \end{vmatrix} + 2\begin{vmatrix} 4 & 0 \\ 0 & 4 \end{vmatrix} = 2(0-12) + 4(8-0) + 2(16-0) = -24 + 32 + 32 = 40$$

Since $D = 0$ but not all other determinants are 0, there is no solution (inconsistent).

43. Let a represent the adult tickets and c represent the children's tickets. The system of equations is:

$$a + c = 127$$
$$2a + 1.5c = 214$$

Multiply the first equation by –2:

$$-2a - 2c = -254$$
$$2a + 1.5c = 214$$

Adding yields:

$$-0.5c = -40$$
$$c = 80$$
$$a = 47$$

There were 47 adult tickets and 80 children's tickets sold.

44.   Let d represent the number of dimes and q represent the number of quarters. The system of equations is:

$$d + q = 20$$
$$0.10d + 0.25q = 3.20$$

Multiply the first equation by –0.10:

$$-0.10d - 0.10q = -2.00$$
$$0.10d + 0.25q = 3.20$$

Adding yields:

$$0.15q = 1.20$$
$$q = 8$$
$$d = 12$$

John has 12 dimes and 8 quarters.

45.   Let x represent the amount invested at 12% and y represent the amount invested at 15%. The system of equations is:

$$x + y = 12000$$
$$0.12x + 0.15y = 1650$$

Multiplying the first equation by –0.12:

$$-0.12x - 0.12y = -1440$$
$$0.12x + 0.15y = 1650$$

Adding yields:

$$0.03y = 210$$
$$y = 7000$$
$$x = 5000$$

Ms. Jones invested $5,000 at 12% and $7,000 at 15%.

46.   Let b and c represent the rates of the boat and current. The system of equations is:

$$2(b + c) = 28$$
$$3(b - c) = 30$$

Simplifying the system:

$$b + c = 14$$
$$b - c = 10$$

Adding yields:

$$2b = 24$$
$$b = 12$$
$$c = 2$$

The boat's rate is 12 mph and the current's rate is 2 mph.

47.   Graphing the solution set:

48.   Graphing the solution set:

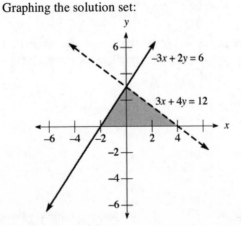

49. Graphing the solution set:

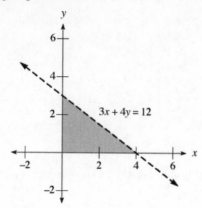

50. Graphing the solution set:

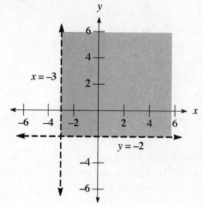

Chapter 5
Exponents and Polynomials

5.1 Properties of Exponents

1. Evaluating: $4^2 = 4 \cdot 4 = 16$

3. Evaluating: $-4^2 = -4 \cdot 4 = -16$

5. Evaluating: $-0.3^3 = -0.3 \cdot 0.3 \cdot 0.3 = -0.027$

7. Evaluating: $2^5 = 2 \cdot 2 \cdot 2 \cdot 2 \cdot 2 = 32$

9. Evaluating: $\left(\frac{1}{2}\right)^3 = \frac{1}{2} \cdot \frac{1}{2} \cdot \frac{1}{2} = \frac{1}{8}$

11. Evaluating: $\left(-\frac{5}{6}\right)^2 = \left(-\frac{5}{6}\right) \cdot \left(-\frac{5}{6}\right) = \frac{25}{36}$

13. Using properties of exponents: $x^5 \cdot x^4 = x^{5+4} = x^9$

15. Using properties of exponents: $\left(2^3\right)^2 = 2^{3 \cdot 2} = 2^6 = 64$

17. Using properties of exponents: $\left(-\frac{2}{3}x^2\right)^3 = \left(-\frac{2}{3}x^2\right)\left(-\frac{2}{3}x^2\right)\left(-\frac{2}{3}x^2\right) = -\frac{8}{27}x^6$

19. Using properties of exponents: $-3a^2\left(2a^4\right) = -6a^{2+4} = -6a^6$

21. Writing with positive exponents: $3^{-2} = \frac{1}{3^2} = \frac{1}{9}$

23. Writing with positive exponents: $(-2)^{-5} = \frac{1}{(-2)^5} = -\frac{1}{32}$

25. Writing with positive exponents: $\left(\frac{3}{4}\right)^{-2} = \left(\frac{4}{3}\right)^2 = \frac{16}{9}$

27. Writing with positive exponents: $\left(\frac{1}{3}\right)^{-2} + \left(\frac{1}{2}\right)^{-3} = 3^2 + 2^3 = 9 + 8 = 17$

29. Simplifying: $x^{-4}x^7 = x^{-4+7} = x^3$

31. Simplifying: $\left(a^2b^{-5}\right)^3 = a^6b^{-15} = \frac{a^6}{b^{15}}$

33. Simplifying: $\left(5y^4\right)^{-3}\left(2y^{-2}\right)^3 = 5^{-3}y^{-12}2^3y^{-6} = \frac{2^3}{5^3}y^{-18} = \frac{8}{125y^{18}}$

35. Simplifying: $\left(\frac{1}{2}x^3\right)\left(\frac{2}{3}x^4\right)\left(\frac{3}{5}x^{-7}\right) = \frac{1}{2} \cdot \frac{2}{3} \cdot \frac{3}{5}x^0 = \frac{1}{5}$

37. Simplifying: $\left(4a^5b^2\right)\left(2b^{-5}c^2\right)\left(3a^7c^4\right) = 24a^{5+7}b^{2-5}c^{2+4} = 24a^{12}b^{-3}c^6 = \frac{24a^{12}c^6}{b^3}$

39. Simplifying: $\left(2x^2y^{-5}\right)^3\left(3x^{-4}y^2\right)^{-4} = 2^3x^6y^{-15} \cdot 3^{-4}x^{16}y^{-8} = \frac{2^3}{3^4}x^{22}y^{-23} = \frac{8x^{22}}{81y^{23}}$

41. Simplifying: $\dfrac{x^{-1}}{x^9} = x^{-1-9} = x^{-10} = \dfrac{1}{x^{10}}$

43. Simplifying: $\dfrac{a^4}{a^{-6}} = a^{4-(-6)} = a^{4+6} = a^{10}$

45. Simplifying: $\dfrac{t^{-10}}{t^{-4}} = t^{-10-(-4)} = t^{-10+4} = t^{-6} = \dfrac{1}{t^6}$

47. Simplifying: $\left(\dfrac{x^5}{x^3}\right)^6 = \left(x^{5-3}\right)^6 = \left(x^2\right)^6 = x^{12}$

49. Simplifying: $\dfrac{\left(x^5\right)^6}{\left(x^3\right)^4} = \dfrac{x^{30}}{x^{12}} = x^{30-12} = x^{18}$

51. Simplifying: $\dfrac{\left(x^{-2}\right)^3 \left(x^3\right)^{-2}}{x^{10}} = \dfrac{x^{-6} x^{-6}}{x^{10}} = \dfrac{x^{-12}}{x^{10}} = x^{-12-10} = x^{-22} = \dfrac{1}{x^{22}}$

53. Simplifying: $\dfrac{5a^8 b^3}{20 a^5 b^{-4}} = \dfrac{5}{20} a^{8-5} b^{3-(-4)} = \dfrac{1}{4} a^3 b^7 = \dfrac{a^3 b^7}{4}$

55. Simplifying: $\dfrac{\left(3x^{-2} y^8\right)^4}{\left(9x^4 y^{-3}\right)^2} = \dfrac{81 x^{-8} y^{32}}{81 x^8 y^{-6}} = x^{-8-8} y^{32+6} = x^{-16} y^{38} = \dfrac{y^{38}}{x^{16}}$

57. Simplifying: $\left(\dfrac{8x^2 y}{4x^4 y^{-3}}\right)^4 = \left(2x^{2-4} y^{1+3}\right)^4 = \left(2x^{-2} y^4\right)^4 = 16 x^{-8} y^{16} = \dfrac{16 y^{16}}{x^8}$

59. Simplifying: $\left(\dfrac{x^{-5} y^2}{x^{-3} y^5}\right)^{-2} = \left(x^{-5+3} y^{2-5}\right)^{-2} = \left(x^{-2} y^{-3}\right)^{-2} = x^4 y^6$

61. Writing as a perfect square: $x^4 y^2 = \left(x^2 y\right)^2$

63. Writing as a perfect square: $9a^2 b^4 = \left(3ab^2\right)^2$

65. Writing as a perfect cube: $8a^3 = \left(2a\right)^3$

67. Writing as a perfect cube: $64 x^3 y^{12} = \left(4xy^4\right)^3$

69. **a.** Substituting $x = 2$: $x^3 x^2 = 2^3 2^2 = 8 \bullet 4 = 32$

b. Substituting $x = 2$: $\left(x^3\right)^2 = \left(2^3\right)^2 = (8)^2 = 64$

c. Substituting $x = 2$: $x^5 = 2^5 = 32$

d. Substituting $x = 2$: $x^5 = 2^6 = 64$

71. **a.** Substituting $x = 2$: $\dfrac{x^5}{x^2} = \dfrac{2^5}{2^2} = \dfrac{32}{4} = 8$

b. Substituting $x = 2$: $x^3 = 2^3 = 8$

c. Substituting $x = 2$: $\dfrac{x^2}{x^6} = \dfrac{2^2}{2^6} = \dfrac{4}{64} = \dfrac{1}{16}$

d. Substituting $x = 2$: $x^{-4} = 2^{-4} = \dfrac{1}{2^4} = \dfrac{1}{16}$

73. **a.** Writing as a perfect square: $\dfrac{1}{49} = \left(\dfrac{1}{7}\right)^2$

b. Writing as a perfect square: $\dfrac{1}{121} = \left(\dfrac{1}{11}\right)^2$

c. Writing as a perfect square: $\dfrac{1}{4x^2} = \left(\dfrac{1}{2x}\right)^2$

d. Writing as a perfect square: $\dfrac{1}{64 x^4} = \left(\dfrac{1}{8x^2}\right)^2$

75. Simplifying: $2 \bullet 2^{n-1} = 2^{1+n-1} = 2^n$

77. Simplifying: $\dfrac{ar^6}{ar^3} = r^{6-3} = r^3$

79. Writing in scientific notation: $378{,}000 = 3.78 \times 10^5$

81. Writing in scientific notation: $4{,}900 = 4.9 \times 10^3$

83. Writing in scientific notation: $0.00037 = 3.7 \times 10^{-4}$

85. Writing in scientific notation: $0.00495 = 4.95 \times 10^{-3}$

87. Writing in expanded form: $5.34 \times 10^3 = 5{,}340$

89. Writing in expanded form: $7.8 \times 10^6 = 7{,}800{,}000$

91. Writing in expanded form: $3.44 \times 10^{-3} = 0.00344$

93. Writing in expanded form: $4.9 \times 10^{-1} = 0.49$

95. Simplifying: $\left(4 \times 10^{10}\right)\left(2 \times 10^{-6}\right) = 8 \times 10^{10-6} = 8 \times 10^4$

97. Simplifying: $\dfrac{8 \times 10^{14}}{4 \times 10^5} = 2 \times 10^{14-5} = 2 \times 10^9$

99. Simplifying: $\dfrac{\left(5 \times 10^6\right)\left(4 \times 10^{-8}\right)}{8 \times 10^4} = \dfrac{20 \times 10^{-2}}{8 \times 10^4} = \dfrac{20}{8} \times 10^{-2-4} = 2.5 \times 10^{-6}$

101. Multiplying: $8x^3 \bullet 10y^6 = 80x^3y^6$

103. Multiplying: $8x^3 \bullet 9y^3 = 72x^3y^3$

105. Multiplying: $3x \bullet 5y = 15xy$

107. Multiplying: $4x^6y^6 \bullet 3x = 12x^7y^6$

109. Multiplying: $27a^6c^3 \bullet 2b^2c = 54a^6b^2c^4$

111. Dividing: $\dfrac{10x^5}{5x^2} = 2x^{5-2} = 2x^3$

113. Dividing: $\dfrac{20x^3}{5x^2} = 4x^{3-2} = 4x$

115. Dividing: $\dfrac{8x^3y^5}{-2x^2y} = -4x^{3-2}y^{5-1} = -4xy^4$

117. Dividing: $\dfrac{4x^4y^3}{-2x^2y} = -2x^{4-2}y^{3-1} = -2x^2y^2$

119. Simplifying: $\dfrac{2.00 \times 10^8}{3.98 \times 10^6} \approx 0.5025 \times 10^2 \approx 50$

121. Using a calculator: $10^{-4.1} \approx 7.9 \times 10^{-5}$

123. Writing in scientific notation: $630,000,000 = 6.3 \times 10^8$ seconds

125. **a.** Writing in scientific notation: $73,000 = 7.3 \times 10^4$

b. Writing in scientific notation: $194,000 = 1.94 \times 10^5$

c. Writing in scientific notation: $1,900,000 = 1.9 \times 10^6$

127. Multiplying to find the distance: $\left(1.7 \times 10^6 \text{ light-years}\right)\left(5.9 \times 10^{12} \text{ miles/light-year}\right) \approx 1.003 \times 10^{19}$ miles

129. Writing in scientific notation: $\$4.22 \times 10^{11}$

131. Writing in scientific notation: $15,000,000,000,000 = 1.5 \times 10^{13}$

133. Multiply the second equation by -2:
$$4x + 3y = 10$$
$$-4x - 2y = -8$$
Adding yields $y = 2$. Substituting into the first equation:
$$4x + 3(2) = 10$$
$$4x + 6 = 10$$
$$4x = 4$$
$$x = 1$$
The solution is $(1,2)$.

135. Multiply the second equation by -5:
$$4x + 5y = 5$$
$$-6x - 5y = -10$$
Adding yields:
$$-2x = -5$$
$$x = \frac{5}{2}$$
Substituting into the first equation:
$$4\left(\frac{5}{2}\right) + 5y = 5$$
$$10 + 5y = 5$$
$$5y = -5$$
$$y = -1$$
The solution is $\left(\frac{5}{2}, -1\right)$.

137. Substituting into the first equation:

$$x + (x + 3) = 3$$
$$2x + 3 = 3$$
$$2x = 0$$
$$x = 0$$

The solution is $(0,3)$.

139. Substituting into the first equation:

$$2x - 3(3x - 5) = -6$$
$$2x - 9x + 15 = -6$$
$$-7x + 15 = -6$$
$$-7x = -21$$
$$x = 3$$

The solution is $(3,4)$.

141. Simplifying: $-4x + 9x = 5x$

143. Simplifying: $5x^2 + 3x^2 = 8x^2$

145. Simplifying: $-8x^3 + 10x^3 = 2x^3$

147. Simplifying: $2x + 3 - 2x - 8 = -5$

149. Simplifying: $-1(2x - 3) = -2x + 3$

151. Simplifying: $-3(-3x - 2) = 9x + 6$

153. Simplifying: $-500 + 27(100) - 0.1(100)^2 ! -500 + 2,700 - 1,000 ! 1,200$

155. Simplifying: $x^{m+2} \bullet x^{-2m} \bullet x^{m-5} = x^{m+2-2m+m-5} = x^{-3} = \dfrac{1}{x^3}$

157. Simplifying: $\left(y^m\right)^2 \left(y^{-3}\right)^m \left(y^{m+3}\right) = y^{2m} \bullet y^{-3m} \bullet y^{m+3} = y^{2m-3m+m+3} = y^3$

159. Simplifying: $\dfrac{x^{n+2}}{x^{n-3}} = x^{n+2-(n-3)} = x^{n+2-n+3} = x^5$

5.2 Polynomials, Sums, and Differences

1. This is a trinomial. The degree is 2 and the leading coefficient is 5.

3. This is a binomial. The degree is 1 and the leading coefficient is 3.

5. This is a trinomial. The degree is 2 and the leading coefficient is 8.

7. This is a polynomial. The degree is 3 and the leading coefficient is 4.

9. This is a monomial. The degree is 0 and the leading coefficient is $-\dfrac{3}{4}$.

11. This is a trinomial. The degree is 3 and the leading coefficient is 6.

13. Simplifying: $(4x + 2) + (3x - 1) ! 7x + 1$

15. Simplifying: $2x^2 - 3x + 10x - 15 = 2x^2 + 7x - 15$

17. Simplifying: $12a^2 + 8ab - 15ab - 10b^2 = 12a^2 - 7ab - 10b^2$

19. Simplifying: $\left(5x^2 - 6x + 1\right) - \left(4x^2 + 7x - 2\right) = 5x^2 - 6x + 1 - 4x^2 - 7x + 2 = x^2 - 13x + 3$

21. Simplifying: $\left(\frac{1}{2}x^2 - \frac{1}{3}x - \frac{1}{6}\right) - \left(\frac{1}{4}x^2 + \frac{7}{12}x\right) + \left(\frac{1}{3}x - \frac{1}{12}\right) = \frac{1}{2}x^2 - \frac{1}{3}x - \frac{1}{6} - \frac{1}{4}x^2 - \frac{7}{12}x + \frac{1}{3}x - \frac{1}{12} = \frac{1}{4}x^2 - \frac{7}{12}x - \frac{1}{4}$

23. Simplifying: $\left(y^3 - 2y^2 - 3y + 4\right) - \left(2y^3 - y^2 + y - 3\right) = y^3 - 2y^2 - 3y + 4 - 2y^3 + y^2 - y + 3 = -y^3 - y^2 - 4y + 7$

25. Simplifying:

$$\left(5x^3 - 4x^2\right) - (3x + 4) + \left(5x^2 - 7\right) - \left(3x^3 + 6\right) = 5x^3 - 4x^2 - 3x - 4 + 5x^2 - 7 - 3x^3 - 6 = 2x^3 + x^2 - 3x - 17$$

27. Simplifying:

$$\left(\frac{4}{7}x^2 - \frac{1}{7}xy + \frac{1}{14}y^2\right) - \left(\frac{1}{2}x^2 - \frac{2}{7}xy - \frac{9}{14}y^2\right) = \frac{4}{7}x^2 - \frac{1}{7}xy + \frac{1}{14}y^2 - \frac{1}{2}x^2 + \frac{2}{7}xy + \frac{9}{14}y^2$$
$$= \frac{8}{14}x^2 - \frac{7}{14}x^2 - \frac{1}{7}xy + \frac{2}{7}xy + \frac{1}{14}y^2 + \frac{9}{14}y^2$$
$$= \frac{1}{14}x^2 + \frac{1}{7}xy + \frac{5}{7}y^2$$

29. Simplifying:

$$\left(3a^3 + 2a^2b + ab^2 - b^3\right) - \left(6a^3 - 4a^2b + 6ab^2 - b^3\right) = 3a^3 + 2a^2b + ab^2 - b^3 - 6a^3 + 4a^2b - 6ab^2 + b^3$$
$$= -3a^3 + 6a^2b - 5ab^2$$

31. Subtracting: $\left(2x^2 - 7x\right) - \left(2x^2 - 4x\right) = 2x^2 - 7x - 2x^2 + 4x = -3x$

33. Adding: $\left(x^2 - 6xy + y^2\right) + \left(2x^2 - 6xy - y^2\right) = x^2 - 6xy + y^2 + 2x^2 - 6xy - y^2 = 3x^2 - 12xy$

35. Subtracting: $\left(9x^5 - 4x^3 - 6\right) - \left(-8x^5 - 4x^3 + 6\right) = 9x^5 - 4x^3 - 6 + 8x^5 + 4x^3 - 6 = 17x^5 - 12$

37. Adding:

$$\left(11a^2 + 3ab + 2b^2\right) + \left(9a^2 - 2ab + b^2\right) + \left(-6a^2 - 3ab + 5b^2\right)$$
$$= 11a^2 + 3ab + 2b^2 + 9a^2 - 2ab + b^2 - 6a^2 - 3ab + 5b^2$$
$$= 14a^2 - 2ab + 8b^2$$

39. Simplifying: $-\left[2 - (4 - x)\right] = -(2 - 4 + x) = -(-2 + x) = 2 - x$

41. Simplifying: $-5\left[-(x - 3) - (x + 2)\right] = -5\left(-x + 3 - x - 2\right) = -5\left(-2x + 1\right) = 10x - 5$

43. Simplifying: $4x - 5\left[3 - (x - 4)\right] = 4x - 5(3 - x + 4) = 4x - 5(7 - x) = 4x - 35 + 5x = 9x - 35$

45. Simplifying:

$$-\left(3x - 4y\right) - \left[\left(4x + 2y\right) - \left(3x + 7y\right)\right] = -(3x - 4y) - (4x + 2y - 3x - 7y)$$
$$= -(3x - 4y) - (x - 5y)$$
$$= -3x + 4y - x + 5y$$
$$= -4x + 9y$$

47. Simplifying:

$$4a - \left\{3a + 2\left[a - 5(a + 1) + 4\right]\right\} = 4a - \left[3a + 2(a - 5a - 5 + 4)\right]$$
$$= 4a - \left[3a + 2(-4a - 1)\right]$$
$$= 4a - (3a - 8a - 2)$$
$$= 4a - (-5a - 2)$$
$$= 4a + 5a + 2$$
$$= 9a + 2$$

49. Evaluating when $x = 2$: $2(2)^2 - 3(2) - 4 = 2(4) - 3(2) - 4 = 8 - 6 - 4 = -2$

51. **a.** Evaluating when $x = 12$: $P(12) = \frac{3}{2}(12)^2 - \frac{3}{4}(12) + 1 = \frac{3}{2}(144) - \frac{3}{4}(12) + 1 = 216 - 9 + 1 = 208$

   b. Evaluating when $x = -8$: $P(-8) = \frac{3}{2}(-8)^2 - \frac{3}{4}(-8) + 1 = \frac{3}{2}(64) - \frac{3}{4}(-8) + 1 = 96 + 6 + 1 = 103$

53. **a.** Evaluating when $x = 5$: $Q(5) = (5)^3 - (5)^2 + (5) - 1 = 125 - 25 + 5 - 1 = 104$

   b. Evaluating when $x = -2$: $Q(-2) = (-2)^3 - (-2)^2 + (-2) - 1 = -8 - 4 - 2 - 1 = -15$

55. **a.** Evaluating when $x = 10$: $R(10) = 11.5(10) - 0.05(10)^2 = 115 - 5 = 110$

   b. Evaluating when $x = -10$: $R(-10) = 11.5(-10) - 0.05(-10)^2 = -115 - 5 = -120$

57. **a.** Evaluating when $x = -4$: $P(-4) = 600 + 1,000(-4) - 100(-4)^2 = 600 - 4,000 - 1,600 = -5,000$

   b. Evaluating when $x = 4$: $P(4) = 600 + 1,000(4) - 100(4)^2 = 600 + 4,000 - 1,600 = 3,000$

59. **a.** Simplifying: $(3x - 5) - (3a - 5) = 3x - 5 - 3a + 5 = 3x - 3a$

   b. Simplifying: $(2x + 3) - (2a + 3) = 2x + 3 - 2a - 3 = 2x - 2a$

61. Substituting $t = 3$: $h = -16(3)^2 + 128(3) = 240$ feet   Substituting $t = 5$: $h = -16(5)^2 + 128(5) = 240$ feet

63. The weekly profit is given by:

$$P = R - C = \left(100x - 0.5x^2\right) - (60x + 300) = 100x - 0.5x^2 - 60x - 300 = -0.5x^2 + 40x - 300$$

Substituting $x = 60$: $P = -0.5(60)^2 + 40(60) - 300 = \$300$

65. The weekly profit is given by:

$$P = R - C = \left(10x - 0.002x^2\right) - (800 + 6.5x) = 10x - 0.002x^2 - 800 - 6.5x = -0.002x^2 + 3.5x - 800$$

Substituting $x = 1000$: $P = -0.002(1000)^2 + 3.5(1000) - 800 = \$700$

67. Simplifying: $-1(5-x)!\ -5+x!\ x-5$

69. Simplifying: $-1(7-x)!\ -7+x!\ x-7$

71. Simplifying: $5\left(x-\frac{1}{5}\right)=5\bullet x-5\bullet\frac{1}{5}=5x-1$

73. Simplifying: $x\left(1-\frac{1}{x}\right)=x\bullet1-x\bullet\frac{1}{x}=x-1$

75. Simplifying: $12\left(\frac{1}{4}x+\frac{2}{3}y\right)=12\bullet\frac{1}{4}x+12\bullet\frac{2}{3}y=3x+8y$

77. Simplifying: $2x^2-3x+10x-15=2x^2+7x-15$

79. Simplifying: $\left(6x^3-2x^2y+8xy^2\right)+\left(-9x^2y+3xy^2-12y^3\right)=6x^3-11x^2y+11xy^2-12y^3$

81. Simplifying: $4x^3(-3x)=-12x^{3+1}=-12x^4$

83. Simplifying: $4x^3\left(5x^2\right)=20x^{3+2}=20x^5$

85. Simplifying: $\left(a^3\right)^2=a^{3\bullet2}=a^6$

87. Simplifying: $11.5(130)-0.05(130)^2=1,495-845=650$

89. a.   From the graph: $f(-3)=5$
    b.   From the graph: $f(0)=-4$
    c.   From the graph: $f(1)=-3$
    d.   From the graph: $g(-1)=3$
    e.   From the graph: $g(0)=4$
    f.   From the graph: $g(2)=0$
    g.   From the graph: $f\left[g(2)\right]=f(0)=-4$
    h.   From the graph: $g\left[f(2)\right]=g(0)=4$

5.3 Multiplication of Polynomials

1.   Multiplying: $2x\left(6x^2-5x+4\right)=2x\bullet6x^2-2x\bullet5x+2x\bullet4=12x^3-10x^2+8x$

3.   Multiplying: $-3a^2\left(a^3-6a^2+7\right)=-3a^2\bullet a^3-\left(-3a^2\right)\bullet6a^2+\left(-3a^2\right)\bullet7=-3a^5+18a^4-21a^2$

5.   Multiplying: $2a^2b\left(a^3-ab+b^3\right)=2a^2b\bullet a^3-2a^2b\bullet ab+2a^2b\bullet b^3=2a^5b-2a^3b^2+2a^2b^4$

7.   Multiplying using the vertical format:

$$\begin{array}{rr} x & -5 \\ x & +3 \\ \hline x^2 & -5x \\ +3x & -15 \\ \hline x^2\quad -2x & -15 \end{array}$$

The product is $x^2-2x-15$.

9.   Multiplying using the vertical format:

$$\begin{array}{rr} 2x^2 & -3 \\ 3x^2 & -5 \\ \hline 6x^4 & -9x^2 \\ -10x^2 & +15 \\ \hline 6x^4\quad -19x^2 & +15 \end{array}$$

The product is $6x^4-19x^2+15$.

11. Multiplying using the vertical format:

$$\begin{array}{rrr} x^2 & +6x & +5 \\ & x & +3 \\ \hline x^3 & +6x^2 & +5x \\ & +3x^2\quad +18x & +15 \\ \hline x^3 & +9x^2\quad +23x & +15 \end{array}$$

The product is $x^3+9x^2+23x+15$.

13. Multiplying using the vertical format:

$$\begin{array}{rrr} a^2 & +ab & +b^2 \\ & a & -b \\ \hline a^3 & +a^2b & +ab^2 \\ & -a^2b\quad -ab^2 & -b^3 \\ \hline a^3 & & -b^3 \end{array}$$

The product is a^3-b^3 .

15. Multiplying using the vertical format:

$$
\begin{array}{rrr}
4x^2 & -2xy & +y^2 \\
 & 2x & +y \\
\hline
8x^3 & -4x^2y & +2xy^2 \\
 & +4x^2y & -2xy^2 & +y^3 \\
\hline
8x^3 & & & +y^3
\end{array}
$$

The product is $8x^3 + y^3$.

17. Multiplying using the vertical format:

$$
\begin{array}{rrr}
a^2 & +ab & +b^2 \\
 & 2a & -3b \\
\hline
2a^3 & +2a^2b & +2ab^2 \\
 & -3a^2b & -3ab^2 & -3b^3 \\
\hline
2a^3 & -a^2b & -ab^2 & -3b^3
\end{array}
$$

The product is $2a^3 - a^2b - ab^2 - 3b^3$.

19. Multiplying using FOIL: $(x-2)(x+3) = x^2 + 3x - 2x - 6 = x^2 + x - 6$

21. Multiplying using FOIL: $(2a+3)(3a+2) = 6a^2 + 4a + 9a + 6 = 6a^2 + 13a + 6$

23. Multiplying using FOIL: $(5-3t)(4+2t) = 20 + 10t - 12t - 6t^2 = 20 - 2t - 6t^2$

25. Multiplying using FOIL: $(x^3+3)(x^3-5) = x^6 - 5x^3 + 3x^3 - 15 = x^6 - 2x^3 - 15$

27. Multiplying using FOIL: $(5x-6y)(4x+3y) = 20x^2 + 15xy - 24xy - 18y^2 = 20x^2 - 9xy - 18y^2$

29. Multiplying using FOIL: $\left(3t+\frac{1}{3}\right)\left(6t-\frac{2}{3}\right) = 18t^2 - 2t + 2t - \frac{2}{9} = 18t^2 - \frac{2}{9}$

31. Finding the product: $(5x+2y)^2 = (5x)^2 + 2(5x)(2y) + (2y)^2 = 25x^2 + 20xy + 4y^2$

33. Finding the product: $\left(5-3t^3\right)^2 = (5)^2 - 2(5)\left(3t^3\right) + \left(3t^3\right)^2 = 25 - 30t^3 + 9t^6$

35. Finding the product: $(2a+3b)(2a-3b) = (2a)^2 - (3b)^2 = 4a^2 - 9b^2$

37. Finding the product: $\left(3r^2+7s\right)\left(3r^2-7s\right) = \left(3r^2\right)^2 - (7s)^2 = 9r^4 - 49s^2$

39. Finding the product: $\left(y+\frac{3}{2}\right)^2 = (y)^2 + 2(y)\left(\frac{3}{2}\right) + \left(\frac{3}{2}\right)^2 = y^2 + 3y + \frac{9}{4}$

41. Finding the product: $\left(a-\frac{1}{2}\right)^2 = (a)^2 - 2(a)\left(\frac{1}{2}\right) + \left(\frac{1}{2}\right)^2 = a^2 - a + \frac{1}{4}$

43. Finding the product: $\left(x+\frac{1}{4}\right)^2 = (x)^2 + 2(x)\left(\frac{1}{4}\right) + \left(\frac{1}{4}\right)^2 = x^2 + \frac{1}{2}x + \frac{1}{16}$

45. Finding the product: $\left(t+\frac{1}{3}\right)^2 = (t)^2 + 2(t)\left(\frac{1}{3}\right) + \left(\frac{1}{3}\right)^2 = t^2 + \frac{2}{3}t + \frac{1}{9}$

47. Finding the product: $\left(\frac{1}{3}x-\frac{2}{5}\right)\left(\frac{1}{3}x+\frac{2}{5}\right) = \left(\frac{1}{3}x\right)^2 - \left(\frac{2}{5}\right)^2 = \frac{1}{9}x^2 - \frac{4}{25}$

49. Expanding and simplifying:

$$
\begin{aligned}
(x-2)^3 &= (x-2)(x-2)^2 \\
&= (x-2)\left(x^2-4x+4\right) \\
&= x^3 - 4x^2 + 4x - 2x^2 + 8x - 8 \\
&= x^3 - 6x^2 + 12x - 8
\end{aligned}
$$

51. Expanding and simplifying:

$$
\begin{aligned}
\left(x-\frac{1}{2}\right)^3 &= \left(x-\frac{1}{2}\right)\left(x-\frac{1}{2}\right)^2 \\
&= \left(x-\frac{1}{2}\right)\left(x^2-x+\frac{1}{4}\right) \\
&= x^3 - x^2 + \frac{1}{4}x - \frac{1}{2}x^2 + \frac{1}{2}x - \frac{1}{8} \\
&= x^3 - \frac{3}{2}x^2 + \frac{3}{4}x - \frac{1}{8}
\end{aligned}
$$

53. Expanding and simplifying:

$$
\begin{aligned}
3(x-1)(x-2)(x-3) &= 3(x-1)\left(x^2-5x+6\right) \\
&= 3\left(x^3 - 5x^2 + 6x - x^2 + 5x - 6\right) \\
&= 3\left(x^3 - 6x^2 + 11x - 6\right) \\
&= 3x^3 - 18x^2 + 33x - 18
\end{aligned}
$$

55. Expanding and simplifying: $\left(b^2+8\right)\left(a^2+1\right) = a^2b^2 + b^2 + 8a^2 + 8$

57.   Expanding and simplifying:
$$(x+1)^2 + (x+2)^2 + (x+3)^2 = x^2 + 2x + 1 + x^2 + 4x + 4 + x^2 + 6x + 9 = 3x^2 + 12x + 14$$

59.   Expanding and simplifying:
$$(2x+3)^2 - (2x-3)^2 = (4x^2 + 12x + 9) - (4x^2 - 12x + 9) = 4x^2 + 12x + 9 - 4x^2 + 12x - 9 = 24x$$

61.   Simplifying: $(x+3)^2 - 2(x+3) - 8 = x^2 + 6x + 9 - 2x - 6 - 8 = x^2 + 4x - 5$

63.   Simplifying: $(2a-3)^2 - 9(2a-3) + 20 = 4a^2 - 12a + 9 - 18a + 27 + 20 = 4a^2 - 30a + 56$

65.   Simplifying:
$$2(4a+2)^2 - 3(4a+2) - 20 = 2(16a^2 + 16a + 4) - 3(4a+2) - 20$$
$$= 32a^2 + 32a + 8 - 12a - 6 - 20$$
$$= 32a^2 + 20a - 18$$

67.   Multiplying using FOIL: $\left[(x+y) - 4\right]\left[(x+y) + 5\right] = (x+y)^2 + (x+y) - 20 = x^2 + 2xy + y^2 + x + y - 20$

69.   Evaluating when $a = 2$ and $b = 3$:
$$a^4 - b^4 = 2^4 - 3^4 = 16 - 81 = -65$$
$$(a-b)^4 = (2-3)^4 = (-1)^4 = 1$$
$$(a^2 + b^2)(a+b)(a-b) = (2^2 + 3^2)(2+3)(2-3) = (13)(5)(-1) = -65$$

71.   Since $R = xp$, substitute $x = 900 - 300p$ to obtain: $R(p) = (900 - 300p)p = 900p - 300p^2$

   To find $R(x)$, we first solve for p:
$$x = 900 - 300p$$
$$300p = 900 - x$$
$$p = 3 - \frac{x}{300}$$

   Now substitute to obtain: $R(x) = x\left(3 - \dfrac{x}{300}\right) = 3x - \dfrac{x^2}{300}$

   Substituting $p = \$1.60$: $R(1.60) = 900(1.60) - 300(1.60)^2 = \$672$. The revenue is $\$672$.

73.   Since $R = xp$, substitute $x = 350 - 10p$ to obtain: $R(p) = (350 - 10p)p = 350p - 10p^2$

   To find $R(x)$, we first solve for p:
$$x = 350 - 10p$$
$$10p = 350 - x$$
$$p = 35 - \frac{x}{10}$$

   Now substitute to obtain: $R(x) = x\left(35 - \dfrac{x}{10}\right) = 35x - \dfrac{x^2}{10}$

   Substituting $x = 65$: $R(65) = 35(65) - \dfrac{(65)^2}{10} = \$1,852.50$. The revenue is $\$1,852.50$.

75.   Since $R(x) = 35x - \dfrac{x^2}{10}$ and $C(x) = 5x + 500$, then:

$$P(x) = R(x) - C(x) = 35x - \frac{x^2}{10} - (5x + 500) = -\frac{x^2}{10} + 30x - 500$$

$$P(60) = -\frac{(60)^2}{10} + 30(60) - 500 = \$940$$

77. Expanding the formula:

$$A = 100(1+r)^4$$
$$= 100(1+r)^2(1+r)^2$$
$$= 100(1+2r+r^2)(1+2r+r^2)$$
$$= 100(1+2r+r^2+2r+4r^2+2r^3+r^2+2r^3+r^4)$$
$$= 100(1+4r+6r^2+4r^3+r^4)$$
$$= 100+400r+600r^2+400r^3+100r^4$$

79. Multiply the first equation by –1 and add it to the second equation:

$$-x-y-z=-6$$
$$2x-y+z=3$$

Adding yields the equation $x-2y=-3$. Multiply the first equation by 3 and add it to the third equation:

$$3x+3y+3z=18$$
$$x+2y-3z=-4$$

Adding yields the equation $4x+5y=14$. So the system becomes:

$$x-2y=-3$$
$$4x+5y=14$$

Multiply the first equation by –4 and add it to the second equation:

$$-4x+8y=12$$
$$4x+5y=14$$

Adding yields:

$$13y=26$$
$$y=2$$

Substituting to find x:

$$4x+5(2) ! \ 14$$
$$4x+10 ! \ 14$$
$$4x ! \ 4$$
$$x ! \ 1$$

Substituting into the original first equation:

$$1+2+z=6$$
$$z+3=6$$
$$z=3$$

The solution is (1,2,3).

81. Multiply the third equation by –1 and add it to the first equation:

$$3x+4y=15$$
$$-4y+3z=-9$$

Adding yields the equation $3x+3z=6$, or $x+z=2$. So the system becomes:

$$2x-5z=-3$$
$$x+z=2$$

Multiply the second equation by –2:

$$2x-5z=-3$$
$$-2x-2z=-4$$

Adding yields:

$$-7z=-7$$
$$z=1$$

Substituting to find x:

$$2x-5(1)=-3$$
$$2x-5=-3$$
$$2x=2$$
$$x=1$$

Substituting into the original first equation:

$$3(1)+4y=15$$
$$4y+3=15$$
$$4y=12$$
$$y=3$$

The solution is (1,3,1).

83.   Simplifying: $\dfrac{8a^3}{a}=8a^{3-1}=8a^2$

85.   Simplifying: $\dfrac{-48a}{a}=-48$

87.   Simplifying: $\dfrac{16a^5b^4}{8a^2b^3}=2a^{5-2}b^{4-3}=2a^3b$

89.   Simplifying: $\dfrac{-24a^5b^5}{8a^5b^3}=-3a^{5-5}b^{5-3}=-3b^2$

91.   Simplifying: $\dfrac{x^3y^4}{-x^3}=-x^{3-3}y^4=-y^4$

93.   Multiplying: $(x^n-2)(x^n-3)=x^{2n}-3x^n-2x^n+6=x^{2n}-5x^n+6$

95.   Multiplying: $(2x^n+3)(5x^n-1)=10x^{2n}-2x^n+15x^n-3=10x^{2n}+13x^n-3$

97.   Multiplying: $(x^n+5)^2=(x^n)^2+2(x^n)(5)+(5)^2=x^{2n}+10x^n+25$

99.   Multiplying: $(x^n+1)(x^{2n}-x^n+1)=x^{3n}-x^{2n}+x^n+x^{2n}-x^n+1=x^{3n}+1$

5.4 The Greatest Common Factor and Factoring by Grouping

1.   Factoring the expression: $10x^3-15x^2=5x^2(2x-3)$

3.   Factoring the expression: $9y^6+18y^3=9y^3(y^3+2)$

5.   Factoring the expression: $9a^2b-6ab^2=3ab(3a-2b)$

7.   Factoring the expression: $21xy^4+7x^2y^2=7xy^2(3y^2+x)$

9.   Factoring the expression: $3a^2-21a+30\,!\,3(a^2-7a+10)$

11.   Factoring the expression: $4x^3-16x^2-20x=4x(x^2-4x-5)$

13.   Factoring the expression: $10x^4y^2+20x^3y^3-30x^2y^4=10x^2y^2(x^2+2xy-3y^2)$

15.   Factoring the expression: $-x^2y+xy^2-x^2y^2\,!\,xy(-x+y-xy)$

17.   Factoring the expression: $4x^3y^2z-8x^2y^2z^2+6xy^2z^3=2xy^2z(2x^2-4xz+3z^2)$

19.   Factoring the expression: $20a^2b^2c^2-30ab^2c+25a^2bc^2=5abc(4abc-6b+5ac)$

21.   Factoring the expression: $5x(a-2b)-3y(a-2b)=(a-2b)(5x-3y)$

23.   Factoring the expression: $3x^2(x+y)^2-6y^2(x+y)^2=3(x+y)^2(x^2-2y^2)$

25.   Factoring the expression: $2x^2(x+5)+7x(x+5)+6(x+5)=(x+5)(2x^2+7x+6)$

27.   Factoring by grouping: $3xy+3y+2ax+2a=3y(x+1)+2a(x+1)=(x+1)(3y+2a)$

29.   Factoring by grouping: $x^2y+x+3xy+3=x(xy+1)+3(xy+1)=(x+3)(xy+1)$

31.   Factoring by grouping: $3xy^2-6y^2+4x-8=3y^2(x-2)+4(x-2)=(x-2)(3y^2+4)$

33.   Factoring by grouping: $x^2-ax-bx+ab=x(x-a)-b(x-a)=(x-a)(x-b)$

35.   Factoring by grouping: $ab+5a-b-5=a(b+5)-1(b+5)=(b+5)(a-1)$

37. Factoring by grouping: $a^4 b^2 + a^4 - 5b^2 - 5 = a^4 \left(b^2 + 1\right) - 5\left(b^2 + 1\right) = \left(b^2 + 1\right)\left(a^4 - 5\right)$

39. Factoring by grouping: $x^3 + 3x^2 - 4x - 12 = x^2 \left(x + 3\right) - 4\left(x + 3\right) = \left(x + 3\right)\left(x^2 - 4\right)$

41. Factoring by grouping: $x^3 + 2x^2 - 25x - 50 = x^2 \left(x + 2\right) - 25\left(x + 2\right) = \left(x + 2\right)\left(x^2 - 25\right)$

43. Factoring by grouping: $2x^3 + 3x^2 - 8x - 12 = x^2 \left(2x + 3\right) - 4\left(2x + 3\right) = \left(2x + 3\right)\left(x^2 - 4\right)$

45. Factoring by grouping: $4x^3 + 12x^2 - 9x - 27 = 4x^2 \left(x + 3\right) - 9\left(x + 3\right) = \left(x + 3\right)\left(4x^2 - 9\right)$

47. It will be $3 \bullet 2 = 6$.

49. Using factoring by grouping:
$$P + Pr + \left(P + Pr\right)r = \left(P + Pr\right) + \left(P + Pr\right)r = \left(P + Pr\right)\left(1 + r\right) = P\left(1 + r\right)\left(1 + r\right) = P\left(1 + r\right)^2$$

51. Factoring the revenue: $R(x) = 11.5x - 0.05x^2 = x\left(11.5 - 0.05x\right)$. So the price is $p = 11.5 - 0.05x$.
Substituting $x = 125$: $p = 11.5 - 0.05\left(125\right) = \$5.25$. The price is $\$5.25$.

53. Factoring the revenue: $R(x) = 35x - 0.1x^2 = x\left(35 - 0.1x\right)$. So the price is $p = 35 - 0.1x$.
Substituting $x = 65$: $p = 35 - 0.1\left(65\right) = \$28.50$. The price is $\$28.50$.

55. Evaluating the determinant: $\begin{vmatrix} 3 & 5 \\ -6 & 2 \end{vmatrix} = 3(2) - 5(-6) = 6 + 30 = 36$

57. Expanding along the first column:
$$\begin{vmatrix} 1 & -2 & 3 \\ 0 & 4 & -1 \\ 2 & -4 & 6 \end{vmatrix} = 1\begin{vmatrix} 4 & -1 \\ -4 & 6 \end{vmatrix} - 0\begin{vmatrix} -2 & 3 \\ -4 & 6 \end{vmatrix} + 2\begin{vmatrix} -2 & 3 \\ 4 & -1 \end{vmatrix} = 1(24 - 4) - 0 + 2(2 - 12) = 20 - 20 = 0$$

59. Factoring out the greatest common factor: $3x^4 - 9x^3 y - 18x^2 y^2 = 3x^2 \left(x^2 - 3xy - 6y^2\right)$

61. Factoring out the greatest common factor: $2x^2 \left(x - 3\right) - 4x\left(x - 3\right) - 3\left(x - 3\right) = \left(x - 3\right)\left(2x^2 - 4x - 3\right)$

63. Multiplying using FOIL: $\left(x + 2\right)\left(3x - 1\right) = 3x^2 - x + 6x - 2 = 3x^2 + 5x - 2$

65. Multiplying using FOIL: $\left(x - 1\right)\left(3x - 2\right) = 3x^2 - 2x - 3x + 2 = 3x^2 - 5x + 2$

67. Multiplying using FOIL: $\left(x + 2\right)\left(x + 3\right) = x^2 + 3x + 2x + 6 = x^2 + 5x + 6$

69. Multiplying using FOIL: $\left(2y + 5\right)\left(3y - 7\right) = 6y^2 - 14y + 15y - 35 = 6y^2 + y - 35$

71. Multiplying using FOIL: $\left(4 - 3a\right)\left(5 - a\right) = 20 - 4a - 15a + 3a^2 = 20 - 19a + 3a^2$

73. Completing the table:

Two Numbers a and b	Their Product ab	Their Sum $a + b$
$1, -24$	-24	-23
$-1, 24$	-24	23
$2, -12$	-24	-10
$-2, 12$	-24	10
$3, -8$	-24	-5
$-3, 8$	-24	5
$4, -6$	-24	-2
$-4, 6$	-24	2

5.5 Factoring Trinomials

1. Factoring the trinomial: $x^2 + 7x + 12 = (x+3)(x+4)$ **3.**   Factoring the trinomial: $x^2 - x - 12 = (x+3)(x-4)$

5. Factoring the trinomial: $y^2 + y - 6 = (y+3)(y-2)$ **7.**   Factoring the trinomial: $16 - 6x - x^2 = (2-x)(8+x)$

9. Factoring the trinomial: $12 + 8x + x^2 = (2+x)(6+x)$

11. Factoring the trinomial: $3a^2 - 21a + 30 = 3(a^2 - 7a + 10) = 3(a-5)(a-2)$

13. Factoring the trinomial: $4x^3 - 16x^2 - 20x = 4x(x^2 - 4x - 5) = 4x(x-5)(x+1)$

15. Factoring the trinomial: $x^2 + 3xy + 2y^2 = (x+2y)(x+y)$

17. Factoring the trinomial: $a^2 + 3ab - 18b^2 = (a+6b)(a-3b)$

19. Factoring the trinomial: $x^2 - 2xa - 48a^2 = (x-8a)(x+6a)$

21. Factoring the trinomial: $x^2 - 12xb + 36b^2 = (x-6b)^2$

23. Factoring the trinomial: $3x^2 - 6xy - 9y^2 = 3(x^2 - 2xy - 3y^2) = 3(x-3y)(x+y)$

25. Factoring the trinomial: $2a^5 + 4a^4b + 4a^3b^2 = 2a^3(a^2 + 2ab + 2b^2)$

27. Factoring the trinomial: $10x^4y^2 + 20x^3y^3 - 30x^2y^4 = 10x^2y^2(x^2 + 2xy - 3y^2) = 10x^2y^2(x+3y)(x-y)$

29. Factoring completely: $2x^2 + 7x - 15 = (2x-3)(x+5)$ **31.**   Factoring completely: $2x^2 + x - 15 = (2x-5)(x+3)$

33. Factoring completely: $2x^2 - 13x + 15 = (2x-3)(x-5)$

35. Factoring completely: $2x^2 - 11x + 15 = (2x-5)(x-3)$

37. The trinomial $2x^2 + 7x + 15$ does not factor (prime).   **39.**   Factoring completely: $2 + 7a + 6a^2 = (2+3a)(1+2a)$

41. Factoring completely: $60y^2 - 15y - 45 = 15(4y^2 - y - 3) = 15(4y+3)(y-1)$

43. Factoring completely: $6x^4 - x^3 - 2x^2 \;!\; x^2(6x^2 - x - 2) \;!\; x^2(3x-2)(2x+1)$

45. Factoring completely: $40r^3 - 120r^2 + 90r = 10r(4r^2 - 12r + 9) = 10r(2r-3)^2$

47. Factoring completely: $4x^2 - 11xy - 3y^2 \;!\; (4x+y)(x-3y)$

49. Factoring completely: $10x^2 - 3xa - 18a^2 \;!\; (2x-3a)(5x+6a)$

51. Factoring completely: $18a^2 + 3ab - 28b^2 \;!\; (3a+4b)(6a-7b)$

53. Factoring completely: $600 + 800t - 800t^2 = 200(3 + 4t - 4t^2) = 200(1+2t)(3-2t)$

55. Factoring completely: $9y^4 + 9y^3 - 10y^2 \;!\; y^2(9y^2 + 9y - 10) \;!\; y^2(3y-2)(3y+5)$

57. Factoring completely: $24a^2 - 2a^3 - 12a^4 \;!\; 2a^2(12 - a - 6a^2) \;!\; 2a^2(3+2a)(4-3a)$

59. Factoring completely: $8x^4y^2 - 2x^3y^3 - 6x^2y^4 = 2x^2y^2(4x^2 - xy - 3y^2) = 2x^2y^2(4x+3y)(x-y)$

61. Factoring completely: $300x^4 + 1000x^2 + 300 = 100(3x^4 + 10x^2 + 3) = 100(3x^2 + 1)(x^2 + 3)$

63. Factoring completely: $20a^4 + 37a^2 + 15 = (5a^2 + 3)(4a^2 + 5)$

65. Factoring completely: $9 + 3r^2 - 12r^4 = 3(3 + r^2 - 4r^4) = 3(3 + 4r^2)(1 - r^2)$

67. Factoring completely: $2x^2(x+5) + 7x(x+5) + 6(x+5) = (x+5)(2x^2 + 7x + 6) = (x+5)(2x+3)(x+2)$

69. Factoring completely: $x^2(2x+3)+7x(2x+3)+10(2x+3)=(2x+3)\left(x^2+7x+10\right)=(2x+3)(x+5)(x+2)$

71. Factoring completely: $3x^2(x-3)+7x(x-3)-20(x-3)=(x-3)\left(3x^2+7x-20\right)=(x-3)(3x-5)(x+4)$

73. Factoring completely: $6x^2(x-2)-17x(x-2)+12(x-2)=(x-2)\left(6x^2-17x+12\right)=(x-2)(3x-4)(2x-3)$

75. Factoring completely: $12x^2(x+3)+7x(x+3)-45(x+3)$! $(x+3)\left(12x^2+7x-45\right)$! $(x+3)(4x+9)(3x-5)$

77. Factoring completely:
$$6x^2(5x-2)-11x(5x-2)-10(5x-2)=(5x-2)\left(6x^2-11x-10\right)=(5x-2)(3x+2)(2x-5)$$

79. Factoring completely:
$$20x^2(2x+3)+47x(2x+3)+21(2x+3) ! (2x+3)\left(20x^2+47x+21\right) ! (2x+3)(5x+3)(4x+7)$$

81. Multiplying out, the polynomial is: $(3x+5y)(3x-5y)=9x^2-25y^2$

83. The polynomial factors as $a^2+260a+2500=(a+10)(a+250)$, so the other factor is $a+250$.

85. The polynomial factors as $12x^2-107x+210=(x-6)(12x-35)$, so the other factor is $12x-35$.

87. The polynomial factors as $54x^2+111x+56$! $(6x+7)(9x+8)$, so the other factor is $9x+8$.

89. The polynomial factors as $35x^2+19x-24=(5x-3)(7x+8)$, so the other factor is $7x+8$.

91. Factoring the right side: $y=4x^2+18x-10=2\left(2x^2+9x-5\right)=2(2x-1)(x+5)$

Evaluating when $x=\frac{1}{2}$: $y=2\left(2\bullet\frac{1}{2}-1\right)\left(\frac{1}{2}+5\right)=2(0)\left(\frac{11}{2}\right)=0$

Evaluating when $x=-5$: $y=2\left(2\bullet(-5)-1\right)(-5+5)=2(-11)(0)=0$

Evaluating when $x=2$: $y=2(2\bullet 2-1)(2+5)=2(3)(7)=42$

93. Multiplying: $(2x-3)(2x+3)=(2x)^2-(3)^2=4x^2-9$

95. Multiplying: $(2x-3)^2$! $(2x-3)(2x-3)$! $4x^2-12x+9$

97. Multiplying: $(2x-3)\left(4x^2+6x+9\right)=8x^3+12x^2+18x-12x^2-18x-27=8x^3-27$

99. Writing as a square: $\frac{25}{64}=\left(\frac{5}{8}\right)^2$

101. Writing as a square: $x^6=\left(x^3\right)^2$

103. Writing as a square: $16x^4=\left(4x^2\right)^2$

105. Writing as a cube: $\frac{1}{8}=\left(\frac{1}{2}\right)^3$

107. Writing as a cube: $x^6=\left(x^2\right)^3$

109. Writing as a cube: $27x^3=(3x)^3$

111. Writing as a cube: $8y^3=(2y)^3$

113. Factoring completely: $8x^6+26x^3y^2+15y^4=\left(2x^3+5y^2\right)\left(4x^3+3y^2\right)$

115. Factoring completely: $3x^2+295x-500=(3x-5)(x+100)$

117. Factoring completely: $\frac{1}{8}x^2+x+2=\left(\frac{1}{4}x+1\right)\left(\frac{1}{2}x+2\right)$

119. Factoring completely: $2x^2+1.5x+0.25=(2x+0.5)(x+0.5)$

5.6 Special Factoring

1. Factoring the trinomial: $x^2 - 6x + 9 = (x-3)^2$

3. Factoring the trinomial: $a^2 - 12a + 36$! $(a-6)^2$

5. Factoring the trinomial: $25 - 10t + t^2$! $(5-t)^2$

7. Factoring the trinomial: $\frac{1}{9}x^2 + 2x + 9 = \left(\frac{1}{3}x + 3\right)^2$

9. Factoring the trinomial: $4y^4 - 12y^2 + 9 = \left(2y^2 - 3\right)^2$

11. Factoring the trinomial: $16a^2 + 40ab + 25b^2 = (4a + 5b)^2$

13. Factoring the trinomial: $\frac{1}{25} + \frac{1}{10}t^2 + \frac{1}{16}t^4 = \left(\frac{1}{5} + \frac{1}{4}t^2\right)^2$

15. Factoring the trinomial: $y^2 + 3y + \frac{9}{4} = \left(y + \frac{3}{2}\right)^2$

17. Factoring the trinomial: $a^2 - a + \frac{1}{4}$! $\left(a - \frac{1}{2}\right)^2$

19. Factoring the trinomial: $x^2 - \frac{1}{2}x + \frac{1}{16} = \left(x - \frac{1}{4}\right)^2$

21. Factoring the trinomial: $t^2 + \frac{2}{3}t + \frac{1}{9} = \left(t + \frac{1}{3}\right)^2$

23. Factoring the trinomial: $16x^2 - 48x + 36$! $4\left(4x^2 - 12x + 9\right)$! $4(2x - 3)^2$

25. Factoring the trinomial: $75a^3 + 30a^2 + 3a = 3a\left(25a^2 + 10a + 1\right) = 3a(5a + 1)^2$

27. Factoring the trinomial: $(x+2)^2 + 6(x+2) + 9 = (x + 2 + 3)^2 = (x+5)^2$

29. Factoring: $x^2 - 9$! $(x+3)(x-3)$

31. Factoring: $49x^2 - 64y^2 = (7x + 8y)(7x - 8y)$

33. Factoring: $4a^2 - \frac{1}{4}$! $\left(2a + \frac{1}{2}\right)\left(2a - \frac{1}{2}\right)$

35. Factoring: $x^2 - \frac{9}{25} = \left(x + \frac{3}{5}\right)\left(x - \frac{3}{5}\right)$

37. Factoring: $9x^2 - 16y^2$! $(3x + 4y)(3x - 4y)$

39. Factoring: $250 - 10t^2 = 10\left(25 - t^2\right) = 10(5 + t)(5 - t)$

41. Factoring: $x^4 - 81 = \left(x^2 + 9\right)\left(x^2 - 9\right) = \left(x^2 + 9\right)(x+3)(x-3)$

43. Factoring: $9x^6 - 1$! $\left(3x^3 + 1\right)\left(3x^3 - 1\right)$

45. Factoring: $16a^4 - 81 = \left(4a^2 + 9\right)\left(4a^2 - 9\right) = \left(4a^2 + 9\right)(2a+3)(2a-3)$

47. Factoring: $\dfrac{1}{81} - \dfrac{y^4}{16} = \left(\dfrac{1}{9} + \dfrac{y^2}{4}\right)\left(\dfrac{1}{9} - \dfrac{y^2}{4}\right) = \left(\dfrac{1}{9} + \dfrac{y^2}{4}\right)\left(\dfrac{1}{3} + \dfrac{y}{2}\right)\left(\dfrac{1}{3} - \dfrac{y}{2}\right)$

49. Factoring: $x^6 - y^6 = \left(x^3 + y^3\right)\left(x^3 - y^3\right) = (x+y)\left(x^2 - xy + y^2\right)(x-y)\left(x^2 + xy + y^2\right)$

51. Factoring: $2a^7 - 128a$! $2a\left(a^6 - 64\right)$! $2a\left(a^3 + 8\right)\left(a^3 - 8\right)$! $2a(a+2)\left(a^2 - 2a + 4\right)(a-2)\left(a^2 + 2a + 4\right)$

53. Factoring: $(x-2)^2 - 9$! $(x - 2 + 3)(x - 2 - 3)$! $(x+1)(x-5)$

55. Factoring: $(y+4)^2 - 16 = (y + 4 + 4)(y + 4 - 4) = y(y + 8)$

57. Factoring: $x^2 - 10x + 25 - y^2$! $(x-5)^2 - y^2$! $(x - 5 + y)(x - 5 - y)$

59. Factoring: $a^2 + 8a + 16 - b^2 = (a+4)^2 - b^2 = (a + 4 + b)(a + 4 - b)$

61. Factoring: $x^2 + 2xy + y^2 - a^2 = (x+y)^2 - a^2 = (x + y + a)(x + y - a)$

63. Factoring: $x^3 + 3x^2 - 4x - 12 = x^2(x+3) - 4(x+3) = (x+3)\left(x^2 - 4\right) = (x+3)(x+2)(x-2)$

65. Factoring: $x^3 + 2x^2 - 25x - 50 = x^2(x+2) - 25(x+2) = (x+2)\left(x^2 - 25\right) = (x+2)(x+5)(x-5)$

67. Factoring: $2x^3 + 3x^2 - 8x - 12 = x^2(2x+3) - 4(2x+3) = (2x+3)\left(x^2 - 4\right) = (2x+3)(x+2)(x-2)$

69. Factoring: $4x^3 + 12x^2 - 9x - 27$! $4x^2(x+3) - 9(x+3)$! $(x+3)\left(4x^2 - 9\right)$! $(x+3)(2x+3)(2x-3)$

71. Factoring: $(2x-5)^2 - 100 = (2x-5-10)(2x-5+10) = (2x-15)(2x+5)$

73. Factoring: $(a-3)^2 - (4b)^2 = (a-3-4b)(a-3+4b)$

75. Factoring: $a^2 - 6a + 9 - 16b^2 = (a-3)^2 - (4b)^2 = (a-3-4b)(a-3+4b)$

77. Factoring: $x^2(x+4) - 6x(x+4) + 9(x+4) = (x+4)(x^2 - 6x + 9) = (x+4)(x-3)^2$

79. Factoring: $x^3 - y^3 = (x-y)(x^2 + xy + y^2)$ **81.** Factoring: $a^3 + 8 = (a+2)(a^2 - 2a + 4)$

83. Factoring: $27 + x^3 = (3+x)(9 - 3x + x^2)$ **85.** Factoring: $y^3 - 1 = (y-1)(y^2 + y + 1)$

87. Factoring: $10r^3 - 1,250 = 10(r^3 - 125) = 10(r-5)(r^2 + 5r + 25)$

89. Factoring: $64 + 27a^3 ! (4+3a)(16 - 12a + 9a^2)$ **91.** Factoring: $8x^3 - 27y^3 = (2x-3y)(4x^2 + 6xy + 9y^2)$

93. Factoring: $t^3 + \frac{1}{27} = \left(t + \frac{1}{3}\right)\left(t^2 - \frac{1}{3}t + \frac{1}{9}\right)$ **95.** Factoring: $27x^3 - \frac{1}{27} = \left(3x - \frac{1}{3}\right)\left(9x^2 + x + \frac{1}{9}\right)$

97. Factoring: $64a^3 + 125b^3 = (4a+5b)(16a^2 - 20ab + 25b^2)$

99. Since $9x^2 + 30x + 25 = (3x+5)^2$ and $9x^2 - 30x + 25 = (3x-5)^2$, two values of b are $b = 30$ and $b = -30$.

101. Evaluating the determinants:

$$D = \begin{vmatrix} 4 & -7 \\ 5 & 2 \end{vmatrix} = 4(2) - (-7)(5) = 8 + 35 = 43$$

$$D_x = \begin{vmatrix} 3 & -7 \\ -3 & 2 \end{vmatrix} = 3(2) - (-7)(-3) = 6 - 21 = -15$$

$$D_y = \begin{vmatrix} 4 & 3 \\ 5 & -3 \end{vmatrix} = 4(-3) - 3(5) = -12 - 15 = -27$$

Using Cramer's rule: $x = \dfrac{D_x}{D} = -\dfrac{15}{43}, y = \dfrac{D_y}{D} = -\dfrac{27}{43}$. The solution is $\left(-\dfrac{15}{43}, -\dfrac{27}{43}\right)$.

103. Evaluating the determinants:

$$D = \begin{vmatrix} 3 & 4 & 0 \\ 2 & 0 & -5 \\ 0 & 4 & -3 \end{vmatrix} = 3\begin{vmatrix} 0 & -5 \\ 4 & -3 \end{vmatrix} - 4\begin{vmatrix} 2 & -5 \\ 0 & -3 \end{vmatrix} = 3(0+20) - 4(-6-0) = 60 + 24 = 84$$

$$D_x = \begin{vmatrix} 15 & 4 & 0 \\ -3 & 0 & -5 \\ 9 & 4 & -3 \end{vmatrix} = 15\begin{vmatrix} 0 & -5 \\ 4 & -3 \end{vmatrix} - 4\begin{vmatrix} -3 & -5 \\ 9 & -3 \end{vmatrix} = 15(0+20) - 4(9+45) = 300 - 216 = 84$$

$$D_y = \begin{vmatrix} 3 & 15 & 0 \\ 2 & -3 & -5 \\ 0 & 9 & -3 \end{vmatrix} = 3\begin{vmatrix} -3 & -5 \\ 9 & -3 \end{vmatrix} - 15\begin{vmatrix} 2 & -5 \\ 0 & -3 \end{vmatrix} = 3(9+45) - 15(-6-0) = 162 + 90 = 252$$

$$D_z = \begin{vmatrix} 3 & 4 & 15 \\ 2 & 0 & -3 \\ 0 & 4 & 9 \end{vmatrix} = 3\begin{vmatrix} 0 & -3 \\ 4 & 9 \end{vmatrix} - 2\begin{vmatrix} 4 & 15 \\ 4 & 9 \end{vmatrix} = 3(0+12) - 2(36-60) = 36 + 48 = 84$$

Using Cramer's rule: $x = \dfrac{D_x}{D} = \dfrac{84}{84} = 1, y = \dfrac{D_y}{D} = \dfrac{252}{84} = 3, z = \dfrac{D_z}{D} = \dfrac{84}{84} = 1$. The solution is $(1,3,1)$.

105. Factoring the greatest common factor: $y^3 + 25y = y(y^2 + 25)$

107. Factoring the greatest common factor: $2ab^5 + 8ab^4 + 2ab^3 = 2ab^3(b^2 + 4b + 1)$

109. Factoring by grouping: $4x^2 - 6x + 2ax - 3a = 2x(2x - 3) + a(2x - 3) = (2x - 3)(2x + a)$

111. Factoring the difference of squares: $x^2 - 4 = (x + 2)(x - 2)$

113. Factoring the perfect square trinomial: $x^2 - 6x + 9 = (x - 3)^2$

115. Factoring: $6a^2 - 11a + 4 = (3a - 4)(2a - 1)$

117. Factoring the sum of cubes: $x^3 + 8 \ ! \ (x + 2)(x^2 - 2x + 4)$

119. Factoring completely: $a^2 - b^2 + 6b - 9 = a^2 - (b^2 - 6b + 9) = a^2 - (b - 3)^2 = (a - b + 3)(a + b - 3)$

121. Factoring completely: $(x - 3)^2 - (y + 5)^2 = ((x - 3) - (y + 5))((x - 3) + (y + 5)) = (x - y - 8)(x + y + 2)$

123. Since $144x^2 - 168xy + 49y^2 = (12x - 7y)^2$, a value of k is $k = 144$.

125. Since $49x^2 + 126x + 81 = (7x + 9)^2$ and $49x^2 - 126x + 81 = (7x - 9)^2$, two values of k are $k = \pm 126$.

5.7 Factoring: A General Review

1. Factoring: $x^2 - 81 = (x + 9)(x - 9)$ **3.** Factoring: $x^2 + 2x - 15 = (x - 3)(x + 5)$

5. Factoring: $x^2(x + 2) + 6x(x + 2) + 9(x + 2) = (x + 2)(x^2 + 6x + 9) = (x + 2)(x + 3)^2$

7. Factoring: $x^2y^2 + 2y^2 + x^2 + 2 = y^2(x^2 + 2) + 1(x^2 + 2) = (x^2 + 2)(y^2 + 1)$

9. Factoring: $2a^3b + 6a^2b + 2ab = 2ab(a^2 + 3a + 1)$ **11.** The polynomial $x^2 + x + 1$ does not factor (prime).

13. Factoring: $12a^2 - 75 = 3(4a^2 - 25) = 3(2a + 5)(2a - 5)$

15. Factoring: $9x^2 - 12xy + 4y^2 = (3x - 2y)^2$ **17.** Factoring: $25 - 10t + t^2 = (5 - t)^2$

19. Factoring: $4x^3 + 16xy^2 = 4x(x^2 + 4y^2)$

21. Factoring: $2y^3 + 20y^2 + 50y = 2y(y^2 + 10y + 25) = 2y(y + 5)^2$

23. Factoring: $a^7 + 8a^4b^3 \ ! \ a^4(a^3 + 8b^3) \ ! \ a^4(a + 2b)(a^2 - 2ab + 4b^2)$

25. Factoring: $t^2 + 6t + 9 - x^2 = (t + 3)^2 - x^2 = (t + 3 + x)(t + 3 - x)$

27. Factoring: $x^3 + 5x^2 - 9x - 45 = x^2(x + 5) - 9(x + 5) = (x + 5)(x^2 - 9) = (x + 5)(x + 3)(x - 3)$

29. Factoring: $5a^2 + 10ab + 5b^2 = 5(a^2 + 2ab + b^2) = 5(a + b)^2$

31. The polynomial $x^2 + 49$ does not factor (prime).

33. Factoring: $3x^2 + 15xy + 18y^2 \ ! \ 3(x^2 + 5xy + 6y^2) \ ! \ 3(x + 2y)(x + 3y)$

35. Factoring: $9a^2 + 2a + \frac{1}{9} = \left(3a + \frac{1}{3}\right)^2$

37. Factoring: $x^2(x - 3) - 14x(x - 3) + 49(x - 3) = (x - 3)(x^2 - 14x + 49) = (x - 3)(x - 7)^2$

39. Factoring: $x^2 - 64 = (x + 8)(x - 8)$ **41.** Factoring: $8 - 14x - 15x^2 = (2 - 5x)(4 + 3x)$

43. Factoring: $49a^7 - 9a^5 = a^5(49a^2 - 9) = a^5(7a + 3)(7a - 3)$

45. Factoring: $r^2 - \frac{1}{25} = \left(r + \frac{1}{5}\right)\left(r - \frac{1}{5}\right)$ **47.** The polynomial $49x^2 + 9y^2$ does not factor (prime).

49. Factoring: $100x^2 - 100x - 600 = 100(x^2 - x - 6) = 100(x - 3)(x + 2)$

51. Factoring: $25a^3 + 20a^2 + 3a = a\left(25a^2 + 20a + 3\right) = a\left(5a + 3\right)\left(5a + 1\right)$

53. Factoring: $3x^4 - 14x^2 - 5 = \left(3x^2 + 1\right)\left(x^2 - 5\right)$

55. Factoring: $24a^5 b - 3a^2 b = 3a^2 b\left(8a^3 - 1\right) = 3a^2 b\left(2a - 1\right)\left(4a^2 + 2a + 1\right)$

57. Factoring: $64 - r^3 = \left(4 - r\right)\left(16 + 4r + r^2\right)$

59. Factoring: $20x^4 - 45x^2 = 5x^2\left(4x^2 - 9\right) = 5x^2\left(2x + 3\right)\left(2x - 3\right)$

61. Factoring: $400t^2 - 900 = 100\left(4t^2 - 9\right) = 100\left(2t + 3\right)\left(2t - 3\right)$

63. Factoring: $16x^5 - 44x^4 + 30x^3 = 2x^3\left(8x^2 - 22x + 15\right) = 2x^3\left(4x - 5\right)\left(2x - 3\right)$

65. Factoring: $y^6 - 1 = \left(y^3 + 1\right)\left(y^3 - 1\right) = \left(y + 1\right)\left(y^2 - y + 1\right)\left(y - 1\right)\left(y^2 + y + 1\right)$

67. Factoring: $50 - 2a^2 = 2\left(25 - a^2\right) = 2\left(5 + a\right)\left(5 - a\right)$

69. Factoring: $12x^4 y^2 + 36x^3 y^3 + 27x^2 y^4 = 3x^2 y^2\left(4x^2 + 12xy + 9y^2\right) = 3x^2 y^2\left(2x + 3y\right)^2$

71. Factoring: $x^2 - 4x + 4 - y^2 = \left(x - 2\right)^2 - y^2 = \left(x - 2 + y\right)\left(x - 2 - y\right)$

73. Factoring: $a^2 - \frac{4}{3}ab + \frac{4}{9}b^2 = \left(a - \frac{2}{3}b\right)^2$

75. Factoring: $x^2 - \frac{4}{5}xy + \frac{4}{25}y^2 = \left(x - \frac{2}{5}y\right)^2$

77. Factoring: $a^2 - \frac{5}{3}ab + \frac{25}{36}b^2 = \left(a - \frac{5}{6}b\right)^2$

79. Factoring: $x^2 - \frac{8}{5}xy + \frac{16}{25}y^2 = \left(x - \frac{4}{5}y\right)^2$

81. Factoring: $2x^2\left(x + 2\right) - 13x\left(x + 2\right) + 15\left(x + 2\right) = \left(x + 2\right)\left(2x^2 - 13x + 15\right) = \left(x + 2\right)\left(2x - 3\right)\left(x - 5\right)$

83. Factoring: $\left(x - 4\right)^3 + \left(x - 4\right)^4 = \left(x - 4\right)^3\left(1 + x - 4\right) = \left(x - 4\right)^3\left(x - 3\right)$

85. Factoring: $2y^3 - 54 = 2\left(y^3 - 27\right) = 2\left(y - 3\right)\left(y^2 + 3y + 9\right)$

87. Factoring: $2a^3 - 128b^3 = 2\left(a^3 - 64b^3\right) = 2\left(a - 4b\right)\left(a^2 + 4ab + 16b^2\right)$

89. Factoring: $2x^3 + 432y^3 = 2\left(x^3 + 216y^3\right) = 2\left(x + 6y\right)\left(x^2 - 6xy + 36y^2\right)$

91. Let g and d represent the number of geese and ducks. The system of equations is:
$$g + d = 108$$
$$1.4g + 0.6d = 112.80$$
Substituting $d = 108 - g$ into the second equation:
$$1.4g + 0.6\left(108 - g\right) = 112.80$$
$$1.4g + 64.8 - 0.6g = 112.80$$
$$0.8g = 48$$
$$g = 60$$
$$d = 48$$
He bought 60 geese and 48 ducks.

93. Let o represent the number of oranges and a represent the number of apples. The system of equations is:
$$\frac{o}{3}\left(0.10\right) + \frac{a}{12}\left(0.15\right) = 6.80$$
$$\frac{5o}{3}\left(0.10\right) + \frac{a}{48}\left(0.15\right) = 25.45$$
Clearing each equation of fractions:
$$12 \cdot \frac{o}{3}\left(0.10\right) + 12 \cdot \frac{a}{12}\left(0.15\right) = 12 \cdot 6.80 \qquad 48 \cdot \frac{5o}{3}\left(0.10\right) + 48 \cdot \frac{a}{48}\left(0.15\right) = 48 \cdot 25.45$$
$$0.4o + 0.15a = 81.6 \qquad\qquad 8o + 0.15a = 1221.6$$

So the system becomes:
$$0.4o + 0.15a = 81.6$$
$$8o + 0.15a = 1221.6$$
Multiply the first equation by –20:
$$-8o - 3a = -1632$$
$$8o + 0.15a = 1221.6$$
Adding yields:
$$-2.85a = -410.4$$
$$a = 144$$
Substituting to find o:
$$8o + 0.15(144) = 1221.6$$
$$8o + 21.6 = 1221.6$$
$$8a = 1200$$
$$a = 150$$
So 150 oranges and 144 apples were bought.

95. Simplifying: $x^2 + (x+1)^2 = x^2 + x^2 + 2x + 1 = 2x^2 + 2x + 1$

97. Simplifying: $\dfrac{16t^2 - 64t + 48}{16} = \dfrac{16\left(t^2 - 4t + 3\right)}{16} = t^2 - 4t + 3$

99. Factoring: $x^2 - 2x - 24 = (x-6)(x+4)$

101. Factoring: $2x^3 - 5x^2 - 3x = x\left(2x^2 - 5x - 3\right) = x(2x+1)(x-3)$

103. Factoring: $x^3 + 2x^2 - 9x - 18 = x^2(x+2) - 9(x+2) = (x+2)\left(x^2 - 9\right) = (x+2)(x+3)(x-3)$

105. Solving the equation:
$$x - 6 = 0$$
$$x = 6$$

107. Solving the equation:
$$2x + 1 = 0$$
$$2x = -1$$
$$x = -\tfrac{1}{2}$$

5.8 Solving Equations by Factoring

1. Solving the equation:
$$x^2 - 5x - 6 = 0$$
$$(x+1)(x-6) = 0$$
$$x = -1, 6$$

3. Solving the equation:
$$x^3 - 5x^2 + 6x = 0$$
$$x\left(x^2 - 5x + 6\right) = 0$$
$$x(x-2)(x-3) = 0$$
$$x = 0, 2, 3$$

5. Solving the equation:
$$3y^2 + 11y - 4 = 0$$
$$(3y-1)(y+4) = 0$$
$$y = -4, \tfrac{1}{3}$$

7. Solving the equation:
$$60x^2 - 130x + 60 = 0$$
$$10\left(6x^2 - 13x + 6\right) = 0$$
$$10(3x-2)(2x-3) = 0$$
$$x = \tfrac{2}{3}, \tfrac{3}{2}$$

9. Solving the equation:
$$\tfrac{1}{10}t^2 - \tfrac{5}{2} = 0$$
$$10\left(\tfrac{1}{10}t^2 - \tfrac{5}{2}\right) = 10(0)$$
$$t^2 - 25 = 0$$
$$(t+5)(t-5) = 0$$
$$t = -5, 5$$

11. Solving the equation:
$$100x^4 = 400x^3 + 2100x^2$$
$$100x^4 - 400x^3 - 2100x^2 = 0$$
$$100x^2\left(x^2 - 4x - 21\right) = 0$$
$$100x^2(x-7)(x+3) = 0$$
$$x = -3, 0, 7$$

13.   Solving the equation:

$$\frac{1}{5}y^2 - 2 = -\frac{3}{10}y$$
$$10\left(\frac{1}{5}y^2 - 2\right) = 10\left(-\frac{3}{10}y\right)$$
$$2y^2 - 20 = -3y$$
$$2y^2 + 3y - 20 = 0$$
$$(y+4)(2y-5) = 0$$
$$y = -4, \frac{5}{2}$$

15.   Solving the equation:

$$9x^2 - 12x = 0$$
$$3x(3x-4) = 0$$
$$x = 0, \frac{4}{3}$$

17.   Solving the equation:

$$0.02r + 0.01 = 0.15r^2$$
$$2r + 1 = 15r^2$$
$$15r^2 - 2r - 1 = 0$$
$$(5r+1)(3r-1) = 0$$
$$r = -\frac{1}{5}, \frac{1}{3}$$

19.   Solving the equation:

$$9a^3 = 16a$$
$$9a^3 - 16a = 0$$
$$a\left(9a^2 - 16\right) = 0$$
$$a(3a+4)(3a-4) = 0$$
$$a = -\frac{4}{3}, 0, \frac{4}{3}$$

21.   Solving the equation:

$$-100x = 10x^2$$
$$0 = 10x^2 + 100x$$
$$0 = 10x(x+10)$$
$$x = -10, 0$$

23.   Solving the equation:

$$(x+6)(x-2) = -7$$
$$x^2 + 4x - 12 = -7$$
$$x^2 + 4x - 5 = 0$$
$$(x+5)(x-1) = 0$$
$$x = -5, 1$$

25.   Solving the equation:

$$(y-4)(y+1) = -6$$
$$y^2 - 3y - 4 = -6$$
$$y^2 - 3y + 2 = 0$$
$$(y-2)(y-1) = 0$$
$$y = 1, 2$$

27.   Solving the equation:

$$(x+1)^2 ! \ 3x+7$$
$$x^2 + 2x + 1 ! \ 3x+7$$
$$x^2 - x - 6 ! \ 0$$
$$(x+2)(x-3) ! \ 0$$
$$x ! \ -2, 3$$

29.   Solving the equation:

$$(2r+3)(2r-1) = -(3r+1)$$
$$4r^2 + 4r - 3 = -3r - 1$$
$$4r^2 + 7r - 2 = 0$$
$$(r+2)(4r-1) = 0$$
$$r = -2, \frac{1}{4}$$

31.   Solving the equation:

$$x^3 + 3x^2 - 4x - 12 = 0$$
$$x^2(x+3) - 4(x+3) = 0$$
$$(x+3)\left(x^2 - 4\right) = 0$$
$$(x+3)(x+2)(x-2) = 0$$
$$x = -3, -2, 2$$

33.   Solving the equation:

$$x^3 + 2x^2 - 25x - 50 = 0$$
$$x^2(x+2) - 25(x+2) = 0$$
$$(x+2)\left(x^2 - 25\right) = 0$$
$$(x+2)(x+5)(x-5) = 0$$
$$x = -5, -2, 5$$

35.   Solving the equation:

$$2x^3 + 3x^2 - 8x - 12 = 0$$
$$x^2(2x+3) - 4(2x+3) = 0$$
$$(2x+3)\left(x^2 - 4\right) = 0$$
$$(2x+3)(x+2)(x-2) = 0$$
$$x = -2, -\frac{3}{2}, 2$$

37.   Solving the equation:

$$4x^3 + 12x^2 - 9x - 27 = 0$$
$$4x^2(x+3) - 9(x+3) = 0$$
$$(x+3)\left(4x^2 - 9\right) = 0$$
$$(x+3)(2x+3)(2x-3) = 0$$
$$x = -3, -\frac{3}{2}, \frac{3}{2}$$

39.   Solving the equation:

$$3x^2 + x = 10$$
$$3x^2 + x - 10 = 0$$
$$(3x-5)(x+2) = 0$$
$$x = -2, \frac{5}{3}$$

41.　Solving the equation:

$$12(x+3)+12(x-3)=3(x^2-9)$$
$$12x+36+12x-36=3x^2-27$$
$$24x=3x^2-27$$
$$3x^2-24x-27=0$$
$$3(x^2-8x-9)=0$$
$$3(x-9)(x+1)=0$$
$$x=-1,9$$

43.　Solving the equation:

$$(y+3)^2+y^2=9$$
$$y^2+6y+9+y^2=9$$
$$2y^2+6y=0$$
$$2y(y+3)=0$$
$$y=-3,0$$

45.　Solving the equation:

$$(x+3)^2+1^2=2$$
$$x^2+6x+9+1=2$$
$$x^2+6x+8=0$$
$$(x+4)(x+2)=0$$
$$x=-4,-2$$

47.　Solving the equation:

$$(x+2)(x)=2^3$$
$$x^2+2x=8$$
$$x^2+2x-8=0$$
$$(x+4)(x-2)=0$$
$$x=-4,2$$

49.　Solving $f(x)=0$:

$$\left(x+\frac{3}{2}\right)^2=0$$
$$x+\frac{3}{2}=0$$
$$x=-\frac{3}{2}$$

51.　Solving $f(x)=0$:

$$(x-3)^2-25=0$$
$$x^2-6x+9-25=0$$
$$x^2-6x-16=0$$
$$(x+2)(x-8)=0$$
$$x=-2,8$$

53.　Solving $f(x)=g(x)$:

$$x^2+6x+3=-6$$
$$x^2+6x+9=0$$
$$(x+3)^2=0$$
$$x+3=0$$
$$x=-3$$

55.　Solving $f(x)=g(x)$:

$$x^2+6x+3=10$$
$$x^2+6x-7=0$$
$$(x+7)(x-1)=0$$
$$x=-7,1$$

57.　Solving $h(x)=f(x)$:

$$x^2-5x=0$$
$$x(x-5)=0$$
$$x=0,5$$

59.　Solving $h(x)=f(x)$:

$$x^2-5x=2x+8$$
$$x^2-7x-8=0$$
$$(x-8)(x+1)=0$$
$$x=-1,8$$

61.　**a.**　Solving the equation:

$$9x-25=0$$
$$9x=25$$
$$x=\frac{25}{9}$$

b.　Solving the equation:

$$9x^2-25=0$$
$$(3x+5)(3x-5)=0$$
$$x=-\frac{5}{3},\frac{5}{3}$$

c.　Solving the equation:

$$9x^2-25=56$$
$$9x^2-81=0$$
$$9(x+3)(x-3)=0$$
$$x=-3,3$$

d.　Solving the equation:

$$9x^2-25=30x-50$$
$$9x^2-30x+25=0$$
$$(3x-5)^2=0$$
$$3x-5=0$$
$$3x=5$$
$$x=\frac{5}{3}$$

63. Let t represent the required time. Using the Pythagorean theorem:
$$(15t)^2 + (20t)^2 = 75^2$$
$$225t^2 + 400t^2 = 5625$$
$$625t^2 = 5625$$
$$625t^2 - 5625 = 0$$
$$625(t^2 - 9) = 0$$
$$t = -3, 3$$

Since the time must be positive, the time is 3 hours.

65. Let x and $x + 1$ represent the two integers. The equation is:
$$(x + x + 1)^2 = 81$$
$$(2x + 1)^2 = 81$$
$$4x^2 + 4x + 1 = 81$$
$$4x^2 + 4x - 80 = 0$$
$$x^2 + x - 20 = 0$$
$$(x + 5)(x - 4) = 0$$
$$x = -5, 4$$
$$x + 1 = -4, 5$$

The integers are either –5 and –4, or 4 and 5.

67. Let h represent the height the ladder makes with the building. Using the Pythagorean theorem:
$$7^2 + h^2 = 25^2$$
$$49 + h^2 = 625$$
$$h^2 = 576$$
$$h = 24$$

The ladder reaches a height of 24 feet along the building.

69. Let x, $x + 2$, and $x + 4$ represent the three sides. Using the Pythagorean theorem:
$$x^2 + (x + 2)^2 = (x + 4)^2$$
$$x^2 + x^2 + 4x + 4 = x^2 + 8x + 16$$
$$2x^2 + 4x + 4 = x^2 + 8x + 16$$
$$x^2 - 4x - 12 = 0$$
$$(x - 6)(x + 2) = 0$$
$$x = 6 \quad (x = -2 \text{ is impossible})$$

The lengths of the three sides are 6, 8, and 10.

71. Let w represent the width and $3w + 2$ represent the length. Using the area formula:
$$w(3w + 2) = 16$$
$$3w^2 + 2w = 16$$
$$3w^2 + 2w - 16 = 0$$
$$(3w + 8)(w - 2) = 0$$
$$w = 2 \quad (w = -\tfrac{8}{3} \text{ is impossible})$$

The dimensions are 2 feet by 8 feet.

73.   Let h represent the height and $4h + 2$ represent the base. Using the area formula:
$$\frac{1}{2}(4h+2)(h) = 36$$
$$4h^2 + 2h = 72$$
$$4h^2 + 2h - 72 = 0$$
$$2(2h^2 + h - 36) = 0$$
$$2(2h+9)(h-4) = 0$$
$$h = 4 \quad (h = -\tfrac{9}{2} \text{ is impossible})$$

The base is 18 inches and the height is 4 inches.

75.   Setting $h = 0$ in the equation:

$$32t - 16t^2 = 0$$
$$16t(2 - t) = 0$$
$$t = 0, 2$$

The object is on the ground at 0 and 2 seconds.

77.   Substituting $v = 48$ and $h = 32$:
$$h = vt - 16t^2$$
$$32 = 48t - 16t^2$$
$$0 = -16t^2 + 48t - 32$$
$$0 = -16(t^2 - 3t + 2)$$
$$0 = -16(t - 1)(t - 2)$$
$$t = 1, 2$$

It will reach a height of 32 feet after 1 sec and 2 sec.

79.   Substituting $v = 24$ and $h = 0$:
$$h = vt - 16t^2$$
$$0 = 24t - 16t^2$$
$$0 = -16t^2 + 24t$$
$$0 = -8t(2t - 3)$$
$$t = 0, \tfrac{3}{2}$$

It will be on the ground after 0 sec and $\frac{3}{2}$ sec.

81.   Substituting $h = 192$:
$$192 = 96 + 80t - 16t^2$$
$$0 = -16t^2 + 80t - 96$$
$$0 = -16(t^2 - 5t + 6)$$
$$0 = -16(t - 2)(t - 3)$$
$$t = 2, 3$$

The bullet will be 192 feet in the air after 2 sec and 3 sec.

83.   Substituting $R = \$3,200$:
$$R = xp$$
$$3200 = (1200 - 100p)p$$
$$3200 = 1200p - 100p^2$$
$$0 = -100p^2 + 1200p - 3200$$
$$0 = -100(p^2 - 12p + 32)$$
$$0 = -100(p - 4)(p - 8)$$
$$p = 4, 8$$

The cartridges should be sold for either $4 or $8.

85.   Substituting $R = \$7,000$:
$$R = xp$$
$$7000 = (1700 - 100p)p$$
$$7000 = 1700p - 100p^2$$
$$0 = -100p^2 + 1700p - 7000$$
$$0 = -100(p^2 - 17p + 70)$$
$$0 = -100(p - 7)(p - 10)$$
$$p = 7, 10$$

The calculators should be sold for either $7 or $10.

87.   Multiply the second equation by 5:
$$2x - 5y = -8$$
$$15x + 5y = 25$$
Adding yields:
$$17x = 17$$
$$x = 1$$
Substituting to find y:
$$3 + y = 5$$
$$y = 2$$
The solution is $(1, 2)$.

89. To clear each equation of fractions, multiply the first equation by 6 and the second equation by 20:

$$2x - y = 18$$
$$-4x + 5y = 0$$

Multiply the first equation by 2:

$$4x - 2y = 36$$
$$-4x + 5y = 0$$

Adding yields:

$$3y = 36$$
$$y = 12$$

Substituting to find x:

$$2x - 12 = 18$$
$$2x = 30$$
$$x = 15$$

The solution is $(15, 12)$.

91. Adding the first and third equations:

$$5x = 15$$
$$x = 3$$

Adding the first and second equations:

$$3x - 2z = 7$$
$$9 - 2z = 7$$
$$-2z = -2$$
$$z = 1$$

Substituting to find y:

$$3 + y - 3 = -2$$
$$y = -2$$

The solution is $(3, -2, 1)$.

93. Let x and $2x$ represent the two investments. The equation is:

$$0.05(x) + 0.06(2x) = 680$$
$$0.17x = 680$$
$$x = 4,000$$
$$2x = 8,000$$

John invested $4,000 at 5% and $8,000 at 6%.

95. Graphing the solution set:

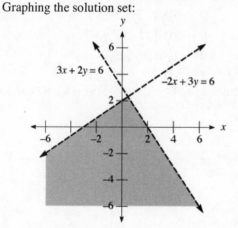

97. Graphing the solution set:

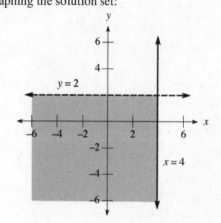

Chapter 5 Review/Test

1. Simplifying: $x^3 \cdot x^7 = x^{3+7} = x^{10}$

2. Simplifying: $\left(5x^3\right)^2 = 5^2 x^6 = 25x^6$

3. Simplifying: $\left(2x^3y\right)^2 \left(-2x^4y^2\right)^3 = 4x^6y^2 \cdot \left(-8x^{12}y^6\right) = -32x^{18}y^8$

4. Writing with positive exponents: $2^{-3} = \dfrac{1}{2^3} = \dfrac{1}{8}$

5. Writing with positive exponents: $\left(\frac{2}{3}\right)^{-2} = \left(\frac{3}{2}\right)^2 = \frac{9}{4}$

6. Writing with positive exponents: $2^{-2} + 4^{-1} = \dfrac{1}{2^2} + \dfrac{1}{4} = \dfrac{1}{4} + \dfrac{1}{4} = \dfrac{1}{2}$

7. Writing in scientific notation: $34,500,000 = 3.45 \times 10^7$

8. Writing in scientific notation: $0.00357 = 3.57 \times 10^{-3}$

9. Writing in expanded form: $4.45 \times 10^4 = 44,500$

10. Writing in expanded form: $4.45 \times 10^{-4} = 0.000445$

11. Simplifying: $\dfrac{a^{-4}}{a^5} = a^{-4-5} = a^{-9} = \dfrac{1}{a^9}$

12. Simplifying: $\dfrac{\left(4x^2\right)\left(-3x^3\right)^2}{\left(12x^{-2}\right)^2} = \dfrac{4x^2\left(9x^6\right)}{144x^{-4}} = \dfrac{36x^8}{144x^{-4}} = \dfrac{36}{144}x^{8+4} = \dfrac{x^{12}}{4}$

13. Simplifying: $\dfrac{x^n x^{3n}}{x^{4n-2}} = \dfrac{x^{4n}}{x^{4n-2}} = x^{4n-(4n-2)} = x^{4n-4n+2} = x^2$

14. Simplifying: $\left(2 \times 10^3\right)\left(4 \times 10^{-5}\right) = 8 \times 10^{3-5} = 8 \times 10^{-2}$

15. Simplifying: $\dfrac{(600,000)(0.000008)}{(4,000)(3,000,000)} = \dfrac{\left(6 \times 10^5\right)\left(8 \times 10^{-6}\right)}{\left(4 \times 10^3\right)\left(3 \times 10^6\right)} = \dfrac{48 \times 10^{-1}}{12 \times 10^9} = \dfrac{48}{12} \times 10^{-1-9} = 4 \times 10^{-10}$

16. Simplifying: $\left(6x^2 - 3x + 2\right) - \left(4x^2 + 2x - 5\right) = 6x^2 - 3x + 2 - 4x^2 - 2x + 5 = 2x^2 - 5x + 7$

17. Simplifying: $\left(x^3 - x\right) - \left(x^2 + x\right) + \left(x^3 - 3\right) - \left(x^2 + 1\right) = x^3 - x - x^2 - x + x^3 - 3 - x^2 - 1 = 2x^3 - 2x^2 - 2x - 4$

18. Subtracting: $\left(3x^2 - 5x - 2\right) - \left(2x^2 - 3x + 1\right) = 3x^2 - 5x - 2 - 2x^2 + 3x - 1 = x^2 - 2x - 3$

19. Simplifying: $-3\left[2x - 4\left(3x + 1\right)\right] = -3\left(2x - 12x - 4\right) = -3\left(-10x - 4\right) = 30x + 12$

20. Evaluating when $x = -2$: $2\left(-2\right)^2 - 3\left(-2\right) + 1 = 2\left(4\right) + 6 + 1 = 8 + 6 + 1 = 15$

21. Multiplying: $3x\left(4x^2 - 2x + 1\right) = 12x^3 - 6x^2 + 3x$

22. Multiplying: $2a^2b^3\left(a^2 + 2ab + b^2\right) = 2a^4b^3 + 4a^3b^4 + 2a^2b^5$

23. Multiplying: $(6 - y)(3 - y) = 18 - 6y - 3y + y^2 = 18 - 9y + y^2$

24. Multiplying: $\left(2x^2 - 1\right)\left(3x^2 + 4\right) = 6x^4 + 8x^2 - 3x^2 - 4 = 6x^4 + 5x^2 - 4$

25. Multiplying: $2t(t + 1)(t - 3) = 2t\left(t^2 - 2t - 3\right) = 2t^3 - 4t^2 - 6t$

26. Multiplying: $(x + 3)\left(x^2 - 3x + 9\right) = x^3 - 3x^2 + 9x + 3x^2 - 9x + 27 = x^3 + 27$

27. Multiplying: $(2x - 3)\left(4x^2 + 6x + 9\right) = 8x^3 + 12x^2 + 18x - 12x^2 - 18x - 27 = 8x^3 - 27$

28. Multiplying: $\left(a^2 - 2\right)^2 = a^4 - 2\left(a^2\right)(2) + 2^2 = a^4 - 4a^2 + 4$

29. Multiplying: $(3x + 5)^2 = (3x)^2 + 2(3x)(5) + 5^2 = 9x^2 + 30x + 25$

30. Multiplying: $(4x - 3y)^2 = (4x)^2 - 2(4x)(3y) + (3y)^2 = 16x^2 - 24xy + 9y^2$

31. Multiplying: $\left(x - \frac{1}{3}\right)\left(x + \frac{1}{3}\right) = (x)^2 - \left(\frac{1}{3}\right)^2 = x^2 - \frac{1}{9}$

32. Multiplying: $(2a + b)(2a - b) = (2a)^2 - (b)^2 = 4a^2 - b^2$

33. Multiplying: $(x - 1)^3 = (x - 1)(x - 1)^2 = (x - 1)(x^2 - 2x + 1) = x^3 - 2x^2 + x - x^2 + 2x - 1 = x^3 - 3x^2 + 3x - 1$

34. Multiplying: $\left(x^m + 2\right)\left(x^m - 2\right) = \left(x^m\right)^2 - (2)^2 = x^{2m} - 4$

35. Factoring: $6x^4 y - 9xy^4 + 18x^3 y^3 = 3xy\left(2x^3 - 3y^3 + 6x^2 y^2\right)$

36. Factoring: $4x^2 (x + y)^2 - 8y^2 (x + y)^2 = 4(x + y)^2 \left(x^2 - 2y^2\right)$

37. Factoring: $8x^2 + 10 - 4x^2 y - 5y = 2\left(4x^2 + 5\right) - y\left(4x^2 + 5\right) = \left(4x^2 + 5\right)(2 - y)$

38. Factoring:
$$\begin{aligned} x^3 + 8b^2 - x^3 y^2 - 8y^2 b^2 &= x^3 - x^3 y^2 + 8b^2 - 8y^2 b^2 \\ &= x^3\left(1 - y^2\right) + 8b^2\left(1 - y^2\right) \\ &= \left(1 - y^2\right)\left(x^3 + 8b^2\right) \\ &= (1 + y)(1 - y)\left(x^3 + 8b^2\right) \end{aligned}$$

39. Factoring: $x^2 - 5x + 6 = (x - 2)(x - 3)$

40. Factoring: $2x^3 + 4x^2 - 30x = 2x\left(x^2 + 2x - 15\right) = 2x(x + 5)(x - 3)$

41. Factoring: $20a^2 - 41ab + 20b^2 = (5a - 4b)(4a - 5b)$

42. Factoring: $6x^4 - 11x^3 - 10x^2 = x^2\left(6x^2 - 11x - 10\right) = x^2 (3x + 2)(2x - 5)$

43. Factoring: $24x^2 y - 6xy - 45y = 3y\left(8x^2 - 2x - 15\right) = 3y(4x + 5)(2x - 3)$

44. Factoring: $x^4 - 16 = \left(x^2 + 4\right)\left(x^2 - 4\right) = \left(x^2 + 4\right)(x + 2)(x - 2)$

45. Factoring: $3a^4 + 18a^2 + 27 = 3\left(a^4 + 6a^2 + 9\right) = 3\left(a^2 + 3\right)^2$

46. Factoring: $a^3 - 8 = (a - 2)\left(a^2 + 2a + 4\right)$

47. Factoring: $5x^3 + 30x^2 y + 45xy^2 = 5x\left(x^2 + 6xy + 9y^2\right) = 5x(x + 3y)^2$

48. Factoring: $3a^3 b - 27ab^3 = 3ab\left(a^2 - 9b^2\right) = 3ab(a - 3b)(a + 3b)$

49. Factoring: $x^2 - 10x + 25 - y^2 = (x - 5)^2 - y^2 = (x - 5 + y)(x - 5 - y)$

50. Factoring: $36 - 25a^2 = (6 + 5a)(6 - 5a)$

51. Factoring: $x^3 + 4x^2 - 9x - 36 = x^2 (x + 4) - 9(x + 4) = (x + 4)\left(x^2 - 9\right) = (x + 4)(x + 3)(x - 3)$

52. Solving the equation:

$$\begin{aligned} x^2 + 5x + 6 &= 0 \\ (x + 3)(x + 2) &= 0 \\ x &= -3, -2 \end{aligned}$$

53. Solving the equation:

$$\begin{aligned} \frac{5}{6} y^2 &= \frac{1}{4} y + \frac{1}{3} \\ 12\left(\frac{5}{6} y^2\right) &= 12\left(\frac{1}{4} y + \frac{1}{3}\right) \\ 10y^2 &= 3y + 4 \\ 10y^2 - 3y - 4 &= 0 \\ (5y - 4)(2y + 1) &= 0 \\ y &= -\frac{1}{2}, \frac{4}{5} \end{aligned}$$

54.   Solving the equation:

$$9x^2 - 25 = 0$$
$$(3x+5)(3x-5) = 0$$
$$x = -\frac{5}{3}, \frac{5}{3}$$

55.   Solving the equation:

$$5x^2 = -10x$$
$$5x^2 + 10x = 0$$
$$5x(x+2) = 0$$
$$x = -2, 0$$

56.   Solving the equation:

$$(x+2)(x-5) = 8$$
$$x^2 - 3x - 10 = 8$$
$$x^2 - 3x - 18 = 0$$
$$(x-6)(x+3) = 0$$
$$x = -3, 6$$

57.   Solving the equation:

$$x^3 + 4x^2 - 9x - 36 = 0$$
$$x^2(x+4) - 9(x+4) = 0$$
$$(x+4)(x^2 - 9) = 0$$
$$(x+4)(x+3)(x-3) = 0$$
$$x = -4, -3, 3$$

58.   Let x and $x + 2$ represent the two even integers. The equation is:

$$x(x+2) = 80$$
$$x^2 + 2x - 80 = 0$$
$$(x+10)(x-8) = 0$$
$$x = -10, 8$$
$$x + 2 = -8, 10$$

The integers are either –10 and –8, or 8 and 10.

59.   Let x and $x + 1$ represent the two integers. The equation is:

$$x^2 + (x+1)^2 = 41$$
$$x^2 + x^2 + 2x + 1 = 41$$
$$2x^2 + 2x - 40 = 0$$
$$x^2 + x - 20 = 0$$
$$(x+5)(x-4) = 0$$
$$x = -5, 4$$
$$x + 1 = -4, 5$$

The integers are either –5 and –4, or 4 and 5.

60.   Let x, $x + 1$, and $x + 2$ represent the three sides. Using the Pythagorean theorem:

$$x^2 + (x+1)^2 = (x+2)^2$$
$$x^2 + x^2 + 2x + 1 = x^2 + 4x + 4$$
$$2x^2 + 2x + 1 = x^2 + 4x + 4$$
$$x^2 - 2x - 3 = 0$$
$$(x-3)(x+1) = 0$$
$$x = 3 \quad (x = -1 \text{ is impossible})$$

The lengths of the three sides are 3, 4, and 5.

61.   Let x, $x + 2$, and $x + 4$ represent the three sides. Using the Pythagorean theorem:

$$x^2 + (x+2)^2 = (x+4)^2$$
$$x^2 + x^2 + 4x + 4 = x^2 + 8x + 16$$
$$2x^2 + 4x + 4 = x^2 + 8x + 16$$
$$x^2 - 4x - 12 = 0$$
$$(x-6)(x+2) = 0$$
$$x = 6 \quad (x = -2 \text{ is impossible})$$

The lengths of the three sides are 6, 8, and 10.

Chapter 6
Rational Expressions and Rational Functions

6.1 Basic Properties and Reducing to Lowest Terms

1. **a.** Simplifying: $\dfrac{6+1}{36-1} = \dfrac{7}{35} = \dfrac{1}{5}$

   b. Simplifying: $\dfrac{x+3}{x^2-9} = \dfrac{x+3}{(x+3)(x-3)} = \dfrac{1}{x-3}$ if $x \neq -3, 3$

   c. Simplifying: $\dfrac{x^2-3x}{x^2-9} = \dfrac{x(x-3)}{(x+3)(x-3)} = \dfrac{x}{x+3}$ if $x \neq -3, 3$

   d. Simplifying: $\dfrac{x^3-27}{x^2-9} = \dfrac{(x-3)(x^2+3x+9)}{(x+3)(x-3)} = \dfrac{x^2+3x+9}{x+3}$ if $x \neq -3, 3$

   e. Simplifying: $\dfrac{x^3-27}{x^3-3x^2} = \dfrac{(x-3)(x^2+3x+9)}{x^2(x-3)} = \dfrac{x^2+3x+9}{x^2}$ if $x \neq 0, 3$

3. Finding each function value:

   $h(0) = \dfrac{0-3}{0+1} = \dfrac{-3}{1} = -3$

   $h(3) = \dfrac{3-3}{3+1} = \dfrac{0}{4} = 0$

   $h(1) = \dfrac{1-3}{1+1} = \dfrac{-2}{2} = -1$

   $h(-3) = \dfrac{-3-3}{-3+1} = \dfrac{-6}{-2} = 3$

   $h(-1) = \dfrac{-1-3}{-1+1} = \dfrac{-4}{0}$, which is undefined

5. The domain is $\{x \mid x \neq 1\}$.

7. Setting the denominator equal to 0:
   $$t^2 - 16 = 0$$
   $$(t+4)(t-4) = 0$$
   $$t = -4, 4$$
   The domain is $\{t \mid t \neq -4, t \neq 4\}$.

9. Setting the denominator equal to 0:
   $$3x - 15 = 0$$
   $$3x = 15$$
   $$x = 5$$
   The domain is $\{x \mid x \neq 5\}$.

11. The domain is all real numbers.

13. The domain is $\{x \mid x \neq 0\}$.

15. Setting the denominator equal to 0:
$$x^2 - x - 20 = 0$$
$$(x+4)(x-5) = 0$$
$$x = -4, 5$$
The domain is $\{x \mid x \neq -4, x \neq 5\}$.

17. Reducing to lowest terms: $\dfrac{x^2 - 16}{6x + 24} = \dfrac{(x+4)(x-4)}{6(x+4)} = \dfrac{x-4}{6}$

19. Reducing to lowest terms: $\dfrac{12x - 9y}{3x^2 + 3xy} = \dfrac{3(4x - 3y)}{3x(x+y)} = \dfrac{4x - 3y}{x(x+y)}$

21. Reducing to lowest terms: $\dfrac{a^4 - 81}{a - 3} = \dfrac{\left(a^2 + 9\right)\left(a^2 - 9\right)}{a - 3} = \dfrac{\left(a^2 + 9\right)(a+3)(a-3)}{a - 3} = \left(a^2 + 9\right)(a + 3)$

23. Reducing to lowest terms: $\dfrac{a^2 - 4a - 12}{a^2 + 8a + 12} = \dfrac{(a-6)(a+2)}{(a+6)(a+2)} = \dfrac{a-6}{a+6}$

25. Reducing to lowest terms: $\dfrac{20y^2 - 45}{10y^2 - 5y - 15} = \dfrac{5\left(4y^2 - 9\right)}{5\left(2y^2 - y - 3\right)} = \dfrac{5(2y+3)(2y-3)}{5(2y-3)(y+1)} = \dfrac{2y+3}{y+1}$

27. Reducing to lowest terms: $\dfrac{a^3 + b^3}{a^2 - b^2} = \dfrac{(a+b)\left(a^2 - ab + b^2\right)}{(a+b)(a-b)} = \dfrac{a^2 - ab + b^2}{a - b}$

29. Reducing to lowest terms: $\dfrac{8x^4 - 8x}{4x^4 + 4x^3 + 4x^2} = \dfrac{8x\left(x^3 - 1\right)}{4x^2\left(x^2 + x + 1\right)} = \dfrac{8x(x-1)\left(x^2 + x + 1\right)}{4x^2\left(x^2 + x + 1\right)} = \dfrac{2(x-1)}{x}$

31. Reducing to lowest terms: $\dfrac{6x^2 + 7xy - 3y^2}{6x^2 + xy - y^2} = \dfrac{(2x+3y)(3x-y)}{(2x+y)(3x-y)} = \dfrac{2x+3y}{2x+y}$

33. Reducing to lowest terms: $\dfrac{ax + 2x + 3a + 6}{ay + 2y - 4a - 8} = \dfrac{x(a+2) + 3(a+2)}{y(a+2) - 4(a+2)} = \dfrac{(a+2)(x+3)}{(a+2)(y-4)} = \dfrac{x+3}{y-4}$

35. Reducing to lowest terms: $\dfrac{x^2 + bx - 3x - 3b}{x^2 - 2bx - 3x + 6b} = \dfrac{x(x+b) - 3(x+b)}{x(x-2b) - 3(x-2b)} = \dfrac{(x+b)(x-3)}{(x-2b)(x-3)} = \dfrac{x+b}{x-2b}$

37. Reducing to lowest terms:

$$\dfrac{x^3 + 3x^2 - 4x - 12}{x^2 + x - 6} = \dfrac{x^2(x+3) - 4(x+3)}{(x+3)(x-2)} = \dfrac{(x+3)\left(x^2 - 4\right)}{(x+3)(x-2)} = \dfrac{(x+3)(x+2)(x-2)}{(x+3)(x-2)} = x + 2$$

39. Reducing to lowest terms: $\dfrac{4x^4 - 25}{6x^3 - 4x^2 + 15x - 10} = \dfrac{\left(2x^2 + 5\right)\left(2x^2 - 5\right)}{2x^2(3x - 2) + 5(3x - 2)} = \dfrac{\left(2x^2 + 5\right)\left(2x^2 - 5\right)}{(3x - 2)\left(2x^2 + 5\right)} = \dfrac{2x^2 - 5}{3x - 2}$

41. Reducing to lowest terms: $\dfrac{x^3 - 8}{x^2 - 4} = \dfrac{(x-2)\left(x^2 + 2x + 4\right)}{(x-2)(x+2)} = \dfrac{x^2 + 2x + 4}{x + 2}$

43. Reducing to lowest terms: $\dfrac{64 + t^3}{16 - 4t + t^2} = \dfrac{(4+t)\left(16 - 4t + t^2\right)}{16 - 4t + t^2} = 4 + t$

45. Reducing to lowest terms: $\dfrac{8x^3 - 27}{4x^2 - 9} = \dfrac{(2x-3)\left(4x^2 + 6x + 9\right)}{(2x-3)(2x+3)} = \dfrac{4x^2 + 6x + 9}{2x + 3}$

47. Reducing to lowest terms: $\dfrac{x - 4}{4 - x} = \dfrac{x - 4}{-1(x - 4)} = -1$

49. Reducing to lowest terms: $\dfrac{y^2 - 36}{6 - y} = \dfrac{(y+6)(y-6)}{-1(y-6)} = -(y+6)$

51. Reducing to lowest terms: $\dfrac{1 - 9a^2}{9a^2 - 6a + 1} = \dfrac{-1(9a^2 - 1)}{(3a-1)^2} = \dfrac{-1(3a+1)(3a-1)}{(3a-1)^2} = -\dfrac{3a+1}{3a-1}$

53. Reducing to lowest terms: $\dfrac{(3x-5)-(3a-5)}{x-a} = \dfrac{3x-3a}{x-a} = \dfrac{3(x-a)}{x-a} = 3$

55. Reducing to lowest terms: $\dfrac{(x^2-4)-(a^2-4)}{x-a} = \dfrac{x^2-a^2}{x-a} = \dfrac{(x+a)(x-a)}{x-a} = x+a$

57. Evaluating the functions:

$$f(0) = \dfrac{0^2 - 4}{0 - 2} = \dfrac{-4}{-2} = 2$$
$$g(0) = 0 + 2 = 2$$

59. Evaluating the functions:

$$f(2) = \dfrac{2^2 - 4}{2 - 2} = \dfrac{0}{0}, \text{ which is undefined}$$
$$g(2) = 2 + 2 = 4$$

61. Evaluating the functions:

$$f(0) = \dfrac{0^2 - 1}{0 - 1} = \dfrac{-1}{-1} = 1$$
$$g(0) = 0 + 1 = 1$$

63. Evaluating the functions:

$$f(2) = \dfrac{2^2 - 1}{2 - 1} = \dfrac{3}{1} = 3$$
$$g(2) = 2 + 1 = 3$$

65. The graph of $y = x + 2$ contains the point (2,4), while the other graph does not.

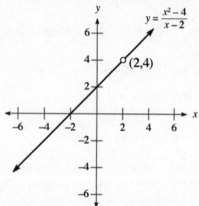

67. **a.** Adding: $(x^2 - 7x) + (7x - 49) = x^2 - 7x + 7x - 49 = x^2 - 49$

b. Subtracting: $(x^2 - 7x) - (7x - 49) = x^2 - 7x - 7x + 49 = x^2 - 14x + 49$

c. Multiplying: $(x^2 - 7x)(7x - 49) = 7x^3 - 49x^2 - 49x^2 + 343x = 7x^3 - 98x^2 + 343x$

d. Reducing: $\dfrac{x^2 - 7x}{7x - 49} = \dfrac{x(x-7)}{7(x-7)} = \dfrac{x}{7}$

69. Completing the table:

Weeks x	Weight (lb) $W(x)$
0	200
1	194
4	184
12	173
24	168

71. Dividing: $\dfrac{\pi \cdot 50 \text{ feet}}{20 \text{ seconds}} \approx 7.9$ feet per second

73. **a.** The domain is $\{t \mid 20 \leq t \leq 50\}$.

   b. Graphing the function:

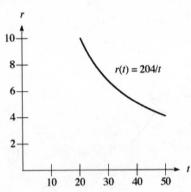

$r(t) = 204/t$

75. **a.** The domain is $\{d \mid 1 \leq d \leq 6\}$.

   b. Graphing the function:

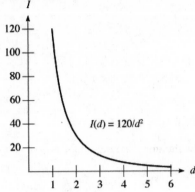

$I(d) = 120/d^2$

77. Subtracting: $\left(4x^2 - 5x + 5\right) - \left(x^2 + 2x + 1\right) = 4x^2 - 5x + 5 - x^2 - 2x - 1 = 3x^2 - 7x + 4$

79. Subtracting: $(10x - 11) - (10x - 20) = 10x - 11 - 10x + 20 = 9$

81. Subtracting: $\left(4x^3\right) - \left(4x^3 - 8x^2\right) = 4x^3 - 4x^3 + 8x^2 = 8x^2$

83. Dividing: $\dfrac{10x^5}{5x^2} = 2x^{5-2} = 2x^3$

85. Dividing: $\dfrac{4x^4 y^3}{-2x^2 y} = -2x^{4-2} y^{3-1} = -2x^2 y^2$

87. Dividing: $4,628 \div 25 = 185.12$

89. Multiplying: $2x^2 (2x - 4) = 4x^3 - 8x^2$

91. Multiplying:

$$(2x - 4)\left(2x^2 + 4x + 5\right) = 2x\left(2x^2 + 4x + 5\right) - 4\left(2x^2 + 4x + 5\right)$$
$$= 4x^3 + 8x^2 + 10x - 8x^2 - 16x - 20$$
$$= 4x^3 - 6x - 20$$

93. Subtracting: $\left(2x^2 - 7x + 9\right) - \left(2x^2 - 4x\right) = 2x^2 - 7x + 9 - 2x^2 + 4x = -3x + 9$

95. Factoring: $x^2 - a^2 = (x + a)(x - a)$

97. Factoring: $x^2 - 6xy - 7y^2 = (x - 7y)(x + y)$

99. **a.** From the graph: $f(2) = 2$

   b. From the graph: $f(-1) = -4$

   c. From the graph: $f(0)$ is undefined

   d. From the graph: $g(3) = 2$

   e. From the graph: $g(6) = 1$

   f. From the graph: $g(-1) = -6$

   g. From the graph: $f(g(6)) = f(1) = 4$

   h. From the graph: $g(f(-2)) = g(-2) = -3$

6.2 Division of Polynomials

1. Dividing: $\dfrac{4x^3 - 8x^2 + 6x}{2x} = \dfrac{4x^3}{2x} - \dfrac{8x^2}{2x} + \dfrac{6x}{2x} = 2x^2 - 4x + 3$

3. Dividing: $\dfrac{10x^4 + 15x^3 - 20x^2}{-5x^2} = \dfrac{10x^4}{-5x^2} + \dfrac{15x^3}{-5x^2} - \dfrac{20x^2}{-5x^2} = -2x^2 - 3x + 4$

5. Dividing: $\dfrac{8y^5 + 10y^3 - 6y}{4y^3} = \dfrac{8y^5}{4y^3} + \dfrac{10y^3}{4y^3} - \dfrac{6y}{4y^3} = 2y^2 + \dfrac{5}{2} - \dfrac{3}{2y^2}$

7. Dividing: $\dfrac{5x^3 - 8x^2 - 6x}{-2x^2} = \dfrac{5x^3}{-2x^2} - \dfrac{8x^2}{-2x^2} - \dfrac{6x}{-2x^2} = -\dfrac{5}{2}x + 4 + \dfrac{3}{x}$

9. Dividing: $\dfrac{28a^3 b^5 + 42a^4 b^3}{7a^2 b^2} = \dfrac{28a^3 b^5}{7a^2 b^2} + \dfrac{42a^4 b^3}{7a^2 b^2} = 4ab^3 + 6a^2 b$

11. Dividing: $\dfrac{10x^3y^2 - 20x^2y^3 - 30x^3y^3}{-10x^2y} = \dfrac{10x^3y^2}{-10x^2y} - \dfrac{20x^2y^3}{-10x^2y} - \dfrac{30x^3y^3}{-10x^2y} = -xy + 2y^2 + 3xy^2$

13. Dividing by factoring: $\dfrac{x^2 - x - 6}{x - 3} = \dfrac{(x-3)(x+2)}{x-3} = x + 2$

15. Dividing by factoring: $\dfrac{2a^2 - 3a - 9}{2a+3} = \dfrac{(2a+3)(a-3)}{2a+3} = a - 3$

17. Dividing by factoring: $\dfrac{5x^2 - 14xy - 24y^2}{x - 4y} = \dfrac{(5x+6y)(x-4y)}{x-4y} = 5x + 6y$

19. Dividing by factoring: $\dfrac{x^3 - y^3}{x - y} = \dfrac{(x-y)(x^2+xy+y^2)}{x-y} = x^2 + xy + y^2$

21. Dividing by factoring: $\dfrac{y^4 - 16}{y-2} = \dfrac{(y^2+4)(y^2-4)}{y-2} = \dfrac{(y^2+4)(y+2)(y-2)}{y-2} = (y^2+4)(y+2)$

23. Dividing by factoring:

$$\frac{x^3 + 2x^2 - 25x - 50}{x - 5} = \frac{x^2(x+2) - 25(x+2)}{x-5} = \frac{(x+2)(x^2-25)}{x-5} = \frac{(x+2)(x+5)(x-5)}{x-5} = (x+2)(x+5)$$

25. Dividing using long division:

$$
\begin{array}{r}
x - 7 \\
x + 2 \overline{\smash{)}\, x^2 - 5x - 7} \\
\underline{x^2 + 2x} \\
-7x - 7 \\
\underline{-7x - 14} \\
7
\end{array}
$$

The quotient is $x - 7 + \dfrac{7}{x+2}$.

27. Dividing using long division:

$$
\begin{array}{r}
2x^2 - 5x + 1 \\
x + 1 \overline{\smash{)}\, 2x^3 - 3x^2 - 4x + 5} \\
\underline{2x^3 + 2x^2} \\
-5x^2 - 4x \\
\underline{-5x^2 - 5x} \\
x + 5 \\
\underline{x + 1} \\
4
\end{array}
$$

The quotient is $2x^2 - 5x + 1 + \dfrac{4}{x+1}$.

29. Dividing using long division:

$$
\begin{array}{r}
y^2 - 3y - 13 \\
2y - 3 \overline{\smash{)}\, 2y^3 - 9y^2 - 17y + 39} \\
\underline{2y^3 - 3y^2} \\
-6y^2 - 17y \\
\underline{-6y^2 + 9y} \\
-26y + 39 \\
\underline{-26y + 39} \\
0
\end{array}
$$

The quotient is $y^2 - 3y - 13$.

31. Dividing using long division:

$$
\begin{array}{r}
3y^2 + 6y + 8 \\
2y - 4 \overline{\smash{)}\, 6y^3 + 0y^2 - 8y + 5} \\
\underline{6y^3 - 12y^2} \\
12y^2 - 8y \\
\underline{12y^2 - 24y} \\
16y + 5 \\
\underline{16y - 32} \\
37
\end{array}
$$

The quotient is $3y^2 + 6y + 8 + \dfrac{37}{2y-4}$.

33. Dividing using long division:

$$a-2\overline{\smash{\big)}\,a^4+0a^3+0a^2-2a+5} \quad \begin{array}{c} a^3+2a^2+4a+6 \end{array}$$

$$\underline{a^4-2a^3}$$
$$2a^3+0a^2$$
$$\underline{2a^3-4a^2}$$
$$4a^2-2a$$
$$\underline{4a^2-8a}$$
$$6a+5$$
$$\underline{6a-12}$$
$$17$$

(handwritten: $a^3+2a^2+4a+6 + \dfrac{17}{a-2}$)

The quotient is $a^3+2a^2+4a+6+\dfrac{17}{a-2}$.

35. Dividing using long division:

$$y-2\overline{\smash{\big)}\,y^4+0y^3+0y^2+0y-16} \quad \begin{array}{c} y^3+2y^2+4y+8 \end{array}$$

$$\underline{y^4-2y^3}$$
$$2y^3+0y^2$$
$$\underline{2y^3-4y^2}$$
$$4y^2+0y$$
$$\underline{4y^2-8y}$$
$$8y-16$$
$$\underline{8y-16}$$
$$0$$

The quotient is y^3+2y^2+4y+8.

37. Dividing: $h(x)=\dfrac{x^2-36}{4x-24}=\dfrac{(x+6)(x-6)}{4(x-6)}=\dfrac{x+6}{4}$. The domain is $\{x\,|\,x\neq 6\}$.

39. Dividing: $h(x)=\dfrac{x^2-16x+64}{x^2-4x-32}=\dfrac{(x-8)^2}{(x-8)(x+4)}=\dfrac{x-8}{x+4}$. The domain is $\{x\,|\,x\neq -4, x\neq 8\}$.

41. Dividing: $h(x)=\dfrac{x^3-27}{x-3}=\dfrac{(x-3)(x^2+3x+9)}{x-3}=x^2+3x+9$. The domain is $\{x\,|\,x\neq 3\}$.

43. **a.** Evaluating the formula: $\dfrac{f(x+h)-f(x)}{h}=\dfrac{4x+4h-4x}{h}=\dfrac{4h}{h}=4$

   b. Evaluating the formula: $\dfrac{f(x)-f(a)}{x-a}=\dfrac{4x-4a}{x-a}=\dfrac{4(x-a)}{x-a}=4$

45. **a.** Evaluating the formula: $\dfrac{f(x+h)-f(x)}{h}=\dfrac{5(x+h)+3-(5x+3)}{h}=\dfrac{5x+5h+3-5x-3}{h}=\dfrac{5h}{h}=5$

   b. Evaluating the formula: $\dfrac{f(x)-f(a)}{x-a}=\dfrac{(5x+3)-(5a+3)}{x-a}=\dfrac{5x-5a}{x-a}=\dfrac{5(x-a)}{x-a}=5$

47. **a.** Evaluating the formula:

$$\dfrac{f(x+h)-f(x)}{h}=\dfrac{(x+h)^2-x^2}{h}=\dfrac{x^2+2xh+h^2-x^2}{h}=\dfrac{2xh+h^2}{h}=\dfrac{h(2x+h)}{h}=2x+h$$

   b. Evaluating the formula: $\dfrac{f(x)-f(a)}{x-a}=\dfrac{x^2-a^2}{x-a}=\dfrac{(x+a)(x-a)}{x-a}=x+a$

49. **a.** Evaluating the formula:

$$\dfrac{f(x+h)-f(x)}{h}=\dfrac{(x+h)^2+1-(x^2+1)}{h}=\dfrac{x^2+2xh+h^2+1-x^2-1}{h}=\dfrac{2xh+h^2}{h}=\dfrac{h(2x+h)}{h}=2x+h$$

   b. Evaluating the formula: $\dfrac{f(x)-f(a)}{x-a}=\dfrac{(x^2+1)-(a^2+1)}{x-a}=\dfrac{x^2-a^2}{x-a}=\dfrac{(x+a)(x-a)}{x-a}=x+a$

51.   **a.**   Evaluating the formula:

$$\frac{f(x+h)-f(x)}{h} = \frac{(x+h)^2 - 3(x+h) + 4 - \left(x^2 - 3x + 4\right)}{h}$$

$$= \frac{x^2 + 2xh + h^2 - 3x - 3h + 4 - x^2 + 3x - 4}{h}$$

$$= \frac{2xh + h^2 - 3h}{h}$$

$$= \frac{h(2x + h - 3)}{h}$$

$$= 2x + h - 3$$

b.   Evaluating the formula:

$$\frac{f(x)-f(a)}{x-a} = \frac{\left(x^2 - 3x + 4\right) - \left(a^2 - 3a + 4\right)}{x-a}$$

$$= \frac{x^2 - a^2 - 3x + 3a}{x-a}$$

$$= \frac{(x+a)(x-a) - 3(x-a)}{x-a}$$

$$= \frac{(x-a)(x+a-3)}{x-a}$$

$$= x + a - 3$$

53.   **a.**   Evaluating the formula:

$$\frac{f(x+h)-f(x)}{h} = \frac{2(x+h)^2 + 3(x+h) - 4 - \left(2x^2 + 3x - 4\right)}{h}$$

$$= \frac{2x^2 + 4xh + 2h^2 + 3x + 3h - 4 - 2x^2 - 3x + 4}{h}$$

$$= \frac{4xh + 2h^2 + 3h}{h}$$

$$= \frac{h(4x + 2h + 3)}{h}$$

$$= 4x + 2h + 3$$

b.   Evaluating the formula:

$$\frac{f(x)-f(a)}{x-a} = \frac{\left(2x^2 + 3x - 4\right) - \left(2a^2 + 3a - 4\right)}{x-a}$$

$$= \frac{2x^2 - 2a^2 + 3x - 3a}{x-a}$$

$$= \frac{2(x+a)(x-a) + 3(x-a)}{x-a}$$

$$= \frac{(x-a)(2x + 2a + 3)}{x-a}$$

$$= 2x + 2a + 3$$

55. **a.** Using long division:

$$x - 2 \overline{)x^3 - 3x^2 + 5x - 6} \quad \begin{array}{c} x^2 - x + 3 \end{array}$$

$$\begin{array}{r} x^2 - x + 3 \\ x - 2 \overline{\smash{)}x^3 - 3x^2 + 5x - 6} \\ \underline{x^3 - 2x^2} \\ -x^2 + 5x \\ \underline{-x^2 + 2x} \\ 3x - 6 \\ \underline{3x - 6} \\ 0 \end{array}$$

Since the remainder is 0, $x - 2$ is a factor of $x^3 - 3x^2 + 5x - 6$.

Also note that: $P(2) = (2)^3 - 3(2)^2 + 5(2) - 6 = 8 - 12 + 10 - 6 = 0$

b. Using long division:

$$\begin{array}{r} x^3 - x + 1 \\ x - 5 \overline{\smash{)}x^4 - 5x^3 - x^2 + 6x - 5} \\ \underline{x^4 - 5x^3} \\ -x^2 + 6x \\ \underline{-x^2 + 5x} \\ x - 5 \\ \underline{x - 5} \\ 0 \end{array}$$

Since the remainder is 0, $x - 5$ is a factor of $x^4 - 5x^3 - x^2 + 6x - 5$.

Also note that: $P(5) = (5)^4 - 5(5)^3 - (5)^2 + 6(5) - 5 = 625 - 625 - 25 + 30 - 5 = 0$

57. **a.** Using long division:

$$\begin{array}{r} x^2 + 6x + 5 \\ x + 4 \overline{\smash{)}x^3 + 10x^2 + 29x + 20} \\ \underline{x^3 + 4x^2} \\ 6x^2 + 29x \\ \underline{6x^2 + 24x} \\ 5x + 20 \\ \underline{5x + 20} \\ 0 \end{array}$$

So $x^3 + 10x^2 + 29x + 20 = (x + 4)(x^2 + 6x + 5) = (x + 4)(x + 5)(x + 1)$.

b. Reducing the fraction: $\dfrac{x^3 + 10x^2 + 29x + 20}{x + 4} = \dfrac{(x + 4)(x + 5)(x + 1)}{x + 4} = (x + 5)(x + 1)$

c. The answers to the problems are the same. Long division produces the same result as factoring and simplifying.

59. **a.** Using long division:

$$
\begin{array}{r}
x^2 + x - 12 \\
x+2 \overline{\smash{\big)}\, x^3 + 3x^2 - 10x - 24} \\
\underline{x^3 + 2x^2} \\
x^2 - 10x \\
\underline{x^2 + 2x} \\
-12x - 24 \\
\underline{-12x - 24} \\
0
\end{array}
$$

So $x^3 + 3x^2 - 10x - 24 = (x+2)(x^2 + x - 12) = (x+2)(x+4)(x-3)$.

b. Reducing the fraction: $\dfrac{x^3 + 3x^2 - 10x - 24}{x+2} = \dfrac{(x+2)(x+4)(x-3)}{x+2} = (x+4)(x-3)$

c. The answers to the problems are the same. Long division produces the same result as factoring and simplifying.

61. Evaluating the function: $P(3) = (3)^2 + 4(3) - 8 = 9 + 12 - 8 = 13$

The remainder is the same (13).

63. **a.** Completing the table:

x	1	5	10	15	20
$C(x)$	2.15	2.75	3.50	4.25	5.00

b. The average cost function is $\bar{C}(x) = \dfrac{2}{x} + 0.15$.

c. Completing the table:

x	1	5	10	15	20
$\bar{C}(x)$	2.15	0.55	0.35	0.28	0.25

d. The average cost function decreases.

65. Dividing: $\dfrac{3}{5} \div \dfrac{2}{7} = \dfrac{3}{5} \cdot \dfrac{7}{2} = \dfrac{21}{10}$

67. Dividing: $\dfrac{3}{4} \div \dfrac{6}{11} = \dfrac{3}{4} \cdot \dfrac{11}{6} = \dfrac{11}{8}$

69. Dividing: $\dfrac{4}{9} \div 8 = \dfrac{4}{9} \cdot \dfrac{1}{8} = \dfrac{1}{18}$

71. Dividing: $8 \div \dfrac{1}{4} = 8 \cdot 4 = 32$

73. Simplifying: $\left(\dfrac{1}{3}\right)^{-2} + \left(\dfrac{1}{2}\right)^{-3} = 3^2 + 2^3 = 9 + 8 = 17$

75. Simplifying: $\left(9x^{-4}y^9\right)^{-2}\left(3x^2y^{-1}\right)^4 = 9^{-2}x^8y^{-18} \cdot 3^4 x^8 y^{-4} = \dfrac{1}{81} \cdot 81x^{16}y^{-22} = \dfrac{x^{16}}{y^{22}}$

77. Multiplying: $\dfrac{6}{7} \cdot \dfrac{14}{18} = \dfrac{6}{7} \cdot \dfrac{2\cdot 7}{3\cdot 6} = \dfrac{2}{3}$

79. Multiplying: $5y^2 \cdot 4x^2 = 20x^2y^2$

81. Multiplying: $9x^4 \cdot 8y^5 = 72x^4y^5$

83. Factoring: $x^2 - 4 = (x+2)(x-2)$

85. Factoring: $x^3 - x^2y = x^2(x-y)$

87. Factoring: $2y^2 - 2 = 2(y^2 - 1) = 2(y+1)(y-1)$

89. Using long division:

$$
\begin{array}{r}
4x^3 - x^2 + 3 \\
x^2 - 5 \overline{\smash{\big)}\, 4x^5 - x^4 - 20x^3 + 8x^2 - 15} \\
\underline{4x^5 \qquad\quad -20x^3} \\
-x^4 \qquad\qquad +8x^2 \\
\underline{-x^4 \qquad\qquad +5x^2} \\
3x^2 - 15 \\
\underline{3x^2 - 15} \\
0
\end{array}
$$

The quotient is $4x^3 - x^2 + 3$.

91. Using long division:

$$
\begin{array}{r}
0.5x^2 - 0.4x + 0.3 \\
x + 0.2 \overline{\smash{\big)}\, 0.5x^3 - 0.3x^2 + 0.22x + 0.06} \\
\underline{0.5x^3 + 0.1x^2} \\
-0.4x^2 + 0.22x \\
\underline{-0.4x^2 - 0.08x} \\
0.3x + 0.06 \\
\underline{0.3x + 0.06} \\
0
\end{array}
$$

The quotient is $0.5x^2 - 0.4x + 0.3$.

93. Using long division:

$$
\begin{array}{r}
\frac{3}{2}x - \frac{5}{2} \\
2x+4{\overline{\smash{\big)}\,3x^2 + x - 9}} \\
\underline{3x^2 + 6x} \\
-5x - 9 \\
\underline{-5x - 10} \\
1
\end{array}
$$

The quotient is $\frac{3}{2}x - \frac{5}{2} + \dfrac{1}{2x+4}$.

95. Using long division:

$$
\begin{array}{r}
\frac{2}{3}x + \frac{1}{3} \\
3x-1{\overline{\smash{\big)}\,2x^2 + \frac{1}{3}x + \frac{5}{3}}} \\
\underline{2x^2 - \frac{2}{3}x} \\
x + \frac{5}{3} \\
\underline{x - \frac{1}{3}} \\
2
\end{array}
$$

The quotient is $\frac{2}{3}x + \frac{1}{3} + \dfrac{2}{3x-1}$.

6.3 Multiplication and Division of Rational Expressions

1. Performing the operations: $\dfrac{2}{9} \cdot \dfrac{3}{4} = \dfrac{2}{3 \cdot 3} \cdot \dfrac{3}{2 \cdot 2} = \dfrac{1}{2 \cdot 3} = \dfrac{1}{6}$

3. Performing the operations: $\dfrac{3}{4} \div \dfrac{1}{3} = \dfrac{3}{4} \cdot \dfrac{3}{1} = \dfrac{9}{4}$

5. Performing the operations: $\dfrac{3}{7} \cdot \dfrac{14}{24} \div \dfrac{1}{2} = \dfrac{1}{4} \div \dfrac{1}{2} = \dfrac{1}{4} \cdot \dfrac{2}{1} = \dfrac{2}{4} = \dfrac{1}{2}$

7. Performing the operations: $\dfrac{10x^2}{5y^2} \cdot \dfrac{15y^3}{2x^4} = \dfrac{150x^2y^3}{10x^4y^2} = \dfrac{15y}{x^2}$

9. Performing the operations: $\dfrac{11a^2b}{5ab^2} \div \dfrac{22a^3b^2}{10ab^4} = \dfrac{11a^2b}{5ab^2} \cdot \dfrac{10ab^4}{22a^3b^2} = \dfrac{110a^3b^5}{110a^4b^4} = \dfrac{b}{a}$

11. Performing the operations: $\dfrac{6x^2}{5y^3} \cdot \dfrac{11z^2}{2x^2} \div \dfrac{33z^5}{10y^8} = \dfrac{33z^2}{5y^3} \cdot \dfrac{10y^8}{33z^5} = \dfrac{2y^8z^2}{y^3z^5} = \dfrac{2y^5}{z^3}$

13. Performing the operations: $\dfrac{x^2-9}{x^2-4} \cdot \dfrac{x-2}{x-3} = \dfrac{(x+3)(x-3)}{(x+2)(x-2)} \cdot \dfrac{x-2}{x-3} = \dfrac{x+3}{x+2}$

15. Performing the operations: $\dfrac{y^2-1}{y+2} \cdot \dfrac{y^2+5y+6}{y^2+2y-3} = \dfrac{(y+1)(y-1)}{y+2} \cdot \dfrac{(y+2)(y+3)}{(y+3)(y-1)} = y+1$

17. Performing the operations: $\dfrac{3x-12}{x^2-4} \cdot \dfrac{x^2+6x+8}{x-4} = \dfrac{3(x-4)}{(x+2)(x-2)} \cdot \dfrac{(x+4)(x+2)}{x-4} = \dfrac{3(x+4)}{x-2}$

19. Performing the operations: $\dfrac{5x+2y}{25x^2-5xy-6y^2} \cdot \dfrac{20x^2-7xy-3y^2}{4x+y} = \dfrac{5x+2y}{(5x+2y)(5x-3y)} \cdot \dfrac{(5x-3y)(4x+y)}{4x+y} = 1$

21. Performing the operations:

$$\dfrac{a^2-5a+6}{a^2-2a-3} \div \dfrac{a-5}{a^2+3a+2} = \dfrac{a^2-5a+6}{a^2-2a-3} \cdot \dfrac{a^2+3a+2}{a-5} = \dfrac{(a-3)(a-2)}{(a-3)(a+1)} \cdot \dfrac{(a+2)(a+1)}{a-5} = \dfrac{(a-2)(a+2)}{a-5}$$

23. Performing the operations:

$$\dfrac{4t^2-1}{6t^2+t-2} \div \dfrac{8t^3+1}{27t^3+8} = \dfrac{4t^2-1}{6t^2+t-2} \cdot \dfrac{27t^3+8}{8t^3+1} = \dfrac{(2t+1)(2t-1)}{(3t+2)(2t-1)} \cdot \dfrac{(3t+2)(9t^2-6t+4)}{(2t+1)(4t^2-2t+1)} = \dfrac{9t^2-6t+4}{4t^2-2t+1}$$

25. Performing the operations:

$$\dfrac{2x^2-5x-12}{4x^2+8x+3} \div \dfrac{x^2-16}{2x^2+7x+3} = \dfrac{2x^2-5x-12}{4x^2+8x+3} \cdot \dfrac{2x^2+7x+3}{x^2-16} = \dfrac{(2x+3)(x-4)}{(2x+1)(2x+3)} \cdot \dfrac{(2x+1)(x+3)}{(x+4)(x-4)} = \dfrac{x+3}{x+4}$$

27. Performing the operations:

$$\frac{6a^2b+2ab^2-20b^3}{4a^2b-16b^3}\cdot\frac{10a^2-22ab+4b^2}{27a^3-125b^3}=\frac{2b\left(3a^2+ab-10b^2\right)}{4b\left(a^2-4b^2\right)}\cdot\frac{2\left(5a^2-11ab+2b^2\right)}{\left(3a-5b\right)\left(9a^2+15ab+25b^2\right)}$$

$$=\frac{2b\left(3a-5b\right)\left(a+2b\right)}{4b\left(a+2b\right)\left(a-2b\right)}\cdot\frac{2\left(5a-b\right)\left(a-2b\right)}{\left(3a-5b\right)\left(9a^2+15ab+25b^2\right)}$$

$$=\frac{5a-b}{9a^2+15ab+25b^2}$$

29. Performing the operations:

$$\frac{360x^3-490x}{36x^2+84x+49}\cdot\frac{30x^2+83x+56}{150x^3+65x^2-280x}=\frac{10x\left(36x^2-49\right)}{\left(6x+7\right)^2}\cdot\frac{\left(6x+7\right)\left(5x+8\right)}{5x\left(30x^2+13x-56\right)}$$

$$=\frac{10x\left(6x+7\right)\left(6x-7\right)}{\left(6x+7\right)^2}\cdot\frac{\left(6x+7\right)\left(5x+8\right)}{5x\left(6x-7\right)\left(5x+8\right)}$$

$$=2$$

31. Performing the operations:

$$\frac{x^5-x^2}{5x^5-5x}\cdot\frac{10x^4-10x^2}{2x^4+2x^3+2x^2}=\frac{x^2\left(x^3-1\right)}{5x\left(x^4-1\right)}\cdot\frac{10x^2\left(x^2-1\right)}{2x^2\left(x^2+x+1\right)}$$

$$=\frac{x^2\left(x-1\right)\left(x^2+x+1\right)}{5x\left(x^2+1\right)\left(x-1\right)\left(x-1\right)}\cdot\frac{10x^2\left(x+1\right)\left(x-1\right)}{2x^2\left(x^2+x+1\right)}$$

$$=\frac{x\left(x-1\right)}{x^2+1}$$

33. Performing the operations:

$$\frac{a^2-16b^2}{a^2-8ab+16b^2}\cdot\frac{a^2-9ab+20b^2}{a^2-7ab+12b^2}\div\frac{a^2-25b^2}{a^2-6ab+9b^2}$$

$$=\frac{a^2-16b^2}{a^2-8ab+16b^2}\cdot\frac{a^2-9ab+20b^2}{a^2-7ab+12b^2}\cdot\frac{a^2-6ab+9b^2}{a^2-25b^2}$$

$$=\frac{\left(a+4b\right)\left(a-4b\right)}{\left(a-4b\right)^2}\cdot\frac{\left(a-5b\right)\left(a-4b\right)}{\left(a-3b\right)\left(a-4b\right)}\cdot\frac{\left(a-3b\right)^2}{\left(a+5b\right)\left(a-5b\right)}$$

$$=\frac{\left(a+4b\right)\left(a-3b\right)}{\left(a-4b\right)\left(a+5b\right)}$$

35. Performing the operations:

$$\frac{2y^2-7y-15}{42y^2-29y-5}\cdot\frac{12y^2-16y+5}{7y^2-36y+5}\div\frac{4y^2-9}{49y^2-1}=\frac{2y^2-7y-15}{42y^2-29y-5}\cdot\frac{12y^2-16y+5}{7y^2-36y+5}\cdot\frac{49y^2-1}{4y^2-9}$$

$$=\frac{\left(2y+3\right)\left(y-5\right)}{\left(6y-5\right)\left(7y+1\right)}\cdot\frac{\left(6y-5\right)\left(2y-1\right)}{\left(7y-1\right)\left(y-5\right)}\cdot\frac{\left(7y+1\right)\left(7y-1\right)}{\left(2y+3\right)\left(2y-3\right)}$$

$$=\frac{2y-1}{2y-3}$$

37. Performing the operations:

$$\frac{xy-2x+3y-6}{xy+2x-4y-8} \cdot \frac{xy+x-4y-4}{xy-x+3y-3} = \frac{x(y-2)+3(y-2)}{x(y+2)-4(y+2)} \cdot \frac{x(y+1)-4(y+1)}{x(y-1)+3(y-1)}$$

$$= \frac{(y-2)(x+3)}{(y+2)(x-4)} \cdot \frac{(y+1)(x-4)}{(y-1)(x+3)}$$

$$= \frac{(y-2)(y+1)}{(y+2)(y-1)}$$

39. Performing the operations:

$$\frac{xy^2-y^2+4xy-4y}{xy-3y+4x-12} \div \frac{xy^3+2xy^2+y^3+2y^2}{xy^2-3y^2+2xy-6y} = \frac{xy^2-y^2+4xy-4y}{xy-3y+4x-12} \cdot \frac{xy^2-3y^2+2xy-6y}{xy^3+2xy^2+y^3+2y^2}$$

$$= \frac{y^2(x-1)+4y(x-1)}{y(x-3)+4(x-3)} \cdot \frac{y^2(x-3)+2y(x-3)}{xy^2(y+2)+y^2(y+2)}$$

$$= \frac{y(x-1)(y+4)}{(x-3)(y+4)} \cdot \frac{y(x-3)(y+2)}{y^2(y+2)(x+1)}$$

$$= \frac{x-1}{x+1}$$

41. Performing the operations:

$$\frac{2x^3+10x^2-8x-40}{x^3+4x^2-9x-36} \cdot \frac{x^2+x-12}{2x^2+14x+20} = \frac{2x^2(x+5)-8(x+5)}{x^2(x+4)-9(x+4)} \cdot \frac{(x+4)(x-3)}{2(x^2+7x+10)}$$

$$= \frac{2(x+5)(x^2-4)}{(x+4)(x^2-9)} \cdot \frac{(x+4)(x-3)}{2(x+5)(x+2)}$$

$$= \frac{2(x+5)(x+2)(x-2)}{(x+4)(x+3)(x-3)} \cdot \frac{(x+4)(x-3)}{2(x+5)(x+2)}$$

$$= \frac{x-2}{x+3}$$

43. a. Simplifying: $\dfrac{16-1}{64-1} = \dfrac{15}{63} = \dfrac{5}{21}$

    b. Reducing: $\dfrac{25x^2-9}{125x^3-27} = \dfrac{(5x-3)(5x+3)}{(5x-3)(25x^2+15x+9)} = \dfrac{5x+3}{25x^2+15x+9}$

    c. Multiplying: $\dfrac{25x^2-9}{125x^3-27} \cdot \dfrac{5x-3}{5x+3} = \dfrac{(5x-3)(5x+3)}{(5x-3)(25x^2+15x+9)} \cdot \dfrac{5x-3}{5x+3} = \dfrac{5x-3}{25x^2+15x+9}$

    d. Dividing: $\dfrac{25x^2-9}{125x^3-27} \div \dfrac{5x-3}{25x^2+15x+9} = \dfrac{(5x-3)(5x+3)}{(5x-3)(25x^2+15x+9)} \cdot \dfrac{25x^2+15x+9}{5x-3} = \dfrac{5x+3}{5x-3}$

45. a. Multiplying: $f(x) \bullet g(x) = \dfrac{x^2-x-6}{x-1} \cdot \dfrac{x+2}{x^2-4x+3} = \dfrac{(x-3)(x+2)}{x-1} \cdot \dfrac{x+2}{(x-3)(x-1)} = \dfrac{(x+2)^2}{(x-1)^2}$

    b. Dividing: $f(x) \div g(x) = \dfrac{x^2-x-6}{x-1} \div \dfrac{x+2}{x^2-4x+3} = \dfrac{(x-3)(x+2)}{x-1} \cdot \dfrac{(x-3)(x-1)}{x+2} = (x-3)^2$

47. Multiplying:

$$f(x) \bullet g(x) = \frac{x^3 - 9x^2 - 3x + 27}{4x^2 - 12} \bullet \frac{x^2 - 2x - 8}{x^2 - 81}$$

$$= \frac{x^2(x-9) - 3(x-9)}{4(x^2 - 3)} \bullet \frac{(x-4)(x+2)}{(x-9)(x+9)}$$

$$= \frac{(x-9)(x^2 - 3)}{4(x^2 - 3)} \bullet \frac{(x-4)(x+2)}{(x-9)(x+9)}$$

$$= \frac{(x-4)(x+2)}{4(x+9)}$$

49. Multiplying:

$$f(x) \bullet g(x) = \frac{x^3 - 3x^2 - 4x + 12}{x+2} \bullet \frac{x^2 + 7x + 12}{x^2 - 5x + 6}$$

$$= \frac{x^2(x-3) - 4(x-3)}{x+2} \bullet \frac{(x+4)(x+3)}{(x-2)(x-3)}$$

$$= \frac{(x-3)(x+2)(x-2)}{x+2} \bullet \frac{(x+4)(x+3)}{(x-2)(x-3)}$$

$$= (x+4)(x+3)$$

51. Finding the product: $(3x - 6) \bullet \dfrac{x}{x-2} = \dfrac{3(x-2)}{1} \bullet \dfrac{x}{x-2} = 3x$

53. Finding the product: $\left(x^2 - 25\right) \bullet \dfrac{2}{x-5} = \dfrac{(x+5)(x-5)}{1} \bullet \dfrac{2}{x-5} = 2(x+5)$

55. Finding the product: $\left(x^2 - 3x + 2\right) \bullet \dfrac{3}{3x-3} = \dfrac{(x-2)(x-1)}{1} \bullet \dfrac{3}{3(x-1)} = x - 2$

57. Finding the product: $(y-3)(y-4)(y+3) \bullet \dfrac{-1}{y^2 - 9} = \dfrac{(y-3)(y-4)(y+3)}{1} \bullet \dfrac{-1}{(y+3)(y-3)} = -(y-4)$

59. Finding the product: $a(a+5)(a-5) \bullet \dfrac{a+1}{a^2 + 5a} = \dfrac{a(a+5)(a-5)}{1} \bullet \dfrac{a+1}{a(a+5)} = (a-5)(a+1)$

61. Dividing: $\left(x^2 - 2x - 8\right) \div \dfrac{x^2 - x - 6}{x-4} = \dfrac{(x-4)(x+2)}{1} \bullet \dfrac{x-4}{(x-3)(x+2)} = \dfrac{(x-4)^2}{x-3}$

63. Dividing: $(3-x) \div \dfrac{x^2 - 9}{x-1} = \dfrac{-(x-3)}{1} \bullet \dfrac{x-1}{(x+3)(x-3)} = -\dfrac{x-1}{x+3}$

65. Dividing:

$$(xy - 2x - 3y + 6) \div \frac{x^2 - 2x - 3}{x^2 - 6x - 7} = \frac{x(y-2) - 3(y-2)}{1} \bullet \frac{(x-7)(x+1)}{(x-3)(x+1)}$$

$$= \frac{(y-2)(x-3)}{1} \bullet \frac{(x-7)(x+1)}{(x-3)(x+1)}$$

$$= (y-2)(x-7)$$

67. Simplifying: $\dfrac{x^2(x+2) + 6x(x+2) + 9(x+2)}{x^2 - 2x - 8} = \dfrac{(x+2)\left(x^2 + 6x + 9\right)}{(x-4)(x+2)} = \dfrac{(x+2)(x+3)^2}{(x-4)(x+2)} = \dfrac{(x+3)^2}{x-4}$

69. Simplifying: $\dfrac{2x^2(x+3) - 5x(x+3) + 3(x+3)}{x^2 + x - 6} = \dfrac{(x+3)\left(2x^2 - 5x + 3\right)}{(x+3)(x-2)} = \dfrac{(x+3)(2x-3)(x-1)}{(x+3)(x-2)} = \dfrac{(2x-3)(x-1)}{x-2}$

71.   Completing the table:

Number of Copies x	Price per Copy ($) $p(x)$
1	20.33
10	9.33
20	6.40
50	4.00
100	3.05

73.   Finding the revenue: $R = 100 \cdot \dfrac{2(100+60)}{100+5} = 100 \cdot \dfrac{320}{105} \approx \$305.00$

75.   First use long division to find the length of the square base:

$$
\begin{array}{r}
x - 3 \\
x^2 + x + 1 \overline{\smash{\big)}\, x^3 - 2x^2 - 2x - 3} \\
\underline{x^3 + x^2 + x} \\
-3x^2 - 3x - 3 \\
\underline{-3x^2 - 3x - 3} \\
0
\end{array}
$$

The area of the base is therefore: $A = (x-3)^2 = x^2 - 6x + 9$

77.   Multiplying the polynomials: $2x^2(5x^3 + 4x - 3) = 2x^2 \cdot 5x^3 + 2x^2 \cdot 4x - 2x^2 \cdot 3 = 10x^5 + 8x^3 - 6x^2$

79.   Multiplying the polynomials: $(3a-1)(4a+5) = 12a^2 + 15a - 4a - 5 = 12a^2 + 11a - 5$

81.   Multiplying the polynomials: $(3x+7)(4y-2) = 12xy + 28y - 6x - 14$

83.   Multiplying the polynomials: $\left(3 - t^2\right)^2 = (3)^2 - 2(3)\left(t^2\right) + \left(t^2\right)^2 = 9 - 6t^2 + t^4$

85.   Multiplying the polynomials:

$$3(x+1)(x+2)(x+3) = 3(x+1)\left(x^2 + 5x + 6\right) = 3\left(x^3 + 6x^2 + 11x + 6\right) = 3x^3 + 18x^2 + 33x + 18$$

87.   Combining: $\frac{4}{9} + \frac{2}{9} = \frac{6}{9} = \frac{2 \cdot 3}{3 \cdot 3} = \frac{2}{3}$

89.   Combining: $\frac{3}{14} + \frac{7}{30} = \frac{3}{14} \cdot \frac{15}{15} + \frac{7}{30} \cdot \frac{7}{7} = \frac{45}{210} + \frac{49}{210} = \frac{94}{210} = \frac{47}{105}$

91.   Multiplying: $-1(7-x) = -7 + x = x - 7$ **93.**   Factoring: $x^2 - 1 = (x+1)(x-1)$

95.   Factoring: $2x + 10 = 2(x+5)$ **97.**   Factoring: $a^3 - b^3 = (a-b)\left(a^2 + ab + b^2\right)$

99.   Performing the operations:

$$\frac{x^6 + y^6}{x^4 + 4x^2y^2 + 3y^4} \div \frac{x^4 + 3x^2y^2 + 2y^4}{x^4 + 5x^2y^2 + 6y^4} = \frac{x^6 + y^6}{x^4 + 4x^2y^2 + 3y^4} \cdot \frac{x^4 + 5x^2y^2 + 6y^4}{x^4 + 3x^2y^2 + 2y^4}$$

$$= \frac{\left(x^2 + y^2\right)\left(x^4 - x^2y^2 + y^4\right)}{\left(x^2 + 3y^2\right)\left(x^2 + y^2\right)} \cdot \frac{\left(x^2 + 3y^2\right)\left(x^2 + 2y^2\right)}{\left(x^2 + 2y^2\right)\left(x^2 + y^2\right)}$$

$$= \frac{x^4 - x^2y^2 + y^4}{x^2 + y^2}$$

101. Performing the operations:

$$\frac{a^2(2a+b)+6a(2a+b)+5(2a+b)}{3a^2(2a+b)-2a(2a+b)+(2a+b)} \div \frac{a+1}{a-1} = \frac{a^2(2a+b)+6a(2a+b)+5(2a+b)}{3a^2(2a+b)-2a(2a+b)+(2a+b)} \cdot \frac{a-1}{a+1}$$

$$= \frac{(2a+b)(a^2+6a+5)}{(2a+b)(3a^2-2a+1)} \cdot \frac{a-1}{a+1}$$

$$= \frac{(2a+b)(a+5)(a+1)}{(2a+b)(3a^2-2a+1)} \cdot \frac{a-1}{a+1}$$

$$= \frac{(a+5)(a-1)}{3a^2-2a+1}$$

103. Performing the operations: $\dfrac{a^3-a^2b}{ac-a} \div \left(\dfrac{a-b}{c-1}\right)^2 = \dfrac{a^3-a^2b}{ac-a} \cdot \left(\dfrac{c-1}{a-b}\right)^2 = \dfrac{a^2(a-b)}{a(c-1)} \cdot \dfrac{(c-1)^2}{(a-b)^2} = \dfrac{a(c-1)}{a-b}$

6.4 Addition and Subtraction of Rational Expressions

1. Combining the fractions: $\frac{3}{4}+\frac{1}{2}=\frac{3}{4}+\frac{1}{2}\cdot\frac{2}{2}=\frac{3}{4}+\frac{2}{4}=\frac{5}{4}$

3. Combining the fractions: $\frac{2}{5}-\frac{1}{15}=\frac{2}{5}\cdot\frac{3}{3}-\frac{1}{15}=\frac{6}{15}-\frac{1}{15}=\frac{5}{15}=\frac{1}{3}$

5. Combining the fractions: $\frac{5}{6}+\frac{7}{8}=\frac{5}{6}\cdot\frac{4}{4}+\frac{7}{8}\cdot\frac{3}{3}=\frac{20}{24}+\frac{21}{24}=\frac{41}{24}$

7. Combining the fractions: $\frac{9}{48}-\frac{3}{54}=\frac{9}{48}\cdot\frac{9}{9}-\frac{3}{54}\cdot\frac{8}{8}=\frac{81}{432}-\frac{24}{432}=\frac{57}{432}=\frac{19}{144}$

9. Combining the fractions: $\frac{3}{4}-\frac{1}{8}+\frac{2}{3}=\frac{3}{4}\cdot\frac{6}{6}-\frac{1}{8}\cdot\frac{3}{3}+\frac{2}{3}\cdot\frac{8}{8}=\frac{18}{24}-\frac{3}{24}+\frac{16}{24}=\frac{31}{24}$

11. Combining the rational expressions: $\dfrac{x}{x+3}+\dfrac{3}{x+3}=\dfrac{x+3}{x+3}=1$

13. Combining the rational expressions: $\dfrac{4}{y-4}-\dfrac{y}{y-4}=\dfrac{4-y}{y-4}=\dfrac{-1(y-4)}{y-4}=-1$

15. Combining the rational expressions: $\dfrac{x}{x^2-y^2}-\dfrac{y}{x^2-y^2}=\dfrac{x-y}{x^2-y^2}=\dfrac{x-y}{(x+y)(x-y)}=\dfrac{1}{x+y}$

17. Combining the rational expressions: $\dfrac{2x-3}{x-2}-\dfrac{x-1}{x-2}=\dfrac{2x-3-x+1}{x-2}=\dfrac{x-2}{x-2}=1$

19. Combining the rational expressions: $\dfrac{1}{a}+\dfrac{2}{a^2}-\dfrac{3}{a^3}=\dfrac{1}{a}\cdot\dfrac{a^2}{a^2}+\dfrac{2}{a^2}\cdot\dfrac{a}{a}-\dfrac{3}{a^3}=\dfrac{a^2+2a-3}{a^3}$

21. Combining the rational expressions: $\dfrac{7x-2}{2x+1}-\dfrac{5x-3}{2x+1}=\dfrac{7x-2-5x+3}{2x+1}=\dfrac{2x+1}{2x+1}=1$

23. **a.** Multiplying: $\dfrac{3}{8}\cdot\dfrac{1}{6}=\dfrac{3}{8}\cdot\dfrac{1}{2\cdot3}=\dfrac{1}{16}$

   b. Dividing: $\dfrac{3}{8}\div\dfrac{1}{6}=\dfrac{3}{8}\cdot\dfrac{6}{1}=\dfrac{3}{2\cdot4}\cdot\dfrac{2\cdot3}{1}=\dfrac{9}{4}$

   c. Adding: $\dfrac{3}{8}+\dfrac{1}{6}=\dfrac{3}{8}\cdot\dfrac{3}{3}+\dfrac{1}{6}\cdot\dfrac{4}{4}=\dfrac{9}{24}+\dfrac{4}{24}=\dfrac{13}{24}$

   d. Multiplying: $\dfrac{x+3}{x-3}\cdot\dfrac{5x+15}{x^2-9}=\dfrac{x+3}{x-3}\cdot\dfrac{5(x+3)}{(x+3)(x-3)}=\dfrac{5(x+3)}{(x-3)^2}$

   e. Dividing: $\dfrac{x+3}{x-3}\div\dfrac{5x+15}{x^2-9}=\dfrac{x+3}{x-3}\cdot\dfrac{(x+3)(x-3)}{5(x+3)}=\dfrac{x+3}{5}$

f. Subtracting:

$$\frac{x+3}{x-3} - \frac{5x+15}{x^2-9} = \frac{x+3}{x-3} \cdot \frac{x+3}{x+3} - \frac{5x+15}{(x+3)(x-3)}$$

$$= \frac{x^2+6x+9}{(x+3)(x-3)} - \frac{5x+15}{(x+3)(x-3)}$$

$$= \frac{x^2+x-6}{(x+3)(x-3)}$$

$$= \frac{(x+3)(x-2)}{(x+3)(x-3)}$$

$$= \frac{x-2}{x-3}$$

25. Combining the rational expressions: $\dfrac{2}{t^2} - \dfrac{3}{2t} = \dfrac{2}{t^2} \cdot \dfrac{2}{2} - \dfrac{3}{2t} \cdot \dfrac{t}{t} = \dfrac{4}{2t^2} - \dfrac{3t}{2t^2} = \dfrac{4-3t}{2t^2}$

27. Combining the rational expressions:

$$\frac{3x+1}{2x-6} - \frac{x+2}{x-3} = \frac{3x+1}{2(x-3)} - \frac{x+2}{x-3} \cdot \frac{2}{2} = \frac{3x+1}{2(x-3)} - \frac{2x+4}{2(x-3)} = \frac{3x+1-2x-4}{2(x-3)} = \frac{x-3}{2(x-3)} = \frac{1}{2}$$

29. Combining the rational expressions:

$$\frac{x+1}{2x-2} - \frac{2}{x^2-1} = \frac{x+1}{2(x-1)} \cdot \frac{x+1}{x+1} - \frac{2}{(x+1)(x-1)} \cdot \frac{2}{2}$$

$$= \frac{x^2+2x+1}{2(x+1)(x-1)} - \frac{4}{2(x+1)(x-1)}$$

$$= \frac{x^2+2x-3}{2(x+1)(x-1)}$$

$$= \frac{(x+3)(x-1)}{2(x+1)(x-1)}$$

$$= \frac{x+3}{2(x+1)}$$

31. Combining the rational expressions:

$$\frac{1}{a-b} - \frac{3ab}{a^3-b^3} = \frac{1}{a-b} \cdot \frac{a^2+ab+b^2}{a^2+ab+b^2} - \frac{3ab}{a^3-b^3}$$

$$= \frac{a^2+ab+b^2}{a^3-b^3} - \frac{3ab}{a^3-b^3}$$

$$= \frac{a^2-2ab+b^2}{a^3-b^3}$$

$$= \frac{(a-b)^2}{(a-b)(a^2+ab+b^2)}$$

$$= \frac{a-b}{a^2+ab+b^2}$$

33. Combining the rational expressions:

$$\frac{1}{2y-3} - \frac{18y}{8y^3-27} = \frac{1}{2y-3} \cdot \frac{4y^2+6y+9}{4y^2+6y+9} - \frac{18y}{8y^3-27}$$

$$= \frac{4y^2+6y+9}{8y^3-27} - \frac{18y}{8y^3-27}$$

$$= \frac{4y^2-12y+9}{8y^3-27}$$

$$= \frac{(2y-3)^2}{(2y-3)(4y^2+6y+9)}$$

$$= \frac{2y-3}{4y^2+6y+9}$$

35. Combining the rational expressions:

$$\frac{x}{x^2-5x+6} - \frac{3}{3-x} = \frac{x}{(x-2)(x-3)} + \frac{3}{x-3} \cdot \frac{x-2}{x-2}$$

$$= \frac{x}{(x-2)(x-3)} + \frac{3x-6}{(x-2)(x-3)}$$

$$= \frac{4x-6}{(x-2)(x-3)}$$

$$= \frac{2(2x-3)}{(x-3)(x-2)}$$

37. Combining the rational expressions:

$$\frac{2}{4t-5} + \frac{9}{8t^2-38t+35} = \frac{2}{4t-5} \cdot \frac{2t-7}{2t-7} + \frac{9}{(4t-5)(2t-7)}$$

$$= \frac{4t-14}{(4t-5)(2t-7)} + \frac{9}{(4t-5)(2t-7)}$$

$$= \frac{4t-5}{(4t-5)(2t-7)}$$

$$= \frac{1}{2t-7}$$

39. Combining the rational expressions:

$$\frac{1}{a^2-5a+6} + \frac{3}{a^2-a-2} = \frac{1}{(a-2)(a-3)} \cdot \frac{a+1}{a+1} + \frac{3}{(a-2)(a+1)} \cdot \frac{a-3}{a-3}$$

$$= \frac{a+1}{(a-2)(a-3)(a+1)} + \frac{3a-9}{(a-2)(a-3)(a+1)}$$

$$= \frac{4a-8}{(a-2)(a-3)(a+1)}$$

$$= \frac{4(a-2)}{(a-2)(a-3)(a+1)}$$

$$= \frac{4}{(a-3)(a+1)}$$

41.   Combining the rational expressions:

$$\frac{1}{8x^3-1}-\frac{1}{4x^2-1}=\frac{1}{(2x-1)(4x^2+2x+1)}\cdot\frac{2x+1}{2x+1}-\frac{1}{(2x+1)(2x-1)}\cdot\frac{4x^2+2x+1}{4x^2+2x+1}$$

$$=\frac{2x+1}{(2x+1)(2x-1)(4x^2+2x+1)}-\frac{4x^2+2x+1}{(2x+1)(2x-1)(4x^2+2x+1)}$$

$$=\frac{2x+1-4x^2-2x-1}{(2x+1)(2x-1)(4x^2+2x+1)}$$

$$=\frac{-4x^2}{(2x+1)(2x-1)(4x^2+2x+1)}$$

43.   Combining the rational expressions:

$$\frac{4}{4x^2-9}-\frac{6}{8x^2-6x-9}=\frac{4}{(2x+3)(2x-3)}\cdot\frac{4x+3}{4x+3}-\frac{6}{(2x-3)(4x+3)}\cdot\frac{2x+3}{2x+3}$$

$$=\frac{16x+12}{(2x+3)(2x-3)(4x+3)}-\frac{12x+18}{(2x+3)(2x-3)(4x+3)}$$

$$=\frac{16x+12-12x-18}{(2x+3)(2x-3)(4x+3)}$$

$$=\frac{4x-6}{(2x+3)(2x-3)(4x+3)}$$

$$=\frac{2(2x-3)}{(2x+3)(2x-3)(4x+3)}$$

$$=\frac{2}{(2x+3)(4x+3)}$$

45.   Combining the rational expressions:

$$\frac{4a}{a^2+6a+5}-\frac{3a}{a^2+5a+4}=\frac{4a}{(a+5)(a+1)}\cdot\frac{a+4}{a+4}-\frac{3a}{(a+4)(a+1)}\cdot\frac{a+5}{a+5}$$

$$=\frac{4a^2+16a}{(a+4)(a+5)(a+1)}-\frac{3a^2+15a}{(a+4)(a+5)(a+1)}$$

$$=\frac{4a^2+16a-3a^2-15a}{(a+4)(a+5)(a+1)}$$

$$=\frac{a^2+a}{(a+4)(a+5)(a+1)}$$

$$=\frac{a(a+1)}{(a+4)(a+5)(a+1)}$$

$$=\frac{a}{(a+4)(a+5)}$$

47.   Combining the rational expressions:

$$\frac{2x-1}{x^2+x-6} - \frac{x+2}{x^2+5x+6} = \frac{2x-1}{(x+3)(x-2)} \cdot \frac{x+2}{x+2} - \frac{x+2}{(x+3)(x+2)} \cdot \frac{x-2}{x-2}$$

$$= \frac{2x^2+3x-2}{(x+3)(x+2)(x-2)} - \frac{x^2-4}{(x+3)(x+2)(x-2)}$$

$$= \frac{2x^2+3x-2-x^2+4}{(x+3)(x+2)(x-2)}$$

$$= \frac{x^2+3x+2}{(x+3)(x+2)(x-2)}$$

$$= \frac{(x+2)(x+1)}{(x+3)(x+2)(x-2)}$$

$$= \frac{x+1}{(x-2)(x+3)}$$

49.   Combining the rational expressions:

$$\frac{2x-8}{3x^2+8x+4} + \frac{x+3}{3x^2+5x+2} = \frac{2x-8}{(3x+2)(x+2)} + \frac{x+3}{(3x+2)(x+1)}$$

$$= \frac{2x-8}{(3x+2)(x+2)} \cdot \frac{x+1}{x+1} + \frac{x+3}{(3x+2)(x+1)} \cdot \frac{x+2}{x+2}$$

$$= \frac{2x^2-6x-8}{(3x+2)(x+2)(x+1)} + \frac{x^2+5x+6}{(3x+2)(x+2)(x+1)}$$

$$= \frac{3x^2-x-2}{(3x+2)(x+2)(x+1)}$$

$$= \frac{(3x+2)(x-1)}{(3x+2)(x+2)(x+1)}$$

$$= \frac{x-1}{(x+1)(x+2)}$$

51.   Combining the rational expressions:

$$\frac{2}{x^2+5x+6} - \frac{4}{x^2+4x+3} + \frac{3}{x^2+3x+2} = \frac{2}{(x+3)(x+2)} - \frac{4}{(x+3)(x+1)} + \frac{3}{(x+2)(x+1)}$$

$$= \frac{2}{(x+3)(x+2)} \cdot \frac{x+1}{x+1} - \frac{4}{(x+3)(x+1)} \cdot \frac{x+2}{x+2} + \frac{3}{(x+2)(x+1)} \cdot \frac{x+3}{x+3}$$

$$= \frac{2x+2}{(x+3)(x+2)(x+1)} - \frac{4x+8}{(x+3)(x+2)(x+1)} + \frac{3x+9}{(x+3)(x+2)(x+1)}$$

$$= \frac{2x+2-4x-8+3x+9}{(x+3)(x+2)(x+1)}$$

$$= \frac{x+3}{(x+3)(x+2)(x+1)}$$

$$= \frac{1}{(x+2)(x+1)}$$

53.   Combining the rational expressions:

$$\frac{2x+8}{x^2+5x+6}-\frac{x+5}{x^2+4x+3}-\frac{x-1}{x^2+3x+2}=\frac{2x+8}{(x+3)(x+2)}-\frac{x+5}{(x+3)(x+1)}-\frac{x-1}{(x+2)(x+1)}$$

$$=\frac{2x+8}{(x+3)(x+2)}\cdot\frac{x+1}{x+1}-\frac{x+5}{(x+3)(x+1)}\cdot\frac{x+2}{x+2}-\frac{x-1}{(x+2)(x+1)}\cdot\frac{x+3}{x+3}$$

$$=\frac{2x^2+10x+8}{(x+3)(x+2)(x+1)}-\frac{x^2+7x+10}{(x+3)(x+2)(x+1)}-\frac{x^2+2x-3}{(x+3)(x+2)(x+1)}$$

$$=\frac{2x^2+10x+8-x^2-7x-10-x^2-2x+3}{(x+3)(x+2)(x+1)}$$

$$=\frac{x+1}{(x+3)(x+2)(x+1)}$$

$$=\frac{1}{(x+2)(x+3)}$$

55.   Combining the rational expressions: $2+\dfrac{3}{2x+1}=\dfrac{2}{1}\cdot\dfrac{2x+1}{2x+1}+\dfrac{3}{2x+1}=\dfrac{4x+2}{2x+1}+\dfrac{3}{2x+1}=\dfrac{4x+5}{2x+1}$

57.   Combining the rational expressions: $5+\dfrac{2}{4-t}=\dfrac{5}{1}\cdot\dfrac{4-t}{4-t}+\dfrac{2}{4-t}=\dfrac{20-5t}{4-t}+\dfrac{2}{4-t}=\dfrac{22-5t}{4-t}$

59.   Combining the rational expressions: $x-\dfrac{4}{2x+3}=\dfrac{x}{1}\cdot\dfrac{2x+3}{2x+3}-\dfrac{4}{2x+3}=\dfrac{2x^2+3x}{2x+3}-\dfrac{4}{2x+3}=\dfrac{2x^2+3x-4}{2x+3}$

61.   Combining the rational expressions:

$$\frac{x}{x+2}+\frac{1}{2x+4}-\frac{3}{x^2+2x}=\frac{x}{x+2}\cdot\frac{2x}{2x}+\frac{1}{2(x+2)}\cdot\frac{x}{x}-\frac{3}{x(x+2)}\cdot\frac{2}{2}$$

$$=\frac{2x^2}{2x(x+2)}+\frac{x}{2x(x+2)}-\frac{6}{2x(x+2)}$$

$$=\frac{2x^2+x-6}{2x(x+2)}$$

$$=\frac{(2x-3)(x+2)}{2x(x+2)}$$

$$=\frac{2x-3}{2x}$$

63.   Combining the rational expressions:

$$\frac{1}{x}+\frac{x}{2x+4}-\frac{2}{x^2+2x}=\frac{1}{x}\cdot\frac{2(x+2)}{2(x+2)}+\frac{x}{2(x+2)}\cdot\frac{x}{x}-\frac{2}{x(x+2)}\cdot\frac{2}{2}$$

$$=\frac{2x+4}{2x(x+2)}+\frac{x^2}{2x(x+2)}-\frac{4}{2x(x+2)}$$

$$=\frac{x^2+2x}{2x(x+2)}$$

$$=\frac{x(x+2)}{2x(x+2)}$$

$$=\frac{1}{2}$$

65. Finding the difference:

$$f(x) - g(x) = \frac{2x-1}{4x-16} - \frac{x-3}{x-4}$$

$$= \frac{2x-1}{4(x-4)} - \frac{x-3}{x-4} \cdot \frac{4}{4}$$

$$= \frac{2x-1}{4(x-4)} - \frac{4x-12}{4(x-4)}$$

$$= \frac{2x-1-4x+12}{4(x-4)}$$

$$= \frac{-2x+11}{4(x-4)}$$

$$= -\frac{2x-11}{4(x-4)}$$

67. Finding the sum:

$$f(x) + g(x) = \frac{2}{x+4} + \frac{x-1}{x^2+3x-4}$$

$$= \frac{2}{x+4} \cdot \frac{x-1}{x-1} + \frac{x-1}{(x+4)(x-1)}$$

$$= \frac{2x-2}{(x+4)(x-1)} + \frac{x-1}{(x+4)(x-1)}$$

$$= \frac{3x-3}{(x+4)(x-1)}$$

$$= \frac{3(x-1)}{(x+4)(x-1)}$$

$$= \frac{3}{x+4}$$

69. Finding the sum:

$$f(x) + g(x) = \frac{2x}{x^2-x-2} + \frac{5}{x^2+x-6}$$

$$= \frac{2x}{(x-2)(x+1)} \cdot \frac{x+3}{x+3} + \frac{5}{(x+3)(x-2)} \cdot \frac{x+1}{x+1}$$

$$= \frac{2x^2+6x}{(x+2)(x-1)(x+3)} + \frac{5x+5}{(x+2)(x-1)(x+3)}$$

$$= \frac{2x^2+11x+5}{(x+2)(x-1)(x+3)}$$

$$= \frac{(2x+1)(x+5)}{(x+2)(x-1)(x+3)}$$

71. Finding the sum:

$$f(x) + g(x) = \frac{x}{9x^2-4} + \frac{1}{3x^2-4x-4}$$

$$= \frac{x}{(3x+2)(3x-2)} \cdot \frac{x-2}{x-2} + \frac{1}{(3x+2)(x-2)} \cdot \frac{3x-2}{3x-2}$$

$$= \frac{x^2-2x}{(3x+2)(3x-2)(x-2)} + \frac{3x-2}{(3x+2)(3x-2)(x-2)}$$

$$= \frac{x^2+x-2}{(3x+2)(3x-2)(x-2)}$$

$$= \frac{(x+2)(x-1)}{(3x+2)(3x-2)(x-2)}$$

73.   Finding the sum:

$$\begin{aligned} f(x)+g(x) &= \frac{3x}{2x^2-x-1} + \frac{6}{2x^2-5x-3} \\ &= \frac{3x}{(2x+1)(x-1)} \cdot \frac{x-3}{x-3} + \frac{6}{(2x+1)(x-3)} \cdot \frac{x-1}{x-1} \\ &= \frac{3x^2-9x}{(2x+1)(x-1)(x-3)} + \frac{6x-6}{(2x+1)(x-1)(x-3)} \\ &= \frac{3x^2-3x-6}{(2x+1)(x-1)(x-3)} \\ &= \frac{3(x^2-x-2)}{(2x+1)(x-1)(x-3)} \\ &= \frac{3(x-2)(x+1)}{(2x+1)(x-1)(x-3)} \end{aligned}$$

75.   Finding the difference:

$$\begin{aligned} f(x)-g(x) &= \frac{5x}{x^2-13x+36} - \frac{3x}{x^2-11x+28} \\ &= \frac{5x}{(x-9)(x-4)} \cdot \frac{x-7}{x-7} - \frac{3x}{(x-7)(x-4)} \cdot \frac{x-9}{x-9} \\ &= \frac{5x^2-35x}{(x-9)(x-4)(x-7)} - \frac{3x^2-27x}{(x-9)(x-4)(x-7)} \\ &= \frac{2x^2-8x}{(x-9)(x-4)(x-7)} \\ &= \frac{2x(x-4)}{(x-9)(x-4)(x-7)} \\ &= \frac{2x}{(x-9)(x-7)} \end{aligned}$$

77.   Writing the expression and simplifying: $x + \dfrac{4}{x} = \dfrac{x^2+4}{x}$

79.   Substituting the values: $P = \dfrac{1}{10} + \dfrac{1}{0.2} = 0.1 + 5 = 5.1$

81.   **a.**   Substituting the values:

$$\frac{1}{T} = \frac{1}{24} - \frac{1}{30} = \frac{5}{120} - \frac{4}{120} = \frac{1}{120}$$
$$T = 120$$

The two objects will meet in 120 months.

b.   If $t_A = t_B$, then $\dfrac{1}{T} = \dfrac{1}{t_A} - \dfrac{1}{t_A} = 0$. Since this is impossible, the two objects will never meet.

83.   Writing in scientific notation: $54,000 = 5.4 \times 10^4$ 85.   Writing in scientific notation: $0.00034 = 3.4 \times 10^{-4}$

87.   Writing in expanded form: $6.44 \times 10^3 = 6,440$ 89.   Writing in expanded form: $6.44 \times 10^{-3} = 0.00644$

91.   Simplifying: $(3 \times 10^8)(4 \times 10^{-5}) = 12 \times 10^3 = 1.2 \times 10^4$

93.   Dividing: $\dfrac{3}{4} \div \dfrac{5}{8} = \dfrac{3}{4} \cdot \dfrac{8}{5} = \dfrac{24}{20} = \dfrac{4 \cdot 6}{4 \cdot 5} = \dfrac{6}{5}$ 95.   Multiplying: $x\left(1 + \dfrac{2}{x}\right) = x \cdot 1 + x \cdot \dfrac{2}{x} = x + 2$

97.   Multiplying: $3x\left(\dfrac{1}{x} - \dfrac{1}{3}\right) = 3x \cdot \dfrac{1}{x} - 3x \cdot \dfrac{1}{3} = 3 - x$ 99.   Factoring: $x^2 - 4 = (x+2)(x-2)$

101. Simplifying the expression:

$$\left(1-\frac{1}{x}\right)\left(1-\frac{1}{x+1}\right)\left(1-\frac{1}{x+2}\right)\left(1-\frac{1}{x+3}\right)=\left(\frac{x}{x}-\frac{1}{x}\right)\left(\frac{x+1}{x+1}-\frac{1}{x+1}\right)\left(\frac{x+2}{x+2}-\frac{1}{x+2}\right)\left(\frac{x+3}{x+3}-\frac{1}{x+3}\right)$$

$$=\left(\frac{x-1}{x}\right)\left(\frac{x}{x+1}\right)\left(\frac{x+1}{x+2}\right)\left(\frac{x+2}{x+3}\right)$$

$$=\frac{x-1}{x+3}$$

103. Simplifying the expression:

$$\left(\frac{a^2-b^2}{u^2-v^2}\right)\left(\frac{av-au}{b-a}\right)+\left(\frac{a^2-av}{u+v}\right)\left(\frac{1}{a}\right)=\frac{(a+b)(a-b)}{(u+v)(u-v)}\cdot\frac{a(v-u)}{b-a}+\frac{a(a-v)}{u+v}\cdot\frac{1}{a}$$

$$=\frac{a(a+b)}{u+v}+\frac{a-v}{u+v}$$

$$=\frac{a^2+ab+a-v}{u+v}$$

105. Simplifying the expression:

$$\frac{18x-19}{4x^2+27x-7}-\frac{12x-41}{3x^2+17x-28}=\frac{18x-19}{(4x-1)(x+7)}\cdot\frac{3x-4}{3x-4}-\frac{12x-41}{(3x-4)(x+7)}\cdot\frac{4x-1}{4x-1}$$

$$=\frac{54x^2-129x+76}{(4x-1)(x+7)(3x-4)}-\frac{48x^2-176x+41}{(4x-1)(x+7)(3x-4)}$$

$$=\frac{54x^2-129x+76-48x^2+176x-41}{(4x-1)(x+7)(3x-4)}$$

$$=\frac{6x^2+47x+35}{(4x-1)(x+7)(3x-4)}$$

$$=\frac{(6x+5)(x+7)}{(4x-1)(x+7)(3x-4)}$$

$$=\frac{6x+5}{(4x-1)(3x-4)}$$

107. Simplifying the expression:

$$\left(\frac{1}{y^2-1}\div\frac{1}{y^2+1}\right)\left(\frac{y^3+1}{y^4-1}\right)+\frac{1}{(y+1)^2(y-1)}=\frac{y^2+1}{y^2-1}\cdot\frac{y^3+1}{y^4-1}+\frac{1}{(y+1)^2(y-1)}$$

$$=\frac{y^2+1}{(y+1)(y-1)}\cdot\frac{(y+1)(y^2-y+1)}{(y^2+1)(y+1)(y-1)}+\frac{1}{(y+1)^2(y-1)}$$

$$=\frac{y^2-y+1}{(y+1)(y-1)^2}\cdot\frac{y+1}{y+1}+\frac{1}{(y+1)^2(y-1)}\cdot\frac{y-1}{y-1}$$

$$=\frac{y^3+1+y-1}{(y+1)^2(y-1)^2}$$

$$=\frac{y^3+y}{(y+1)^2(y-1)^2}$$

$$=\frac{y(y^2+1)}{(y+1)^2(y-1)^2}$$

6.5 Complex Fractions

1.   Simplifying the complex fraction: $\dfrac{\frac{3}{4}}{\frac{2}{3}} = \dfrac{\frac{3}{4}\bullet 12}{\frac{2}{3}\bullet 12} = \dfrac{9}{8}$

3.   Simplifying the complex fraction: $\dfrac{\frac{1}{3}-\frac{1}{4}}{\frac{1}{2}+\frac{1}{8}} = \dfrac{\left(\frac{1}{3}-\frac{1}{4}\right)\bullet 24}{\left(\frac{1}{2}+\frac{1}{8}\right)\bullet 24} = \dfrac{8-6}{12+3} = \dfrac{2}{15}$

5.   Simplifying the complex fraction: $\dfrac{3+\frac{2}{5}}{1-\frac{3}{7}} = \dfrac{\left(3+\frac{2}{5}\right)\bullet 35}{\left(1-\frac{3}{7}\right)\bullet 35} = \dfrac{105+14}{35-15} = \dfrac{119}{20}$

7.   Simplifying the complex fraction: $\dfrac{\frac{1}{x}}{1+\frac{1}{x}} = \dfrac{\left(\frac{1}{x}\right)\bullet x}{\left(1+\frac{1}{x}\right)\bullet x} = \dfrac{1}{x+1}$

9.   Simplifying the complex fraction: $\dfrac{1+\frac{1}{a}}{1-\frac{1}{a}} = \dfrac{\left(1+\frac{1}{a}\right)\bullet a}{\left(1-\frac{1}{a}\right)\bullet a} = \dfrac{a+1}{a-1}$

11.   Simplifying the complex fraction: $\dfrac{\frac{1}{x}-\frac{1}{y}}{\frac{1}{x}+\frac{1}{y}} = \dfrac{\left(\frac{1}{x}-\frac{1}{y}\right)\bullet xy}{\left(\frac{1}{x}+\frac{1}{y}\right)\bullet xy} = \dfrac{y-x}{y+x}$

13.   Simplifying the complex fraction:

$\dfrac{\frac{x-5}{x^2-4}}{\frac{x^2-25}{x+2}} = \dfrac{\frac{x-5}{(x+2)(x-2)}\bullet(x+2)(x-2)}{\frac{(x+5)(x-5)}{x+2}\bullet(x+2)(x-2)} = \dfrac{x-5}{(x+5)(x-5)(x-2)} = \dfrac{1}{(x+5)(x-2)}$

15.   Simplifying the complex fraction:

$\dfrac{\frac{4a}{2a^3+2}}{\frac{8a}{4a+4}} = \dfrac{\frac{4a}{2(a+1)\left(a^2-a+1\right)}\bullet 2(a+1)\left(a^2-a+1\right)}{\frac{8a}{4(a+1)}\bullet 2(a+1)\left(a^2-a+1\right)} = \dfrac{4a}{4a\left(a^2-a+1\right)} = \dfrac{1}{a^2-a+1}$

17.   Simplifying the complex fraction: $\dfrac{1-\frac{9}{x^2}}{1-\frac{1}{x}-\frac{6}{x^2}} = \dfrac{\left(1-\frac{9}{x^2}\right)\bullet x^2}{\left(1-\frac{1}{x}-\frac{6}{x^2}\right)\bullet x^2} = \dfrac{x^2-9}{x^2-x-6} = \dfrac{(x+3)(x-3)}{(x+2)(x-3)} = \dfrac{x+3}{x+2}$

19.   Simplifying the complex fraction: $\dfrac{2+\frac{5}{a}-\frac{3}{a^2}}{2-\frac{5}{a}+\frac{2}{a^2}} = \dfrac{\left(2+\frac{5}{a}-\frac{3}{a^2}\right)\bullet a^2}{\left(2-\frac{5}{a}+\frac{2}{a^2}\right)\bullet a^2} = \dfrac{2a^2+5a-3}{2a^2-5a+2} = \dfrac{(2a-1)(a+3)}{(2a-1)(a-2)} = \dfrac{a+3}{a-2}$

21. Simplifying the complex fraction:

$$\frac{27-\dfrac{8}{x^3}}{3+\dfrac{1}{x}-\dfrac{2}{x^2}}=\frac{\left(27-\dfrac{8}{x^3}\right)\bullet x^3}{\left(3+\dfrac{1}{x}-\dfrac{2}{x^2}\right)\bullet x^3}=\frac{27x^3-8}{3x^3+x^2-2x}=\frac{(3x-2)\left(9x^2+6x+4\right)}{x(3x-2)(x+1)}=\frac{9x^2+6x+4}{x(x+1)}$$

23. Simplifying the complex fraction:

$$\frac{1+\dfrac{2}{x}+\dfrac{4}{x^2}+\dfrac{8}{x^3}}{1-\dfrac{16}{x^4}}=\frac{\left(1+\dfrac{2}{x}+\dfrac{4}{x^2}+\dfrac{8}{x^3}\right)\bullet x^4}{\left(1-\dfrac{16}{x^4}\right)\bullet x^4}$$

$$=\frac{x^4+2x^3+4x^2+8x}{x^4-16}$$

$$=\frac{x\left(x^3+2x^2+4x+8\right)}{\left(x^2+4\right)\left(x^2-4\right)}$$

$$=\frac{x(x+2)\left(x^2+4\right)}{\left(x^2+4\right)(x+2)(x-2)}$$

$$=\frac{x}{x-2}$$

25. Simplifying the complex fraction:

$$\frac{2+\dfrac{3}{x}-\dfrac{18}{x^2}-\dfrac{27}{x^3}}{2+\dfrac{9}{x}+\dfrac{9}{x^2}}=\frac{\left(2+\dfrac{3}{x}-\dfrac{18}{x^2}-\dfrac{27}{x^3}\right)\bullet x^3}{\left(2+\dfrac{9}{x}+\dfrac{9}{x^2}\right)\bullet x^3}$$

$$=\frac{2x^3+3x^2-18x-27}{2x^3+9x^2+9x}$$

$$=\frac{x^2(2x+3)-9(2x+3)}{x\left(2x^2+9x+9\right)}$$

$$=\frac{(2x+3)\left(x^2-9\right)}{x(2x+3)(x+3)}$$

$$=\frac{(2x+3)(x+3)(x-3)}{x(2x+3)(x+3)}$$

$$=\frac{x-3}{x}$$

27. Simplifying the complex fraction: $\dfrac{1+\dfrac{1}{x+3}}{1-\dfrac{1}{x+3}}=\dfrac{\left(1+\dfrac{1}{x+3}\right)\bullet(x+3)}{\left(1-\dfrac{1}{x+3}\right)\bullet(x+3)}=\dfrac{x+3+1}{x+3-1}=\dfrac{x+4}{x+2}$

29. Simplifying the complex fraction:

$$\frac{1-\dfrac{1}{a+1}}{1+\dfrac{1}{a-1}}=\frac{\left(1-\dfrac{1}{a+1}\right)\bullet(a+1)(a-1)}{\left(1+\dfrac{1}{a-1}\right)\bullet(a+1)(a-1)}=\frac{(a+1)(a-1)-(a-1)}{(a+1)(a-1)+(a+1)}=\frac{(a-1)(a+1-1)}{(a+1)(a-1+1)}=\frac{a(a-1)}{a(a+1)}=\frac{a-1}{a+1}$$

31. Simplifying the complex fraction: $\dfrac{\dfrac{1}{x+3}+\dfrac{1}{x-3}}{\dfrac{1}{x+3}-\dfrac{1}{x-3}}=\dfrac{\left(\dfrac{1}{x+3}+\dfrac{1}{x-3}\right)\bullet(x+3)(x-3)}{\left(\dfrac{1}{x+3}-\dfrac{1}{x-3}\right)\bullet(x+3)(x-3)}=\dfrac{(x-3)+(x+3)}{(x-3)-(x+3)}=\dfrac{2x}{-6}=-\dfrac{x}{3}$

33. Simplifying the complex fraction:

$$\frac{\dfrac{y+1}{y-1}+\dfrac{y-1}{y+1}}{\dfrac{y+1}{y-1}-\dfrac{y-1}{y+1}}=\frac{\left(\dfrac{y+1}{y-1}+\dfrac{y-1}{y+1}\right)\bullet(y+1)(y-1)}{\left(\dfrac{y+1}{y-1}-\dfrac{y-1}{y+1}\right)\bullet(y+1)(y-1)}$$

$$=\frac{(y+1)^2+(y-1)^2}{(y+1)^2-(y-1)^2}$$

$$=\frac{y^2+2y+1+y^2-2y+1}{y^2+2y+1-y^2+2y-1}$$

$$=\frac{2y^2+2}{4y}$$

$$=\frac{2\left(y^2+1\right)}{4y}$$

$$=\frac{y^2+1}{2y}$$

35. Simplifying the complex fraction: $1-\dfrac{x}{1-\dfrac{1}{x}}=1-\dfrac{x\bullet x}{\left(1-\dfrac{1}{x}\right)\bullet x}=1-\dfrac{x^2}{x-1}=\dfrac{x-1-x^2}{x-1}=\dfrac{-x^2+x-1}{x-1}$

37. Simplifying the complex fraction: $1+\dfrac{1}{1+\dfrac{1}{1+1}}=1+\dfrac{1}{1+\dfrac{1}{2}}=1+\dfrac{1}{\dfrac{3}{2}}=1+\dfrac{2}{3}=\dfrac{5}{3}$

39. Simplifying the complex fraction:

$$\frac{1-\dfrac{1}{x+\dfrac{1}{2}}}{1+\dfrac{1}{x+\dfrac{1}{2}}}=\frac{1-\dfrac{1\bullet 2}{\left(x+\frac{1}{2}\right)\bullet 2}}{1+\dfrac{1\bullet 2}{\left(x+\frac{1}{2}\right)\bullet 2}}=\frac{1-\dfrac{2}{2x+1}}{1+\dfrac{2}{2x+1}}=\frac{\left(1-\dfrac{2}{2x+1}\right)(2x+1)}{\left(1+\dfrac{2}{2x+1}\right)(2x+1)}=\frac{2x+1-2}{2x+1+2}=\frac{2x-1}{2x+3}$$

41. Simplifying the complex fraction:

$$\frac{\dfrac{1}{x+h}-\dfrac{1}{x}}{h}=\frac{\left(\dfrac{1}{x+h}-\dfrac{1}{x}\right)\bullet x(x+h)}{h\bullet x(x+h)}=\frac{x-(x+h)}{hx(x+h)}=\frac{x-x-h}{hx(x+h)}=\frac{-h}{hx(x+h)}=-\frac{1}{x(x+h)}$$

43. Simplifying the complex fraction: $\dfrac{\dfrac{3}{ab}+\dfrac{4}{bc}-\dfrac{2}{ac}}{\dfrac{5}{abc}}=\dfrac{\left(\dfrac{3}{ab}+\dfrac{4}{bc}-\dfrac{2}{ac}\right)\bullet abc}{\left(\dfrac{5}{abc}\right)\bullet abc}=\dfrac{3c+4a-2b}{5}$

45. Simplifying the complex fraction:

$$\frac{\dfrac{t^2-2t-8}{t^2+7t+6}}{\dfrac{t^2-t-6}{t^2+2t+1}}=\frac{\dfrac{(t-4)(t+2)}{(t+6)(t+1)}\bullet(t+6)(t+1)^2}{\dfrac{(t-3)(t+2)}{(t+1)^2}\bullet(t+6)(t+1)^2}=\frac{(t-4)(t+2)(t+1)}{(t-3)(t+2)(t+6)}=\frac{(t-4)(t+1)}{(t+6)(t-3)}$$

47. Simplifying the complex fraction:

$$\frac{5+\dfrac{4}{b-1}}{\dfrac{7}{b+5}-\dfrac{3}{b-1}}=\frac{\left(5+\dfrac{4}{b-1}\right)\bullet(b+5)(b-1)}{\left(\dfrac{7}{b+5}-\dfrac{3}{b-1}\right)\bullet(b+5)(b-1)}$$

$$=\frac{5(b+5)(b-1)+4(b+5)}{7(b-1)-3(b+5)}$$

$$=\frac{(b+5)(5b-5+4)}{7b-7-3b-15}$$

$$=\frac{(b+5)(5b-1)}{4b-22}$$

$$=\frac{(5b-1)(b+5)}{2(2b-11)}$$

49. Simplifying the complex fraction:

$$\frac{\dfrac{3}{x^2-x-6}}{\dfrac{2}{x+2}-\dfrac{4}{x-3}}=\frac{\dfrac{3}{(x-3)(x+2)}\bullet(x-3)(x+2)}{\left(\dfrac{2}{x+2}-\dfrac{4}{x-3}\right)\bullet(x-3)(x+2)}$$

$$=\frac{3}{2(x-3)-4(x+2)}$$

$$=\frac{3}{2x-6-4x-8}$$

$$=\frac{3}{-2x-14}$$

$$=-\frac{3}{2x+14}$$

51. Simplifying the complex fraction:
$$\frac{\dfrac{1}{m-4}+\dfrac{1}{m-5}}{\dfrac{1}{m^2-9m+20}}=\frac{\left(\dfrac{1}{m-4}+\dfrac{1}{m-5}\right)\bullet(m-4)(m-5)}{\dfrac{1}{(m-4)(m-5)}\bullet(m-4)(m-5)}=\frac{(m-5)+(m-4)}{1}=2m-9$$

53.   **a.**   Simplifying the difference quotient:
$$\frac{f(x)-f(a)}{x-a}=\frac{\dfrac{4}{x}-\dfrac{4}{a}}{x-a}=\frac{\left(\dfrac{4}{x}-\dfrac{4}{a}\right)ax}{(x-a)ax}=\frac{4a-4x}{ax(x-a)}=\frac{-4(x-a)}{ax(x-a)}=-\frac{4}{ax}$$

   b.   Simplifying the difference quotient:
$$\frac{f(x)-f(a)}{x-a}=\frac{\dfrac{1}{x+1}-\dfrac{1}{a+1}}{x-a}$$

$$=\frac{\left(\dfrac{1}{x+1}-\dfrac{1}{a+1}\right)(x+1)(a+1)}{(x-a)(x+1)(a+1)}$$

$$=\frac{a+1-x-1}{(x-a)(x+1)(a+1)}$$

$$=\frac{a-x}{(x-a)(x+1)(a+1)}$$

$$=-\frac{1}{(x+1)(a+1)}$$

   c.   Simplifying the difference quotient:
$$\frac{f(x)-f(a)}{x-a}=\frac{\dfrac{1}{x^2}-\dfrac{1}{a^2}}{x-a}=\frac{\left(\dfrac{1}{x^2}-\dfrac{1}{a^2}\right)a^2x^2}{a^2x^2(x-a)}=\frac{a^2-x^2}{a^2x^2(x-a)}=\frac{(a+x)(a-x)}{a^2x^2(x-a)}=-\frac{a+x}{a^2x^2}$$

55.   Rewriting without negative exponents: $f=\left(a^{-1}+b^{-1}\right)^{-1}=\dfrac{1}{a^{-1}+b^{-1}}=\dfrac{1}{\dfrac{1}{a}+\dfrac{1}{b}}=\dfrac{1\bullet ab}{\left(\dfrac{1}{a}+\dfrac{1}{b}\right)\bullet ab}=\dfrac{ab}{a+b}$

57.   **a.**   As v approaches 0, the denominator approaches 1.

    b.   Solving v:

$$h = \frac{f}{1 + \dfrac{v}{s}}$$

$$h = \frac{f \bullet s}{\left(1 + \dfrac{v}{s}\right)s}$$

$$h = \frac{fs}{s + v}$$

$$h(s + v) = fs$$

$$s + v = \frac{fs}{h}$$

$$v = \frac{fs}{h} - s$$

59.   Solving the equation:

$$3x + 60 = 15$$
$$3x = -45$$
$$x = -15$$

61.   Solving the equation:

$$3(y - 3) = 2(y - 2)$$
$$3y - 9 = 2y - 4$$
$$y = 5$$

63.   Solving the equation:

$$10 - 2(x + 3) = x + 1$$
$$10 - 2x - 6 = x + 1$$
$$-2x + 4 = x + 1$$
$$-3x = -3$$
$$x = 1$$

65.   Solving the equation:

$$x^2 - x - 12 = 0$$
$$(x - 4)(x + 3) = 0$$
$$x = -3, 4$$

67.   Solving the equation:

$$(x + 1)(x - 6) = -12$$
$$x^2 - 5x - 6 = -12$$
$$x^2 - 5x + 6 = 0$$
$$(x - 2)(x - 3) = 0$$
$$x = 2, 3$$

69.   Multiplying: $x(y - 2) = xy - 2x$

71.   Multiplying: $6\left(\dfrac{x}{2} - 3\right) = 6 \bullet \dfrac{x}{2} - 6 \bullet 3 = 3x - 18$

73.   Multiplying: $xab \bullet \dfrac{1}{x} = ab$

75.   Factoring: $y^2 - 25 = (y + 5)(y - 5)$

77.   Factoring: $xa + xb = x(a + b)$

79.   Solving the equation:

$$5x - 4 = 6$$
$$5x = 10$$
$$x = 2$$

81.   Simplifying the expression: $\dfrac{\left(\frac{1}{3}\right) - \left(\frac{1}{3}\right)^2}{1 - \frac{1}{3}} = \dfrac{\frac{1}{3} - \frac{1}{9}}{1 - \frac{1}{3}} = \dfrac{\left(\frac{1}{3} - \frac{1}{9}\right) \bullet 9}{\left(1 - \frac{1}{3}\right) \bullet 9} = \dfrac{3 - 1}{9 - 3} = \dfrac{2}{6} = \dfrac{1}{3}$

83.   Simplifying the expression: $\dfrac{\left(\frac{1}{9}\right) - \frac{1}{9}\left(\frac{1}{3}\right)^4}{1 - \frac{1}{3}} = \dfrac{\frac{1}{9} - \frac{1}{729}}{1 - \frac{1}{3}} = \dfrac{\left(\frac{1}{9} - \frac{1}{729}\right) \bullet 729}{\left(1 - \frac{1}{3}\right) \bullet 729} = \dfrac{81 - 1}{729 - 243} = \dfrac{80}{486} = \dfrac{40}{243}$

85. Simplifying the expression:

$$\frac{1+\dfrac{1}{1-\dfrac{a}{b}}}{1-\dfrac{3}{1-\dfrac{a}{b}}} = \frac{1+\dfrac{1\bullet b}{\left(1-\dfrac{a}{b}\right)\bullet b}}{1-\dfrac{3\bullet b}{\left(1-\dfrac{a}{b}\right)\bullet b}} = \frac{1+\dfrac{b}{b-a}}{1-\dfrac{3b}{b-a}} = \frac{\left(1+\dfrac{b}{b-a}\right)(b-a)}{\left(1-\dfrac{3b}{b-a}\right)(b-a)} = \frac{b-a+b}{b-a-3b} = \frac{-a+2b}{-a-2b} = \frac{a-2b}{a+2b}$$

87. Simplifying the expression: $\dfrac{a^{-1}+b^{-1}}{(ab)^{-1}} = \dfrac{\dfrac{1}{a}+\dfrac{1}{b}}{\dfrac{1}{ab}} = \dfrac{\left(\dfrac{1}{a}+\dfrac{1}{b}\right)ab}{\left(\dfrac{1}{ab}\right)ab} = \dfrac{b+a}{1} = a+b$

89. Simplifying the expression:

$$\frac{\left(q^{-2}-t^{-2}\right)^{-1}}{\left(t^{-1}-q^{-1}\right)^{-1}} = \frac{t^{-1}-q^{-1}}{q^{-2}-t^{-2}} = \frac{\dfrac{1}{t}-\dfrac{1}{q}}{\dfrac{1}{q^{2}}-\dfrac{1}{t^{2}}} = \frac{\left(\dfrac{1}{t}-\dfrac{1}{q}\right)q^{2}t^{2}}{\left(\dfrac{1}{q^{2}}-\dfrac{1}{t^{2}}\right)q^{2}t^{2}} = \frac{q^{2}t-qt^{2}}{t^{2}-q^{2}} = \frac{-qt(t-q)}{(t+q)(t-q)} = -\frac{qt}{q+t}$$

6.6 Equations Involving Rational Expressions

1. Solving the equation:

$$\frac{x}{5}+4=\frac{5}{3}$$
$$15\left(\frac{x}{5}+4\right)=15\left(\frac{5}{3}\right)$$
$$3x+60=25$$
$$3x=-35$$
$$x=-\frac{35}{3}$$

3. Solving the equation:

$$\frac{a}{3}+2=\frac{4}{5}$$
$$15\left(\frac{a}{3}+2\right)=15\left(\frac{4}{5}\right)$$
$$5a+30=12$$
$$5a=-18$$
$$a=-\frac{18}{5}$$

5. Solving the equation:

$$\frac{y}{2}+\frac{y}{4}+\frac{y}{6}=3$$
$$12\left(\frac{y}{2}+\frac{y}{4}+\frac{y}{6}\right)=12(3)$$
$$6y+3y+2y=36$$
$$11y=36$$
$$y=\frac{36}{11}$$

7. Solving the equation:

$$\frac{5}{2x}=\frac{1}{x}+\frac{3}{4}$$
$$4x\left(\frac{5}{2x}\right)=4x\left(\frac{1}{x}+\frac{3}{4}\right)$$
$$10=4+3x$$
$$3x=6$$
$$x=2$$

9. Solving the equation:

$$\frac{1}{x}=\frac{1}{3}-\frac{2}{3x}$$
$$3x\left(\frac{1}{x}\right)=3x\left(\frac{1}{3}-\frac{2}{3x}\right)$$
$$3=x-2$$
$$x=5$$

11. Solving the equation:

$$\frac{2x}{x-3}+2=\frac{2}{x-3}$$
$$(x-3)\left(\frac{2x}{x-3}+2\right)=(x-3)\left(\frac{2}{x-3}\right)$$
$$2x+2(x-3)=2$$
$$2x+2x-6=2$$
$$4x=8$$
$$x=2$$

13. Solving the equation:

$$1 - \frac{1}{x} = \frac{12}{x^2}$$

$$x^2\left(1 - \frac{1}{x}\right) = x^2\left(\frac{12}{x^2}\right)$$

$$x^2 - x = 12$$

$$x^2 - x - 12 = 0$$

$$(x+3)(x-4) = 0$$

$$x = -3, 4$$

15. Solving the equation:

$$y - \frac{4}{3y} = -\frac{1}{3}$$

$$3y\left(y - \frac{4}{3y}\right) = 3y\left(-\frac{1}{3}\right)$$

$$3y^2 - 4 = -y$$

$$3y^2 + y - 4 = 0$$

$$(3y+4)(y-1) = 0$$

$$y = -\frac{4}{3}, 1$$

17. Solving the equation:

$$f(x) + g(x) = \frac{5}{8}$$

$$\frac{1}{x-3} + \frac{1}{x+3} = \frac{5}{8}$$

$$8(x-3)(x+3)\left(\frac{1}{x-3} + \frac{1}{x+3}\right) = 8(x-3)(x+3)\left(\frac{5}{8}\right)$$

$$8(x+3) + 8(x-3) = 5(x-3)(x+3)$$

$$8x + 24 + 8x - 24 = 5x^2 - 45$$

$$16x = 5x^2 - 45$$

$$0 = 5x^2 - 16x - 45$$

$$0 = (5x+9)(x-5)$$

$$x = -\frac{9}{5}, 5$$

19. Solving the equation:

$$\frac{f(x)}{g(x)} = 5$$

$$\frac{\frac{1}{x-3}}{\frac{1}{x+3}} = 5$$

$$\frac{x+3}{x-3} = 5$$

$$x+3 = 5(x-3)$$

$$x+3 = 5x - 15$$

$$18 = 4x$$

$$x = \frac{9}{2}$$

21. Solving the equation:

$$f(x) = g(x)$$

$$\frac{1}{x-3} = \frac{1}{x+3}$$

$$x+3 = x-3$$

$$3 = -3 \quad \text{(false)}$$

There is no solution $(\varnothing)$.

23. Solving the equation:

$$f(x) + g(x) = \frac{24}{5}$$

$$\frac{4}{x+2} + \frac{4}{x-2} = \frac{24}{5}$$

$$5(x-2)(x+2)\left(\frac{4}{x+2} + \frac{4}{x-2}\right) = 5(x-2)(x+2)\left(\frac{24}{5}\right)$$

$$20(x-2) + 20(x+2) = 24(x-2)(x+2)$$

$$20x - 40 + 20x + 40 = 24x^2 - 96$$

$$40x = 24x^2 - 96$$

$$0 = 24x^2 - 40x - 96$$

$$0 = 3x^2 - 5x - 12$$

$$0 = (3x+4)(x-3)$$

$$x = -\frac{4}{3}, 3$$

25. Solving the equation:

$$\frac{f(x)}{g(x)} = -5$$

$$\frac{\dfrac{4}{x+2}}{\dfrac{4}{x-2}} = -5$$

$$\frac{x-2}{x+2} = -5$$

$$x - 2 = -5(x+2)$$

$$x - 2 = -5x - 10$$

$$6x = -8$$

$$x = -\frac{4}{3}$$

27. Solving the equation:

$$f(x) = g(x)$$

$$\frac{4}{x+2} = \frac{4}{x-2}$$

$$x - 2 = x + 2$$

$$-2 = 2 \quad \text{(false)}$$

There is no solution $(\varnothing)$.

29. **a.** Solving the equation:

$$6x - 2 = 0$$

$$6x = 2$$

$$x = \frac{1}{3}$$

b. Solving the equation:

$$\frac{6}{x} - 2 = 0$$

$$x\left(\frac{6}{x} - 2\right) = x(0)$$

$$6 - 2x = 0$$

$$6 = 2x$$

$$x = 3$$

c. Solving the equation:

$$\frac{x}{6} - 2 = -\frac{1}{2}$$

$$6\left(\frac{x}{6} - 2\right) = 6\left(-\frac{1}{2}\right)$$

$$x - 12 = -3$$

$$x = 9$$

d. Solving the equation:

$$\frac{6}{x} - 2 = -\frac{1}{2}$$

$$2x\left(\frac{6}{x} - 2\right) = 2x\left(-\frac{1}{2}\right)$$

$$12 - 4x = -x$$

$$12 = 3x$$

$$x = 4$$

e.   Solving the equation:

$$\frac{6}{x^2} + 6 = \frac{20}{x}$$

$$x^2\left(\frac{6}{x^2} + 6\right) = x^2\left(\frac{20}{x}\right)$$

$$6 + 6x^2 = 20x$$

$$6x^2 - 20x + 6 = 0$$

$$3x^2 - 10x + 3 = 0$$

$$(3x - 1)(x - 3) = 0$$

$$x = \tfrac{1}{3}, 3$$

31.   **a.**   Dividing: $\dfrac{6}{x^2 - 2x - 8} \div \dfrac{x + 3}{x + 2} = \dfrac{6}{(x-4)(x+2)} \bullet \dfrac{x+2}{x+3} = \dfrac{6}{(x-4)(x+3)}$

b.   Adding:

$$\frac{6}{x^2 - 2x - 8} + \frac{x + 3}{x + 2} = \frac{6}{(x-4)(x+2)} + \frac{x+3}{x+2} \bullet \frac{x-4}{x-4}$$

$$= \frac{6}{(x-4)(x+2)} + \frac{x^2 - x - 12}{(x-4)(x+2)}$$

$$= \frac{x^2 - x - 6}{(x-4)(x+2)}$$

$$= \frac{(x-3)(x+2)}{(x-4)(x+2)}$$

$$= \frac{x - 3}{x - 4}$$

c.   Solving the equation:

$$\frac{6}{x^2 - 2x - 8} + \frac{x + 3}{x + 2} = 2$$

$$(x-4)(x+2)\left(\frac{6}{(x-4)(x+2)} + \frac{x+3}{x+2}\right) = (x-4)(x+2)(2)$$

$$6 + (x-4)(x+3) = 2(x-4)(x+2)$$

$$6 + x^2 - x - 12 = 2x^2 - 4x - 16$$

$$0 = x^2 - 3x - 10$$

$$0 = (x-5)(x+2)$$

$$x = -2, 5$$

Note that $x = -2$ does not check in the original equation, so the solution is $x = 5$.

33.   Solving the equation:

$$\frac{x+2}{x+1} = \frac{1}{x+1} + 2$$

$$(x+1)\left(\frac{x+2}{x+1}\right) = (x+1)\left(\frac{1}{x+1} + 2\right)$$

$$x + 2 = 1 + 2(x+1)$$

$$x + 2 = 1 + 2x + 2$$

$$x + 2 = 2x + 3$$

$$x = -1 \quad (\text{does not check})$$

There is no solution (–1 does not check), or $\varnothing$.

35.   Solving the equation:

$$\frac{3}{a-2} = \frac{2}{a-3}$$

$$(a-2)(a-3)\left(\frac{3}{a-2}\right) = (a-2)(a-3)\left(\frac{2}{a-3}\right)$$

$$3(a-3) = 2(a-2)$$

$$3a - 9 = 2a - 4$$

$$a = 5$$

37.   Solving the equation:

$$6 - \frac{5}{x^2} = \frac{7}{x}$$

$$x^2\left(6 - \frac{5}{x^2}\right) = x^2\left(\frac{7}{x}\right)$$

$$6x^2 - 5 = 7x$$

$$6x^2 - 7x - 5 = 0$$

$$(2x+1)(3x-5) = 0$$

$$x = -\frac{1}{2}, \frac{5}{3}$$

39.   Solving the equation:

$$\frac{1}{x-1} - \frac{1}{x+1} = \frac{3x}{x^2-1}$$

$$(x+1)(x-1)\left(\frac{1}{x-1} - \frac{1}{x+1}\right) = (x+1)(x-1)\left(\frac{3x}{(x+1)(x-1)}\right)$$

$$(x+1) - (x-1) = 3x$$

$$x + 1 - x + 1 = 3x$$

$$3x = 2$$

$$x = \frac{2}{3}$$

41.   Solving the equation:

$$\frac{2}{x-3} + \frac{x}{x^2-9} = \frac{4}{x+3}$$

$$(x+3)(x-3)\left(\frac{2}{x-3} + \frac{x}{(x+3)(x-3)}\right) = (x+3)(x-3)\left(\frac{4}{x+3}\right)$$

$$2(x+3) + x = 4(x-3)$$

$$2x + 6 + x = 4x - 12$$

$$3x + 6 = 4x - 12$$

$$-x = -18$$

$$x = 18$$

43.   Solving the equation:

$$\frac{3}{2} - \frac{1}{x-4} = \frac{-2}{2x-8}$$

$$2(x-4)\left(\frac{3}{2} - \frac{1}{x-4}\right) = 2(x-4)\left(\frac{-2}{2(x-4)}\right)$$

$$3(x-4) - 2 = -2$$

$$3x - 12 - 2 = -2$$

$$3x - 14 = -2$$

$$3x = 12$$

$$x = 4 \quad \text{(does not check)}$$

There is no solution (4 does not check), or $\varnothing$.

45.   Solving the equation:

$$\frac{t-4}{t^2-3t}=\frac{-2}{t^2-9}$$

$$t(t+3)(t-3)\cdot\frac{t-4}{t(t-3)}=t(t+3)(t-3)\cdot\frac{-2}{(t+3)(t-3)}$$

$$(t+3)(t-4)=-2t$$

$$t^2-t-12=-2t$$

$$t^2+t-12=0$$

$$(t+4)(t-3)=0$$

$$t=-4\quad(t=3\text{ does not check})$$

The solution is –4 (3 does not check).

47.   Solving the equation:

$$\frac{3}{y-4}-\frac{2}{y+1}=\frac{5}{y^2-3y-4}$$

$$(y-4)(y+1)\left(\frac{3}{y-4}-\frac{2}{y+1}\right)=(y-4)(y+1)\left(\frac{5}{(y-4)(y+1)}\right)$$

$$3(y+1)-2(y-4)=5$$

$$3y+3-2y+8=5$$

$$y+11=5$$

$$y=-6$$

49.   Solving the equation:

$$\frac{2}{1+a}=\frac{3}{1-a}+\frac{5}{a}$$

$$a(1+a)(1-a)\left(\frac{2}{1+a}\right)=a(1+a)(1-a)\left(\frac{3}{1-a}+\frac{5}{a}\right)$$

$$2a(1-a)=3a(1+a)+5(1+a)(1-a)$$

$$2a-2a^2=3a+3a^2+5-5a^2$$

$$-2a^2+2a=-2a^2+3a+5$$

$$2a=3a+5$$

$$-a=5$$

$$a=-5$$

51.   Solving the equation:

$$\frac{3}{2x-6}-\frac{x+1}{4x-12}=4$$

$$4(x-3)\left(\frac{3}{2(x-3)}-\frac{x+1}{4(x-3)}\right)=4(x-3)(4)$$

$$6-(x+1)=16x-48$$

$$5-x=16x-48$$

$$-17x=-53$$

$$x=\frac{53}{17}$$

53.   Solving the equation:

$$\frac{y+2}{y^2-y}-\frac{6}{y^2-1}=0$$

$$y(y+1)(y-1)\left(\frac{y+2}{y(y-1)}-\frac{6}{(y+1)(y-1)}\right)=y(y+1)(y-1)(0)$$

$$(y+1)(y+2)-6y=0$$

$$y^2+3y+2-6y=0$$

$$y^2-3y+2=0$$

$$(y-1)(y-2)=0$$

$$y=2\quad(y=1\text{ does not check})$$

The solution is 2 (1 does not check).

55. Solving the equation:

$$\frac{4}{2x-6} - \frac{12}{4x+12} = \frac{12}{x^2-9}$$

$$4(x+3)(x-3)\left(\frac{4}{2(x-3)} - \frac{12}{4(x+3)}\right) = 4(x+3)(x-3)\left(\frac{12}{(x+3)(x-3)}\right)$$

$$8(x+3) - 12(x-3) = 48$$
$$8x+24-12x+36 = 48$$
$$-4x+60 = 48$$
$$-4x = -12$$
$$x = 3 \quad (x=3 \text{ does not check})$$

There is no solution (3 does not check), or $\varnothing$.

57. Solving the equation:

$$\frac{2}{y^2-7y+12} - \frac{1}{y^2-9} = \frac{4}{y^2-y-12}$$

$$(y+3)(y-3)(y-4)\left(\frac{2}{(y-3)(y-4)} - \frac{1}{(y+3)(y-3)}\right) = (y+3)(y-3)(y-4)\left(\frac{4}{(y-4)(y+3)}\right)$$

$$2(y+3) - (y-4) = 4(y-3)$$
$$2y+6-y+4 = 4y-12$$
$$y+10 = 4y-12$$
$$-3y = -22$$
$$y = \frac{22}{3}$$

59. Solving the equation:

$$6x^{-1} + 4 = 7$$
$$x(6x^{-1} + 4) = x(7)$$
$$6 + 4x = 7x$$
$$6 = 3x$$
$$x = 2$$

61. Solving the equation:

$$1 + 5x^{-2} = 6x^{-1}$$
$$x^2(1 + 5x^{-2}) = x^2(6x^{-1})$$
$$x^2 + 5 = 6x$$
$$x^2 - 6x + 5 = 0$$
$$(x-1)(x-5) = 0$$
$$x = 1, 5$$

63. Solving for x:

$$\frac{1}{x} = \frac{1}{b} - \frac{1}{a}$$
$$abx\left(\frac{1}{x}\right) = abx\left(\frac{1}{b} - \frac{1}{a}\right)$$
$$ab = ax - bx$$
$$ab = x(a-b)$$
$$x = \frac{ab}{a-b}$$

65. Solving for R:

$$\frac{1}{R} = \frac{1}{R_1} + \frac{1}{R_2}$$
$$RR_1R_2\left(\frac{1}{R}\right) = RR_1R_2\left(\frac{1}{R_1} + \frac{1}{R_2}\right)$$
$$R_1R_2 = RR_2 + RR_1$$
$$R_1R_2 = R(R_1 + R_2)$$
$$R = \frac{R_1R_2}{R_1 + R_2}$$

67. Solving for y:

$$x = \frac{y-3}{y-1}$$
$$x(y-1) = y-3$$
$$xy - x = y-3$$
$$xy - y = x-3$$
$$y(x-1) = x-3$$
$$y = \frac{x-3}{x-1}$$

69. Solving for y:

$$x = \frac{2y+1}{3y+1}$$
$$x(3y+1) = 2y+1$$
$$3xy + x = 2y+1$$
$$3xy - 2y = -x+1$$
$$y(3x-2) = -x+1$$
$$y = \frac{1-x}{3x-2}$$

71.   Simplifying the left-hand side: $\dfrac{2}{x-y} - \dfrac{1}{y-x} = \dfrac{2}{x-y} - \dfrac{-1}{x-y} = \dfrac{2}{x-y} + \dfrac{1}{x-y} = \dfrac{3}{x-y}$

73.   Completing the table:

Time t (sec)	Speed of Kayak Relative to the Water v (m/sec)	Current of the River c (m/sec)
240	4	1
300	4	2
514	4	3
338	3	1
540	3	2
impossible	3	3

75.   Let x represent the number. The equation is:

$$2(x+3) = 16$$
$$2x + 6 = 16$$
$$2x = 10$$
$$x = 5$$

The number is 5.

77.   Let w represent the width and $2w - 3$ represent the length. The equation is:

$$2w + 2(2w - 3) = 42$$
$$2w + 4w - 6 = 42$$
$$6w = 48$$
$$w = 8$$
$$2w - 3 = 13$$

The width is 8 meters and the length is 13 meters.

79.   Let x and $x + 1$ represent the two integers. The equation is:

$$x^2 + (x+1)^2 = 61$$
$$x^2 + x^2 + 2x + 1 = 61$$
$$2x^2 + 2x - 60 = 0$$
$$x^2 + x - 30 = 0$$
$$(x+6)(x-5) = 0$$
$$x = -6, 5$$
$$x + 1 = -5, 6$$

The integers are either –6 and –5, or 5 and 6.

81.   Let x, $x + 1$, and $x + 2$ represent the three sides. The equation is:

$$x^2 + (x+1)^2 = (x+2)^2$$
$$x^2 + x^2 + 2x + 1 = x^2 + 4x + 4$$
$$x^2 - 2x - 3 = 0$$
$$(x+1)(x-3) = 0$$
$$x = -1, 3$$
$$x + 1 = 0, 4$$
$$x + 2 = 1, 5$$

Since the sides of the triangle must be positive, the sides are 3, 4, and 5.

83.   Multiplying: $39.3 \bullet 60 = 2,358$ **85.**   Dividing: $65,000 \div 5,280 \approx 12.3$

87.   Multiplying: $2x\left(\dfrac{1}{x} + \dfrac{1}{2x}\right) = 2x \bullet \dfrac{1}{x} + 2x \bullet \dfrac{1}{2x} = 2 + 1 = 3$

89. Solving the equation:

$$12(x+3)+12(x-3)=3\left(x^2-9\right)$$
$$12x+36+12x-36=3x^2-27$$
$$24x=3x^2-27$$
$$3x^2-24x-27=0$$
$$3\left(x^2-8x-9\right)=0$$
$$3(x-9)(x+1)=0$$
$$x=-1,9$$

91. Solving the equation:

$$\frac{1}{10}-\frac{1}{12}=\frac{1}{x}$$
$$60x\left(\frac{1}{10}-\frac{1}{12}\right)=60x\left(\frac{1}{x}\right)$$
$$6x-5x=60$$
$$x=60$$

93. Solving the equation:

$$\frac{12}{x}+\frac{8}{x^2}-\frac{75}{x^3}-\frac{50}{x^4}=0$$
$$x^4\left(\frac{12}{x}+\frac{8}{x^2}-\frac{75}{x^3}-\frac{50}{x^4}\right)=x^4(0)$$
$$12x^3+8x^2-75x-50=0$$
$$4x^2(3x+2)-25(3x+2)=0$$
$$(3x+2)\left(4x^2-25\right)=0$$
$$(3x+2)(2x+5)(2x-5)=0$$
$$x=-\frac{5}{2},-\frac{2}{3},\frac{5}{2}$$

95. Solving the equation:

$$\frac{1}{x^3}-\frac{1}{3x^2}-\frac{1}{4x}+\frac{1}{12}=0$$
$$12x^3\left(\frac{1}{x^3}-\frac{1}{3x^2}-\frac{1}{4x}+\frac{1}{12}\right)=12x^3(0)$$
$$12-4x-3x^2+x^3=0$$
$$x^2(x-3)-4(x-3)=0$$
$$(x-3)\left(x^2-4\right)=0$$
$$(x-3)(x+2)(x-2)=0$$
$$x=-2,2,3$$

97. Solving for x:

$$\frac{2}{x}+\frac{4}{x+a}=\frac{-6}{a-x}$$
$$x(x+a)(a-x)\left(\frac{2}{x}+\frac{4}{x+a}\right)=x(x+a)(a-x)\left(\frac{-6}{a-x}\right)$$
$$2(x+a)(a-x)+4x(a-x)=-6x(x+a)$$
$$2a^2-2x^2+4ax-4x^2=-6x^2-6ax$$
$$2a^2=-10ax$$
$$x=-\frac{a}{5}$$

99. Solving for v:

$$\frac{s-vt}{t^2}=-16$$
$$s-vt=-16t^2$$
$$s+16t^2=vt$$
$$v=\frac{16t^2+s}{t}$$

101. Solving for f:

$$\frac{1}{p}=\frac{1}{f}+\frac{1}{g}$$
$$pfg\left(\frac{1}{p}\right)=Pfg\left(\frac{1}{f}+\frac{1}{g}\right)$$
$$fg=pg+pf$$
$$fg-pf=pg$$
$$f(g-p)=pg$$
$$f=\frac{pg}{g-p}$$

6.7 Applications

1.   Let x and $3x$ represent the two numbers. The equation is:

$$\frac{1}{x} + \frac{1}{3x} = \frac{20}{3}$$

$$3x\left(\frac{1}{x} + \frac{1}{3x}\right) = 3x\left(\frac{20}{3}\right)$$

$$3 + 1 = 20x$$

$$20x = 4$$

$$x = \frac{1}{5}$$

The numbers are $\frac{1}{5}$ and $\frac{3}{5}$.

3.   Let x represent the number. The equation is:

$$x + \frac{1}{x} = \frac{10}{3}$$

$$3x\left(x + \frac{1}{x}\right) = 3x\left(\frac{10}{3}\right)$$

$$3x^2 + 3 = 10x$$

$$3x^2 - 10x + 3 = 0$$

$$(3x - 1)(x - 3) = 0$$

$$x = \frac{1}{3}, 3$$

The number is either 3 or $\frac{1}{3}$.

5.   Let x and $x + 1$ represent the two integers. The equation is:

$$\frac{1}{x} + \frac{1}{x+1} = \frac{7}{12}$$

$$12x(x+1)\left(\frac{1}{x} + \frac{1}{x+1}\right) = 12x(x+1)\left(\frac{7}{12}\right)$$

$$12(x+1) + 12x = 7x(x+1)$$

$$12x + 12 + 12x = 7x^2 + 7x$$

$$0 = 7x^2 - 17x - 12$$

$$0 = (7x + 4)(x - 3)$$

$$x = 3 \quad \left(x = -\frac{4}{7} \text{ is not an integer}\right)$$

The two integers are 3 and 4.

7.   Let x represent the number. The equation is:

$$\frac{7 + x}{9 + x} = \frac{5}{6}$$

$$6(9 + x)\left(\frac{7 + x}{9 + x}\right) = 6(9 + x)\left(\frac{5}{6}\right)$$

$$6(7 + x) = 5(9 + x)$$

$$42 + 6x = 45 + 5x$$

$$x = 3$$

The number is 3.

9. Let x represent the speed of the current. Setting the times equal:

$$\frac{3}{5+x} = \frac{1.5}{5-x}$$
$$3(5-x) = 1.5(5+x)$$
$$15 - 3x = 7.5 + 1.5x$$
$$7.5 = 4.5x$$
$$x = \frac{75}{45} = \frac{5}{3}$$

The speed of the current is $\frac{5}{3}$ mph.

11. Let x represent the speed of the boat. Since the total time is 3 hours:

$$\frac{8}{x-2} + \frac{8}{x+2} = 3$$
$$(x+2)(x-2)\left(\frac{8}{x-2} + \frac{8}{x+2}\right) = 3(x+2)(x-2)$$
$$8(x+2) + 8(x-2) = 3x^2 - 12$$
$$16x = 3x^2 - 12$$
$$0 = 3x^2 - 16x - 12$$
$$0 = (3x+2)(x-6)$$
$$x = 6 \quad \left(x = -\frac{2}{3} \text{ is impossible}\right)$$

The speed of the boat is 6 mph.

13. Let r represent the speed of train B and $r + 15$ represent the speed of train A. Since the times are equal:

$$\frac{150}{r+15} = \frac{120}{r}$$
$$150r = 120(r+15)$$
$$150r = 120r + 1800$$
$$30r = 1800$$
$$r = 60$$

The speed of train A is 75 mph and the speed of train B is 60 mph.

15. The smaller plane makes the trip in 3 hours, so the 747 must take $1\frac{1}{2}$ hours to complete the trip. Thus the average

speed is given by: $\dfrac{810 \text{ miles}}{1\frac{1}{2} \text{ hours}} = 540$ miles per hour

17. Let r represent the bus's usual speed. The difference of the two times is $\frac{1}{2}$ hour, therefore:

$$\frac{270}{r} - \frac{270}{r+6} = \frac{1}{2}$$
$$2r(r+6)\left(\frac{270}{r} - \frac{270}{r+6}\right) = 2r(r+6)\left(\frac{1}{2}\right)$$
$$540(r+6) - 540(r) = r(r+6)$$
$$540r + 3240 - 540r = r^2 + 6r$$
$$0 = r^2 + 6r - 3240$$
$$0 = (r-54)(r+60)$$
$$r = 54 \quad (r = -60 \text{ is impossible})$$

The usual speed is 54 mph.

19.   Let x represent the time to fill the tank if both pipes are open. The rate equation is:

$$\frac{1}{8} - \frac{1}{16} = \frac{1}{x}$$

$$16x\left(\frac{1}{8} - \frac{1}{16}\right) = 16x\left(\frac{1}{x}\right)$$

$$2x - x = 16$$

$$x = 16$$

It will take 16 hours to fill the tank if both pipes are open.

21.   Let x represent the time to empty the tank with the drain open. The rate equation is:

$$\frac{1}{3} - \frac{1}{5} = \frac{1}{x}$$

$$15x\left(\frac{1}{3} - \frac{1}{5}\right) = 15x\left(\frac{1}{x}\right)$$

$$5x - 3x = 15$$

$$2x = 15$$

$$x = 7\frac{1}{2}$$

It will take $7\frac{1}{2}$ minutes to empty the tank if both the faucet and drain are open.

23.   Let x represent the time to fill the sink with the hot water faucet. The rate equation is:

$$\frac{1}{3.5} + \frac{1}{x} = \frac{1}{2.1}$$

$$7.35x\left(\frac{1}{3.5} + \frac{1}{x}\right) = 7.35x\left(\frac{1}{2.1}\right)$$

$$2.1x + 7.35 = 3.5x$$

$$7.35 = 1.4x$$

$$x = 5.25$$

It will take $5\frac{1}{4}$ minutes to fill the sink with the hot water faucet.

25.   Converting to acres: $\dfrac{2,224,750 \text{ sq. ft.}}{43,560 \text{ sq. ft./acre}} \approx 51.1 \text{ acres}$

27.   Converting the speed: $\dfrac{5750 \text{ feet}}{11 \text{ minutes}} \cdot \dfrac{1 \text{ mile}}{5280 \text{ feet}} \cdot \dfrac{60 \text{ minutes}}{1 \text{ hour}} \approx 5.9 \text{ mph}$

29.   Converting the speed: $\dfrac{100 \text{ meters}}{10.8 \text{ seconds}} \cdot \dfrac{3.28 \text{ feet}}{1 \text{ meter}} \cdot \dfrac{1 \text{ mile}}{5280 \text{ feet}} \cdot \dfrac{60 \text{ seconds}}{1 \text{ minute}} \cdot \dfrac{60 \text{ minutes}}{1 \text{ hour}} \approx 20.7 \text{ mph}$

31.   Converting the speed: $\dfrac{\pi \bullet 65 \text{ feet}}{30 \text{ seconds}} \cdot \dfrac{1 \text{ mile}}{5280 \text{ feet}} \cdot \dfrac{60 \text{ seconds}}{1 \text{ minute}} \cdot \dfrac{60 \text{ minutes}}{1 \text{ hour}} \approx 4.6 \text{ mph}$

33.   Converting the speed: $\dfrac{2\pi \bullet 2 \text{ inches}}{\frac{1}{300} \text{ minutes}} \cdot \dfrac{1 \text{ foot}}{12 \text{ inches}} \cdot \dfrac{1 \text{ mile}}{5280 \text{ feet}} \cdot \dfrac{60 \text{ minutes}}{1 \text{ hour}} \approx 3.6 \text{ mph}$

35.   Solving the equation:

$$\frac{1}{3}\left[\left(x + \frac{2}{3}x\right) + \frac{1}{3}\left(x + \frac{2}{3}x\right)\right] = 10$$

$$\left(x + \frac{2}{3}x\right) + \frac{1}{3}\left(x + \frac{2}{3}x\right) = 30$$

$$x + \frac{2}{3}x + \frac{1}{3}x + \frac{2}{9}x = 30$$

$$\frac{20}{9}x = 30$$

$$20x = 270$$

$$x = \frac{27}{2}$$

37. Performing the operations: $\dfrac{2a+10}{a^3} \cdot \dfrac{a^2}{3a+15} = \dfrac{2(a+5)}{a^3} \cdot \dfrac{a^2}{3(a+5)} = \dfrac{2}{3a}$

39. Performing the operations: $\left(x^2 - 9\right)\left(\dfrac{x+2}{x+3}\right) = (x+3)(x-3)\left(\dfrac{x+2}{x+3}\right) = (x-3)(x+2)$

41. Performing the operations: $\dfrac{2x-7}{x-2} - \dfrac{x-5}{x-2} = \dfrac{2x-7-x+5}{x-2} = \dfrac{x-2}{x-2} = 1$

43. Simplifying the expression: $\dfrac{\dfrac{1}{x} - \dfrac{1}{3}}{\dfrac{1}{x} + \dfrac{1}{3}} = \dfrac{\left(\dfrac{1}{x} - \dfrac{1}{3}\right) \cdot 3x}{\left(\dfrac{1}{x} + \dfrac{1}{3}\right) \cdot 3x} = \dfrac{3-x}{3+x}$

45. Solving the equation:

$$\frac{x}{x-3} + \frac{3}{2} = \frac{3}{x-3}$$
$$2(x-3)\left(\frac{x}{x-3} + \frac{3}{2}\right) = 2(x-3)\left(\frac{3}{x-3}\right)$$
$$2x + 3(x-3) = 6$$
$$2x + 3x - 9 = 6$$
$$5x = 15$$
$$x = 3 \quad \text{(does not check)}$$

There is no solution (3 does not check).

Chapter 6 Review/Test

1. Reducing the fraction: $\dfrac{125x^4 y z^3}{35x^2 y^4 z^3} = \dfrac{25x^2}{7y^3}$

2. Reducing the fraction: $\dfrac{a^3 - ab^2}{4a+4b} = \dfrac{a\left(a^2 - b^2\right)}{4(a+b)} = \dfrac{a(a+b)(a-b)}{4(a+b)} = \dfrac{a(a-b)}{4}$

3. Reducing the fraction: $\dfrac{x^2 - 25}{x^2 + 10x + 25} = \dfrac{(x+5)(x-5)}{(x+5)^2} = \dfrac{x-5}{x+5}$

4. Reducing the fraction: $\dfrac{ax+x-5a-5}{ax-x-5a+5} = \dfrac{x(a+1)-5(a+1)}{x(a-1)-5(a-1)} = \dfrac{(a+1)(x-5)}{(a-1)(x-5)} = \dfrac{a+1}{a-1}$

5. Dividing: $\dfrac{12x^3 + 8x^2 + 16x}{4x^2} = \dfrac{12x^3}{4x^2} + \dfrac{8x^2}{4x^2} + \dfrac{16x}{4x^2} = 3x + 2 + \dfrac{4}{x}$

6. Dividing: $\dfrac{27a^2 b^3 - 15a^3 b^2 + 21a^4 b^4}{-3a^2 b^2} = \dfrac{27a^2 b^3}{-3a^2 b^2} - \dfrac{15a^3 b^2}{-3a^2 b^2} + \dfrac{21a^4 b^4}{-3a^2 b^2} = -9b + 5a - 7a^2 b^2$

7. Dividing: $\dfrac{x^{6n} - x^{5n}}{x^{3n}} = \dfrac{x^{6n}}{x^{3n}} - \dfrac{x^{5n}}{x^{3n}} = x^{3n} - x^{2n}$

8. Dividing by factoring: $\dfrac{x^2 - x - 6}{x-3} = \dfrac{(x-3)(x+2)}{x-3} = x+2$

9. Dividing by factoring: $\dfrac{5x^2 - 14xy - 24y^2}{x-4y} = \dfrac{(5x+6y)(x-4y)}{x-4y} = 5x+6y$

10. Dividing by factoring: $\dfrac{y^4 - 16}{y-2} = \dfrac{\left(y^2 + 4\right)\left(y^2 - 4\right)}{y-2} = \dfrac{\left(y^2 + 4\right)(y+2)(y-2)}{y-2} = \left(y^2 + 4\right)(y+2) = y^3 + 2y^2 + 4y + 8$

11.   Dividing using long division:

$$2x - 7 \overline{\smash{\big)}\, 8x^2 - 26x - 9} \;\;\; \overset{\displaystyle 4x + 1}{}$$

$$\underline{8x^2 - 28x}$$
$$2x - 9$$
$$\underline{2x - 7}$$
$$-2$$

The quotient is $4x + 1 - \dfrac{2}{2x - 7}$.

12.   Dividing using long division:

$$2y - 3 \overline{\smash{\big)}\, 2y^3 - 9y^2 - 17y + 39} \;\;\; \overset{\displaystyle y^2 - 3y - 13}{}$$

$$\underline{2y^3 - 3y^2}$$
$$-6y^2 - 17y$$
$$\underline{-6y^2 + 9y}$$
$$-26y + 39$$
$$\underline{-26y + 39}$$
$$0$$

The quotient is $y^2 - 3y - 13$.

13.   Performing the operations: $\dfrac{3}{4} \cdot \dfrac{12}{15} \div \dfrac{1}{3} = \dfrac{3}{4} \cdot \dfrac{12}{15} \cdot \dfrac{3}{1} = \dfrac{9}{5}$

14.   Performing the operations: $\dfrac{15x^2 y}{8xy^2} \div \dfrac{10xy}{4x} = \dfrac{15x^2 y}{8xy^2} \cdot \dfrac{4x}{10xy} = \dfrac{60x^3 y}{80x^2 y^3} = \dfrac{3x}{4y^2}$

15.   Performing the operations: $\dfrac{x^3 - 1}{x^4 - 1} \cdot \dfrac{x^2 - 1}{x^2 + x + 1} = \dfrac{(x-1)(x^2 + x + 1)}{(x^2 + 1)(x^2 - 1)} \cdot \dfrac{x^2 - 1}{x^2 + x + 1} = \dfrac{x - 1}{x^2 + 1}$

16.   Performing the operations:

$$\dfrac{a^2 + 5a + 6}{a + 1} \cdot \dfrac{a + 5}{a^2 + 2a - 3} \div \dfrac{a^2 + 7a + 10}{a^2 - 1} = \dfrac{a^2 + 5a + 6}{a + 1} \cdot \dfrac{a + 5}{a^2 + 2a - 3} \cdot \dfrac{a^2 - 1}{a^2 + 7a + 10}$$

$$= \dfrac{(a + 3)(a + 2)}{a + 1} \cdot \dfrac{a + 5}{(a + 3)(a - 1)} \cdot \dfrac{(a + 1)(a - 1)}{(a + 5)(a + 2)}$$

$$= 1$$

17.   Performing the operations:

$$\dfrac{ax + bx + 2a + 2b}{ax - 3a + bx - 3b} \div \dfrac{ax - bx - 2a + 2b}{ax - bx - 3a + 3b} = \dfrac{ax + bx + 2a + 2b}{ax - 3a + bx - 3b} \cdot \dfrac{ax - bx - 3a + 3b}{ax - bx - 2a + 2b}$$

$$= \dfrac{x(a + b) + 2(a + b)}{a(x - 3) + b(x - 3)} \cdot \dfrac{x(a - b) - 3(a - b)}{x(a - b) - 2(a - b)}$$

$$= \dfrac{(a + b)(x + 2)}{(x - 3)(a + b)} \cdot \dfrac{(a - b)(x - 3)}{(a - b)(x - 2)}$$

$$= \dfrac{x + 2}{x - 2}$$

18.   Performing the operations: $\left(4x^2 - 9\right) \cdot \dfrac{x + 3}{2x + 3} = (2x + 3)(2x - 3) \cdot \dfrac{x + 3}{2x + 3} = (2x - 3)(x + 3)$

19.   Performing the operations: $\dfrac{3}{5} - \dfrac{1}{10} + \dfrac{8}{15} = \dfrac{3}{5} \cdot \dfrac{6}{6} - \dfrac{1}{10} \cdot \dfrac{3}{3} + \dfrac{8}{15} \cdot \dfrac{2}{2} = \dfrac{18}{30} - \dfrac{3}{30} + \dfrac{16}{30} = \dfrac{31}{30}$

20.   Performing the operations: $\dfrac{5}{x - 5} - \dfrac{x}{x - 5} = \dfrac{5 - x}{x - 5} = \dfrac{-(x - 5)}{x - 5} = -1$

21.   Performing the operations: $\dfrac{1}{x} + \dfrac{1}{x^2} + \dfrac{1}{x^3} = \dfrac{1}{x} \cdot \dfrac{x^2}{x^2} + \dfrac{1}{x^2} \cdot \dfrac{x}{x} + \dfrac{1}{x^3} = \dfrac{x^2 + x + 1}{x^3}$

22.   Performing the operations:

$$\frac{8}{y^2-16}-\frac{7}{y^2-y-12}=\frac{8}{(y+4)(y-4)}\cdot\frac{y+3}{y+3}-\frac{7}{(y-4)(y+3)}\cdot\frac{y+4}{y+4}$$

$$=\frac{8y+24}{(y+4)(y-4)(y+3)}-\frac{7y+28}{(y+4)(y-4)(y+3)}$$

$$=\frac{8y+24-7y-28}{(y+4)(y-4)(y+3)}$$

$$=\frac{y-4}{(y+4)(y-4)(y+3)}$$

$$=\frac{1}{(y+4)(y+3)}$$

23.   Performing the operations:

$$\frac{x-2}{x^2+5x+4}-\frac{x-4}{2x^2+12x+16}=\frac{x-2}{(x+4)(x+1)}\cdot\frac{2(x+2)}{2(x+2)}-\frac{x-4}{2(x+4)(x+2)}\cdot\frac{x+1}{x+1}$$

$$=\frac{2x^2-8}{2(x+4)(x+1)(x+2)}-\frac{x^2-3x-4}{2(x+4)(x+1)(x+2)}$$

$$=\frac{2x^2-8-x^2+3x+4}{2(x+4)(x+1)(x+2)}$$

$$=\frac{x^2+3x-4}{2(x+4)(x+1)(x+2)}$$

$$=\frac{(x+4)(x-1)}{2(x+4)(x+1)(x+2)}$$

$$=\frac{x-1}{2(x+1)(x+2)}$$

24.   Performing the operations: $3+\dfrac{4}{5x-2}=\dfrac{3(5x-2)+4}{5x-2}=\dfrac{15x-6+4}{5x-2}=\dfrac{15x-2}{5x-2}$

25.   Simplifying the complex fraction: $\dfrac{1+\frac{2}{3}}{1-\frac{2}{3}}=\dfrac{\left(1+\frac{2}{3}\right)\cdot3}{\left(1-\frac{2}{3}\right)\cdot3}=\dfrac{3+2}{3-2}=5$

26.   Simplifying the complex fraction:

$$\frac{\dfrac{4a}{2a^3+2}}{\dfrac{8a}{4a+4}}=\frac{\dfrac{4a}{2(a+1)(a^2-a+1)}\cdot4(a+1)(a^2-a+1)}{\dfrac{8a}{4(a+1)}\cdot4(a+1)(a^2-a+1)}=\frac{8a}{8a(a^2-a+1)}=\frac{1}{a^2-a+1}$$

27.   Simplifying the complex fraction: $1+\dfrac{1}{x+\frac{1}{x}}=1+\dfrac{1\cdot x}{\left(x+\frac{1}{x}\right)\cdot x}=1+\dfrac{x}{x^2+1}=\dfrac{x^2+x+1}{x^2+1}$

28.   Simplifying the complex fraction: $\dfrac{1-\frac{9}{x^2}}{1-\frac{1}{x}-\frac{6}{x^2}}=\dfrac{\left(1-\frac{9}{x^2}\right)\cdot x^2}{\left(1-\frac{1}{x}-\frac{6}{x^2}\right)\cdot x^2}=\dfrac{x^2-9}{x^2-x-6}=\dfrac{(x+3)(x-3)}{(x+2)(x-3)}=\dfrac{x+3}{x+2}$

29. Solving the equation:

$$\frac{3}{x-1} = \frac{3}{5}$$
$$5(x-1)\left(\frac{3}{x-1}\right) = 5(x-1)\left(\frac{3}{5}\right)$$
$$15 = 3x - 3$$
$$18 = 3x$$
$$x = 6$$

30. Solving the equation:

$$\frac{x+1}{3} + \frac{x-3}{4} = \frac{1}{6}$$
$$12\left(\frac{x+1}{3} + \frac{x-3}{4}\right) = 12\left(\frac{1}{6}\right)$$
$$4(x+1) + 3(x-3) = 2$$
$$4x + 4 + 3x - 9 = 2$$
$$7x - 5 = 2$$
$$7x = 7$$
$$x = 1$$

31. Solving the equation:

$$\frac{5}{y+1} = \frac{4}{y+2}$$
$$5(y+2) = 4(y+1)$$
$$5y + 10 = 4y + 4$$
$$y = -6$$

32. Solving the equation:

$$\frac{x+6}{x+3} - 2 = \frac{3}{x+3}$$
$$(x+3)\left(\frac{x+6}{x+3} - 2\right) = (x+3)\left(\frac{3}{x+3}\right)$$
$$x + 6 - 2(x+3) = 3$$
$$x + 6 - 2x - 6 = 3$$
$$-x = 3$$
$$x = -3 \quad (x = -3 \text{ does not check})$$

There is no solution (–3 does not check).

33. Solving the equation:

$$\frac{4}{x^2 - x - 12} + \frac{1}{x^2 - 9} = \frac{2}{x^2 - 7x + 12}$$
$$(x-4)(x+3)(x-3)\left(\frac{4}{(x-4)(x+3)} + \frac{1}{(x+3)(x-3)}\right) = (x-4)(x+3)(x-3)\left(\frac{2}{(x-4)(x-3)}\right)$$
$$4(x-3) + x - 4 = 2(x+3)$$
$$4x - 12 + x - 4 = 2x + 6$$
$$5x - 16 = 2x + 6$$
$$3x = 22$$
$$x = \frac{22}{3}$$

34. Solving the equation:

$$\frac{a+4}{a^2 + 5a} = \frac{-2}{a^2 - 25}$$
$$a(a+5)(a-5) \cdot \frac{a+4}{a(a+5)} = a(a+5)(a-5) \cdot \frac{-2}{(a+5)(a-5)}$$
$$(a-5)(a+4) = -2a$$
$$a^2 - a - 20 = -2a$$
$$a^2 + a - 20 = 0$$
$$(a+5)(a-4) = 0$$
$$a = 4 \quad (a = -5 \text{ does not check})$$

The solution is 4 (–5 does not check).

35.   Let x represent the rate of the truck and $x + 10$ represent the rate of the car. The equation is:

$$\frac{120}{x} - \frac{120}{x+10} = 2$$

$$x(x+10)\left(\frac{120}{x} - \frac{120}{x+10}\right) = 2x(x+10)$$

$$120(x+10) - 120x = 2x^2 + 20x$$

$$1200 = 2x^2 + 20x$$

$$0 = 2\left(x^2 + 10x - 600\right)$$

$$0 = 2(x+30)(x-20)$$

$$x = 20 \quad (x = -30 \text{ is impossible})$$

The car's rate is 30 mph and the truck's rate is 20 mph.

36.   Converting the speed: $\dfrac{3.5 \text{ miles}}{28 \text{ minutes}} \cdot \dfrac{60 \text{ minutes}}{1 \text{ hour}} = 7.5$ mph

37.   Converting the speed: $\dfrac{1088 \text{ feet}}{1 \text{ second}} \cdot \dfrac{1 \text{ mile}}{5280 \text{ feet}} \cdot \dfrac{60 \text{ seconds}}{1 \text{ minute}} \cdot \dfrac{60 \text{ minutes}}{1 \text{ hour}} \approx 742$ mph

Chapter 7
Rational Exponents and Roots

7.1 Rational Exponents

1. Finding the root: $\sqrt{144} = 12$

3. Finding the root: $\sqrt{-144}$ is not a real number

5. Finding the root: $-\sqrt{49} = -7$

7. Finding the root: $\sqrt[3]{-27} = -3$

9. Finding the root: $\sqrt[4]{16} = 2$

11. Finding the root: $\sqrt[4]{-16}$ is not a real number

13. Finding the root: $\sqrt{0.04} = 0.2$

15. Finding the root: $\sqrt[3]{0.008} = 0.2$

17. Finding the root: $\sqrt[3]{125} = 5$

19. Finding the root: $-\sqrt[3]{216} = -6$

21. Finding the root: $\sqrt{\frac{1}{36}} = \frac{1}{6}$

23. Finding the root: $\sqrt[3]{\frac{8}{125}} = \frac{2}{5}$

25. Simplifying: $\sqrt{36a^8} = 6a^4$

27. Simplifying: $\sqrt[3]{27a^{12}} = 3a^4$

29. Simplifying: $\sqrt[5]{32x^{10}y^5} = 2x^2y$

31. Simplifying: $\sqrt[4]{16a^{12}b^{20}} = 2a^3b^5$

33. Writing as a root and simplifying: $36^{1/2} = \sqrt{36} = 6$

35. Writing as a root and simplifying: $-9^{1/2} = -\sqrt{9} = -3$

37. Writing as a root and simplifying: $8^{1/3} = \sqrt[3]{8} = 2$

39. Writing as a root and simplifying: $(-8)^{1/3} = \sqrt[3]{-8} = -2$

41. Writing as a root and simplifying: $32^{1/5} = \sqrt[5]{32} = 2$

43. Writing as a root and simplifying: $\left(\frac{81}{25}\right)^{1/2} = \sqrt{\frac{81}{25}} = \frac{9}{5}$

45. Simplifying: $27^{2/3} = \left(27^{1/3}\right)^2 = 3^2 = 9$

47. Simplifying: $25^{3/2} = \left(25^{1/2}\right)^3 = 5^3 = 125$

49. Simplifying: $27^{-1/3} = \left(27^{1/3}\right)^{-1} = 3^{-1} = \frac{1}{3}$

51. Simplifying: $81^{-3/4} = \left(81^{1/4}\right)^{-3} = 3^{-3} = \frac{1}{3^3} = \frac{1}{27}$

53. Simplifying: $\left(\frac{25}{36}\right)^{-1/2} = \left(\frac{36}{25}\right)^{1/2} = \frac{6}{5}$

55. Simplifying: $\left(\frac{81}{16}\right)^{-3/4} = \left(\frac{16}{81}\right)^{3/4} = \left[\left(\frac{16}{81}\right)^{1/4}\right]^3 = \left(\frac{2}{3}\right)^3 = \frac{8}{27}$

57. Simplifying: $16^{1/2} + 27^{1/3} = 4 + 3 = 7$

59. Simplifying: $8^{-2/3} + 4^{-1/2} = \left(8^{1/3}\right)^{-2} + \left(4^{1/2}\right)^{-1} = 2^{-2} + 2^{-1} = \frac{1}{4} + \frac{1}{2} = \frac{3}{4}$

61. Using properties of exponents: $x^{3/5} \cdot x^{1/5} = x^{3/5+1/5} = x^{4/5}$

63. Using properties of exponents: $\left(a^{3/4}\right)^{4/3} = a^{3/4 \cdot 4/3} = a$

65. Using properties of exponents: $\dfrac{x^{1/5}}{x^{3/5}} = x^{1/5-3/5} = x^{-2/5} = \dfrac{1}{x^{2/5}}$

67. Using properties of exponents: $\dfrac{x^{5/6}}{x^{2/3}} = x^{5/6-2/3} = x^{5/6-4/6} = x^{1/6}$

69. Using properties of exponents: $\left(x^{3/5} y^{5/6} z^{1/3}\right)^{3/5} = x^{3/5 \cdot 3/5} y^{5/6 \cdot 3/5} z^{1/3 \cdot 3/5} = x^{9/25} y^{1/2} z^{1/5}$

71. Using properties of exponents: $\dfrac{a^{3/4} b^2}{a^{7/8} b^{1/4}} = a^{3/4-7/8} b^{2-1/4} = a^{6/8-7/8} b^{8/4-1/4} = a^{-1/8} b^{7/4} = \dfrac{b^{7/4}}{a^{1/8}}$

73. Using properties of exponents: $\dfrac{\left(y^{2/3}\right)^{3/4}}{\left(y^{1/3}\right)^{3/5}} = \dfrac{y^{1/2}}{y^{1/5}} = y^{1/2-1/5} = y^{5/10-2/10} = y^{3/10}$

75. Using properties of exponents: $\left(\dfrac{a^{-1/4}}{b^{1/2}}\right)^{8} = \dfrac{a^{-1/4 \cdot 8}}{b^{1/2 \cdot 8}} = \dfrac{a^{-2}}{b^4} = \dfrac{1}{a^2 b^4}$

77. Using properties of exponents: $\dfrac{\left(r^{-2} s^{1/3}\right)^6}{r^8 s^{3/2}} = \dfrac{r^{-12} s^2}{r^8 s^{3/2}} = r^{-12-8} s^{2-3/2} = r^{-20} s^{1/2} = \dfrac{s^{1/2}}{r^{20}}$

79. Using properties of exponents: $\dfrac{\left(25 a^6 b^4\right)^{1/2}}{\left(8a^{-9} b^3\right)^{-1/3}} = \dfrac{25^{1/2} a^3 b^2}{8^{-1/3} a^3 b^{-1}} = \dfrac{5}{1/2} a^{3-3} b^{2+1} = 10 b^3$

81. Substituting $r = 250$: $v = \left(\dfrac{5 \cdot 250}{2}\right)^{1/2} = 625^{1/2} = 25$. The maximum speed is 25 mph.

83. **a.**   The length of the side is $126 + 86 + 86 + 126 = 424$ pm

   b.   Let d represent the diagonal. Using the Pythagorean theorem:
   $$d^2 = 424^2 + 424^2 = 359{,}552$$
   $$d = \sqrt{359{,}552} \approx 600 \text{ pm}$$

   c.   Converting to meters: $600 \text{ pm} \cdot \dfrac{1 \text{ m}}{10^{12} \text{ pm}} = 6 \times 10^{-10} \text{ m}$

85. **a.**   This graph is B.   **b.**   This graph is A.
   c.   This graph is C.   **d.**   The points of intersection are (0,0) and (1,1).

87. Multiplying: $x^2 \left(x^4 - x\right) = x^2 \cdot x^4 - x^2 \cdot x = x^6 - x^3$

89. Multiplying: $(x-3)(x+5) = x^2 + 5x - 3x - 15 = x^2 + 2x - 15$

91. Multiplying: $\left(x^2 - 5\right)^2 = \left(x^2\right)^2 - 2\left(x^2\right)(5) + 5^2 = x^4 - 10x^2 + 25$

93. Multiplying: $(x-3)\left(x^2 + 3x + 9\right) = x^3 + 3x^2 + 9x - 3x^2 - 9x - 27 = x^3 - 27$

95. Simplifying: $x^2 \left(x^4 - x^3\right) = x^2 \cdot x^4 - x^2 \cdot x^3 = x^6 - x^5$

97. Simplifying: $(3a - 2b)(4a - b) = 12a^2 - 3ab - 8ab + 2b^2 = 12a^2 - 11ab + 2b^2$

99. Simplifying: $\left(x^3 - 2\right)\left(x^3 + 2\right) = \left(x^3\right)^2 - (2)^2 = x^6 - 4$

101. Simplifying: $\dfrac{15x^2 y - 20x^4 y^2}{5xy} = \dfrac{15x^2 y}{5xy} - \dfrac{20x^4 y^2}{5xy} = 3x - 4x^3 y$

103. Factoring: $x^2 - 3x - 10 = (x-5)(x+2)$ **105.** Factoring: $6x^2 + 11x - 10 = (3x-2)(2x+5)$

107. Simplifying: $x^{2/3} \cdot x^{4/3} = x^{2/3+4/3} = x^2$

109. Simplifying: $\left(t^{1/2}\right)^2 = t^{1/2 \cdot 2} = t^1 = t$

111. Simplifying: $\dfrac{x^{2/3}}{x^{1/3}} = x^{2/3-1/3} = x^{1/3}$

113. Simplifying each expression:

$$\left(9^{1/2} + 4^{1/2}\right)^2 = (3+2)^2 = 5^2 = 25$$
$$9 + 4 = 13$$

Note that the values are not equal.

115. Rewriting with exponents: $\sqrt{\sqrt{a}} = \sqrt{a^{1/2}} = \left(a^{1/2}\right)^{1/2} = a^{1/4} = \sqrt[4]{a}$

7.2 More Expressions Involving Rational Exponents

1. Multiplying: $x^{2/3}\left(x^{1/3} + x^{4/3}\right) = x^{2/3} \cdot x^{1/3} + x^{2/3} \cdot x^{4/3} = x + x^2$

3. Multiplying: $a^{1/2}\left(a^{3/2} - a^{1/2}\right) = a^{1/2} \cdot a^{3/2} - a^{1/2} \cdot a^{1/2} = a^2 - a$

5. Multiplying: $2x^{1/3}\left(3x^{8/3} - 4x^{5/3} + 5x^{2/3}\right) = 2x^{1/3} \cdot 3x^{8/3} - 2x^{1/3} \cdot 4x^{5/3} + 2x^{1/3} \cdot 5x^{2/3} = 6x^3 - 8x^2 + 10x$

7. Multiplying:

$$4x^{1/2}y^{3/5}\left(3x^{3/2}y^{-3/5} - 9x^{-1/2}y^{7/5}\right) = 4x^{1/2}y^{3/5} \cdot 3x^{3/2}y^{-3/5} - 4x^{1/2}y^{3/5} \cdot 9x^{-1/2}y^{7/5} = 12x^2 - 36y^2$$

9. Multiplying: $\left(x^{2/3} - 4\right)\left(x^{2/3} + 2\right) = x^{2/3} \cdot x^{2/3} + 2x^{2/3} - 4x^{2/3} - 8 = x^{4/3} - 2x^{2/3} - 8$

11. Multiplying: $\left(a^{1/2} - 3\right)\left(a^{1/2} - 7\right) = a^{1/2} \cdot a^{1/2} - 7a^{1/2} - 3a^{1/2} + 21 = a - 10a^{1/2} + 21$

13. Multiplying: $\left(4y^{1/3} - 3\right)\left(5y^{1/3} + 2\right) = 20y^{2/3} + 8y^{1/3} - 15y^{1/3} - 6 = 20y^{2/3} - 7y^{1/3} - 6$

15. Multiplying: $\left(5x^{2/3} + 3y^{1/2}\right)\left(2x^{2/3} + 3y^{1/2}\right) = 10x^{4/3} + 15x^{2/3}y^{1/2} + 6x^{2/3}y^{1/2} + 9y = 10x^{4/3} + 21x^{2/3}y^{1/2} + 9y$

17. Multiplying: $\left(t^{1/2} + 5\right)^2 = \left(t^{1/2} + 5\right)\left(t^{1/2} + 5\right) = t + 5t^{1/2} + 5t^{1/2} + 25 = t + 10t^{1/2} + 25$

19. Multiplying: $\left(x^{3/2} + 4\right)^2 = \left(x^{3/2} + 4\right)\left(x^{3/2} + 4\right) = x^3 + 4x^{3/2} + 4x^{3/2} + 16 = x^3 + 8x^{3/2} + 16$

21. Multiplying: $\left(a^{1/2} - b^{1/2}\right)^2 = \left(a^{1/2} - b^{1/2}\right)\left(a^{1/2} - b^{1/2}\right) = a - a^{1/2}b^{1/2} - a^{1/2}b^{1/2} + b = a - 2a^{1/2}b^{1/2} + b$

23. Multiplying:

$$\left(2x^{1/2} - 3y^{1/2}\right)^2 = \left(2x^{1/2} - 3y^{1/2}\right)\left(2x^{1/2} - 3y^{1/2}\right)$$
$$= 4x - 6x^{1/2}y^{1/2} - 6x^{1/2}y^{1/2} + 9y$$
$$= 4x - 12x^{1/2}y^{1/2} + 9y$$

25. Multiplying: $\left(a^{1/2} - 3^{1/2}\right)\left(a^{1/2} + 3^{1/2}\right) = \left(a^{1/2}\right)^2 - \left(3^{1/2}\right)^2 = a - 3$

27. Multiplying: $\left(x^{3/2} + y^{3/2}\right)\left(x^{3/2} - y^{3/2}\right) = \left(x^{3/2}\right)^2 - \left(y^{3/2}\right)^2 = x^3 - y^3$

29. Multiplying: $\left(t^{1/2} - 2^{3/2}\right)\left(t^{1/2} + 2^{3/2}\right) = \left(t^{1/2}\right)^2 - \left(2^{3/2}\right)^2 = t - 2^3 = t - 8$

31. Multiplying: $\left(2x^{3/2} + 3^{1/2}\right)\left(2x^{3/2} - 3^{1/2}\right) = \left(2x^{3/2}\right)^2 - \left(3^{1/2}\right)^2 = 4x^3 - 3$

33. Multiplying: $\left(x^{1/3} + y^{1/3}\right)\left(x^{2/3} - x^{1/3}y^{1/3} + y^{2/3}\right) = \left(x^{1/3}\right)^3 + \left(y^{1/3}\right)^3 = x + y$

35. Multiplying: $\left(a^{1/3} - 2\right)\left(a^{2/3} + 2a^{1/3} + 4\right) = \left(a^{1/3}\right)^3 - (2)^3 = a - 8$

37. Multiplying: $\left(2x^{1/3}+1\right)\left(4x^{2/3}-2x^{1/3}+1\right)=\left(2x^{1/3}\right)^3+(1)^3=8x+1$

39. Multiplying: $\left(t^{1/4}-1\right)\left(t^{1/4}+1\right)\left(t^{1/2}+1\right)=\left(t^{1/2}-1\right)\left(t^{1/2}+1\right)=t-1$

41. Dividing: $\dfrac{18x^{3/4}+27x^{1/4}}{9x^{1/4}}=\dfrac{18x^{3/4}}{9x^{1/4}}+\dfrac{27x^{1/4}}{9x^{1/4}}=2x^{1/2}+3$

43. Dividing: $\dfrac{12x^{2/3}y^{1/3}-16x^{1/3}y^{2/3}}{4x^{1/3}y^{1/3}}=\dfrac{12x^{2/3}y^{1/3}}{4x^{1/3}y^{1/3}}-\dfrac{16x^{1/3}y^{2/3}}{4x^{1/3}y^{1/3}}=3x^{1/3}-4y^{1/3}$

45. Dividing: $\dfrac{21a^{7/5}b^{3/5}-14a^{2/5}b^{8/5}}{7a^{2/5}b^{3/5}}=\dfrac{21a^{7/5}b^{3/5}}{7a^{2/5}b^{3/5}}-\dfrac{14a^{2/5}b^{8/5}}{7a^{2/5}b^{3/5}}=3a-2b$

47. Factoring: $12(x-2)^{3/2}-9(x-2)^{1/2}=3(x-2)^{1/2}\left[4(x-2)-3\right]=3(x-2)^{1/2}(4x-8-3)=3(x-2)^{1/2}(4x-11)$

49. Factoring: $5(x-3)^{12/5}-15(x-3)^{7/5}=5(x-3)^{7/5}\left[(x-3)-3\right]=5(x-3)^{7/5}(x-6)$

51. Factoring: $9x(x+1)^{3/2}+6(x+1)^{1/2}=3(x+1)^{1/2}\left[3x(x+1)+2\right]=3(x+1)^{1/2}\left(3x^2+3x+2\right)$

53. Factoring: $x^{2/3}-5x^{1/3}+6=\left(x^{1/3}-2\right)\left(x^{1/3}-3\right)$ 55. Factoring: $a^{2/5}-2a^{1/5}-8=\left(a^{1/5}-4\right)\left(a^{1/5}+2\right)$

57. Factoring: $2y^{2/3}-5y^{1/3}-3=\left(2y^{1/3}+1\right)\left(y^{1/3}-3\right)$ 59. Factoring: $9t^{2/5}-25=\left(3t^{1/5}+5\right)\left(3t^{1/5}-5\right)$

61. Factoring: $4x^{2/7}+20x^{1/7}+25=\left(2x^{1/7}+5\right)^2$ 63. Evaluating: $f(4)=4-2\sqrt{4}-8=4-4-8=-8$

65. Evaluating: $f(25)=2(25)+9\sqrt{25}-5=50+45-5=90$

67. Evaluating: $g\left(\frac{9}{4}\right)=2\left(\frac{9}{4}\right)-\sqrt{\frac{9}{4}}-6=\frac{9}{2}-\frac{3}{2}-6=3-6=-3$

69. Evaluating: $f(-8)=(-8)^{2/3}-2(-8)^{1/3}-8=4+4-8=0$

71. Writing as a single fraction: $\dfrac{3}{x^{1/2}}+x^{1/2}=\dfrac{3}{x^{1/2}}+x^{1/2}\cdot\dfrac{x^{1/2}}{x^{1/2}}=\dfrac{3+x}{x^{1/2}}$

73. Writing as a single fraction: $x^{2/3}+\dfrac{5}{x^{1/3}}=x^{2/3}\cdot\dfrac{x^{1/3}}{x^{1/3}}+\dfrac{5}{x^{1/3}}=\dfrac{x+5}{x^{1/3}}$

75. Writing as a single fraction:

$$\dfrac{3x^2}{\left(x^3+1\right)^{1/2}}+\left(x^3+1\right)^{1/2}=\dfrac{3x^2}{\left(x^3+1\right)^{1/2}}+\left(x^3+1\right)^{1/2}\cdot\dfrac{\left(x^3+1\right)^{1/2}}{\left(x^3+1\right)^{1/2}}=\dfrac{3x^2+x^3+1}{\left(x^3+1\right)^{1/2}}=\dfrac{x^3+3x^2+1}{\left(x^3+1\right)^{1/2}}$$

77. Writing as a single fraction:

$$\dfrac{x^2}{\left(x^2+4\right)^{1/2}}-\left(x^2+4\right)^{1/2}=\dfrac{x^2}{\left(x^2+4\right)^{1/2}}-\left(x^2+4\right)^{1/2}\cdot\dfrac{\left(x^2+4\right)^{1/2}}{\left(x^2+4\right)^{1/2}}$$

$$=\dfrac{x^2-\left(x^2+4\right)}{\left(x^2+4\right)^{1/2}}$$

$$=\dfrac{x^2-x^2-4}{\left(x^2+4\right)^{1/2}}$$

$$=\dfrac{-4}{\left(x^2+4\right)^{1/2}}$$

79. Using the formula $r = \left(\dfrac{A}{P}\right)^{1/t} - 1$ to find the annual rate of return: $r = \left(\dfrac{900}{500}\right)^{1/4} - 1 \approx 0.158$

The annual rate of return is approximately 15.8%.

81. Reducing to lowest terms: $\dfrac{x^2 - 9}{x^4 - 81} = \dfrac{x^2 - 9}{\left(x^2 + 9\right)\left(x^2 - 9\right)} = \dfrac{1}{x^2 + 9}$

83. Dividing: $\dfrac{15x^2 y - 20x^4 y^2}{5xy} = \dfrac{15x^2 y}{5xy} - \dfrac{20x^4 y^2}{5xy} = 3x - 4x^3 y$

85. Dividing using long division:

$$
\begin{array}{r}
5x - 4 \\
2x + 3 \overline{)\,10x^2 + 7x - 12} \\
\underline{10x^2 + 15x} \\
-8x - 12 \\
\underline{-8x - 12} \\
0
\end{array}
$$

87. Dividing using long division:

$$
\begin{array}{r}
x^2 + 5x + 25 \\
x - 5 \overline{)\,x^3 + 0x^2 + 0x - 125} \\
\underline{x^3 - 5x^2} \\
5x^2 + 0x \\
\underline{5x^2 - 25x} \\
25x - 125 \\
\underline{25x - 125} \\
0
\end{array}
$$

89. Simplifying: $\sqrt{25} = 5$

91. Simplifying: $\sqrt{6^2} = 6$

93. Simplifying: $\sqrt{16x^4 y^2} = 4x^2 y$

95. Simplifying: $\sqrt{(5y)^2} = 5y$

97. Simplifying: $\sqrt[3]{27} = 3$

99. Simplifying: $\sqrt[3]{2^3} = 2$

101. Simplifying: $\sqrt[3]{8a^3 b^3} = 2ab$

103. Filling in the blank: $50 = 25 \bullet 2$

105. Filling in the blank: $48x^4 y^3 = 48x^4 y^2 \bullet y$

107. Filling in the blank: $12x^7 y^6 = 4x^6 y^6 \bullet 3x$

7.3 Simplified Form for Radicals

1. Simplifying the radical: $\sqrt{8} = \sqrt{4 \bullet 2} = 2\sqrt{2}$

3. Simplifying the radical: $\sqrt{98} = \sqrt{49 \bullet 2} = 7\sqrt{2}$

5. Simplifying the radical: $\sqrt{288} = \sqrt{144 \bullet 2} = 12\sqrt{2}$

7. Simplifying the radical: $\sqrt{80} = \sqrt{16 \bullet 5} = 4\sqrt{5}$

9. Simplifying the radical: $\sqrt{48} = \sqrt{16 \bullet 3} = 4\sqrt{3}$

11. Simplifying the radical: $\sqrt{675} = \sqrt{225 \bullet 3} = 15\sqrt{3}$

13. Simplifying the radical: $\sqrt[3]{54} = \sqrt[3]{27 \bullet 2} = 3\sqrt[3]{2}$

15. Simplifying the radical: $\sqrt[3]{128} = \sqrt[3]{64 \bullet 2} = 4\sqrt[3]{2}$

17. Simplifying the radical: $\sqrt[3]{432} = \sqrt[3]{216 \bullet 2} = 6\sqrt[3]{2}$

19. Simplifying the radical: $\sqrt[5]{64} = \sqrt[5]{32 \bullet 2} = 2\sqrt[5]{2}$

21. Simplifying the radical: $\sqrt{18x^3} = \sqrt{9x^2 \bullet 2x} = 3x\sqrt{2x}$

23. Simplifying the radical: $\sqrt[4]{32y^7} = \sqrt[4]{16y^4 \bullet 2y^3} = 2y\sqrt[4]{2y^3}$

25. Simplifying the radical: $\sqrt[3]{40x^4 y^7} = \sqrt[3]{8x^3 y^6 \bullet 5xy} = 2xy^2 \sqrt[3]{5xy}$

27. Simplifying the radical: $\sqrt{48a^2 b^3 c^4} = \sqrt{16a^2 b^2 c^4 \bullet 3b} = 4abc^2 \sqrt{3b}$

29. Simplifying the radical: $\sqrt[3]{48a^2 b^3 c^4} = \sqrt[3]{8b^3 c^3 \bullet 6a^2 c} = 2bc\sqrt[3]{6a^2 c}$

31. Simplifying the radical: $\sqrt[5]{64x^8 y^{12}} = \sqrt[5]{32x^5 y^{10} \bullet 2x^3 y^2} = 2xy^2 \sqrt[5]{2x^3 y^2}$

33. Simplifying the radical: $\sqrt[5]{243x^7 y^{10} z^5} = \sqrt[5]{243x^5 y^{10} z^5 \bullet x^2} = 3xy^2 z\sqrt[5]{x^2}$

35. Substituting into the expression: $\sqrt{b^2 - 4ac} = \sqrt{(-6)^2 - 4(2)(3)} = \sqrt{36 - 24} = \sqrt{12} = 2\sqrt{3}$

37. Substituting into the expression: $\sqrt{b^2 - 4ac} = \sqrt{(2)^2 - 4(1)(6)} = \sqrt{4 - 24} = \sqrt{-20}$, which is not a real number

39. Substituting into the expression: $\sqrt{b^2 - 4ac} = \sqrt{\left(-\frac{1}{2}\right)^2 - 4\left(\frac{1}{2}\right)\left(-\frac{5}{4}\right)} = \sqrt{\frac{1}{4} + \frac{5}{2}} = \sqrt{\frac{11}{4}} = \frac{\sqrt{11}}{2}$

41. Rationalizing the denominator: $\frac{2}{\sqrt{3}} = \frac{2}{\sqrt{3}} \cdot \frac{\sqrt{3}}{\sqrt{3}} = \frac{2\sqrt{3}}{3}$ **43.** Rationalizing the denominator: $\frac{5}{\sqrt{6}} = \frac{5}{\sqrt{6}} \cdot \frac{\sqrt{6}}{\sqrt{6}} = \frac{5\sqrt{6}}{6}$

45. Rationalizing the denominator: $\sqrt{\frac{1}{2}} = \frac{1}{\sqrt{2}} \cdot \frac{\sqrt{2}}{\sqrt{2}} = \frac{\sqrt{2}}{2}$ **47.** Rationalizing the denominator: $\sqrt{\frac{1}{5}} = \frac{1}{\sqrt{5}} \cdot \frac{\sqrt{5}}{\sqrt{5}} = \frac{\sqrt{5}}{5}$

49. Rationalizing the denominator: $\frac{4}{\sqrt[3]{2}} = \frac{4}{\sqrt[3]{2}} \cdot \frac{\sqrt[3]{4}}{\sqrt[3]{4}} = \frac{4\sqrt[3]{4}}{2} = 2\sqrt[3]{4}$

51. Rationalizing the denominator: $\frac{2}{\sqrt[3]{9}} = \frac{2}{\sqrt[3]{9}} \cdot \frac{\sqrt[3]{3}}{\sqrt[3]{3}} = \frac{2\sqrt[3]{3}}{3}$

53. Rationalizing the denominator: $\sqrt[4]{\frac{3}{2x^2}} = \frac{\sqrt[4]{3}}{\sqrt[4]{2x^2}} \cdot \frac{\sqrt[4]{8x^2}}{\sqrt[4]{8x^2}} = \frac{\sqrt[4]{24x^2}}{2x}$

55. Rationalizing the denominator: $\sqrt[4]{\frac{8}{y}} = \frac{\sqrt[4]{8}}{\sqrt[4]{y}} \cdot \frac{\sqrt[4]{y^3}}{\sqrt[4]{y^3}} = \frac{\sqrt[4]{8y^3}}{y}$

57. Rationalizing the denominator: $\sqrt[3]{\frac{4x}{3y}} = \frac{\sqrt[3]{4x}}{\sqrt[3]{3y}} \cdot \frac{\sqrt[3]{9y^2}}{\sqrt[3]{9y^2}} = \frac{\sqrt[3]{36xy^2}}{3y}$

59. Rationalizing the denominator: $\sqrt[3]{\frac{2x}{9y}} = \frac{\sqrt[3]{2x}}{\sqrt[3]{9y}} \cdot \frac{\sqrt[3]{3y^2}}{\sqrt[3]{3y^2}} = \frac{\sqrt[3]{6xy^2}}{3y}$

61. Rationalizing the denominator: $\sqrt[4]{\frac{1}{8x^3}} = \frac{1}{\sqrt[4]{8x^3}} \cdot \frac{\sqrt[4]{2x}}{\sqrt[4]{2x}} = \frac{\sqrt[4]{2x}}{2x}$

63. Simplifying: $\sqrt{\frac{27x^3}{5y}} = \frac{\sqrt{27x^3}}{\sqrt{5y}} \cdot \frac{\sqrt{5y}}{\sqrt{5y}} = \frac{\sqrt{135x^3y}}{5y} = \frac{3x\sqrt{15xy}}{5y}$

65. Simplifying: $\sqrt{\frac{75x^3y^2}{2z}} = \frac{\sqrt{75x^3y^2}}{\sqrt{2z}} \cdot \frac{\sqrt{2z}}{\sqrt{2z}} = \frac{\sqrt{150x^3y^2z}}{2z} = \frac{5xy\sqrt{6xz}}{2z}$

67. Simplifying: $\sqrt[3]{\frac{16a^4b^3}{9c}} = \frac{\sqrt[3]{16a^4b^3}}{\sqrt[3]{9c}} \cdot \frac{\sqrt[3]{3c^2}}{\sqrt[3]{3c^2}} = \frac{\sqrt[3]{48a^4b^3c^2}}{3c} = \frac{2ab\sqrt[3]{6ac^2}}{3c}$

69. Simplifying: $\sqrt[3]{\frac{8x^3y^6}{9z}} = \frac{\sqrt[3]{8x^3y^6}}{\sqrt[3]{9z}} \cdot \frac{\sqrt[3]{3z^2}}{\sqrt[3]{3z^2}} = \frac{\sqrt[3]{24x^3y^6z^2}}{3z} = \frac{2xy^2\sqrt[3]{3z^2}}{3z}$

71. Simplifying: $\sqrt{\sqrt{x^2}} = \sqrt{x}$

73. Simplifying: $\sqrt[3]{\sqrt{xy}} = \left((xy)^{1/2}\right)^{1/3} = (xy)^{1/6} = \sqrt[6]{xy}$ **75.** Simplifying: $\sqrt[3]{\sqrt[4]{a}} = \left(a^{1/4}\right)^{1/3} = a^{1/12} = \sqrt[12]{a}$

77. Simplifying: $\sqrt[3]{\sqrt[3]{6x^{10}}} = \left(\left(6x^{10}\right)^{1/3}\right)^{1/3} = \left(6x^{10}\right)^{1/9} = x(6x)^{1/9} = x\sqrt[9]{6x}$

79. Simplifying: $\sqrt[4]{\sqrt[3]{a^{12}b^{24}c^{14}}} = \left(\left(a^{12}b^{24}c^{14}\right)^{1/3}\right)^{1/4} = \left(a^{12}b^{24}c^{14}\right)^{1/12} = ab^2c\left(c^2\right)^{1/12} = ab^2c\left(c^{1/6}\right) = ab^2c\sqrt[6]{c}$

81. Simplifying: $\sqrt[3]{\sqrt[5]{3a^{17}b^{16}c^{30}}} = \left(\left(3a^{17}b^{16}c^{30}\right)^{1/3}\right)^{1/5} = \left(3a^{17}b^{16}c^{30}\right)^{1/15} = abc^2\left(3a^2b\right)^{1/15} = abc^2\sqrt[15]{3a^2b}$

83. Simplifying: $\left(\sqrt{\sqrt[3]{8ab^6}} \right)^2 = \sqrt[3]{8ab^6} = 2b^2 \sqrt[3]{a}$

85. Simplifying: $\sqrt{25x^2} = 5|x|$

87. Simplifying: $\sqrt{27x^3y^2} = \sqrt{9x^2y^2 \cdot 3x} = 3|xy|\sqrt{3x}$

89. Simplifying: $\sqrt{x^2 - 10x + 25} = \sqrt{(x-5)^2} = |x-5|$

91. Simplifying: $\sqrt{4x^2 + 12x + 9} = \sqrt{(2x+3)^2} = |2x+3|$

93. Simplifying: $\sqrt{4a^4 + 16a^3 + 16a^2} = \sqrt{4a^2 (a^2 + 4a + 4)} = \sqrt{4a^2 (a+2)^2} = 2|a(a+2)|$

95. Simplifying: $\sqrt{4x^3 - 8x^2} = \sqrt{4x^2 (x-2)} = 2|x|\sqrt{x-2}$

97. Substituting $a = 9$ and $b = 16$:
$$\sqrt{a+b} = \sqrt{9+16} = \sqrt{25} = 5$$
$$\sqrt{a} + \sqrt{b} = \sqrt{9} + \sqrt{16} = 3 + 4 = 7$$
Thus $\sqrt{a+b} \neq \sqrt{a} + \sqrt{b}$.

99. Substituting $w = 10$ and $l = 15$: $d = \sqrt{l^2 + w^2} = \sqrt{15^2 + 10^2} = \sqrt{225 + 100} = \sqrt{325} = \sqrt{25 \cdot 13} = 5\sqrt{13}$ feet

101. **a.** Substituting $w = 3$, $l = 4$, and $h = 12$:
$$d = \sqrt{l^2 + w^2 + h^2} = \sqrt{4^2 + 3^2 + 12^2} = \sqrt{16 + 9 + 144} = \sqrt{169} = 13 \text{ feet}$$

b. Substituting $w = 2$, $l = 4$, and $h = 6$:
$$d = \sqrt{l^2 + w^2 + h^2} = \sqrt{4^2 + 2^2 + 6^2} = \sqrt{16 + 4 + 36} = \sqrt{56} = 2\sqrt{14} \approx 7.5 \text{ feet}$$

103. Answers will vary.

105. Performing the operations: $\dfrac{8xy^3}{9x^2y} \div \dfrac{16x^2y^2}{18xy^3} = \dfrac{8xy^3}{9x^2y} \cdot \dfrac{18xy^3}{16x^2y^2} = \dfrac{144x^2y^6}{144x^4y^3} = \dfrac{y^3}{x^2}$

107. Performing the operations: $\dfrac{12a^2 - 4a - 5}{2a + 1} \cdot \dfrac{7a + 3}{42a^2 - 17a - 15} = \dfrac{(6a-5)(2a+1)}{2a+1} \cdot \dfrac{7a+3}{(7a+3)(6a-5)} = 1$

109. Performing the operations:
$$\frac{8x^3 + 27}{27x^3 + 1} \div \frac{6x^2 + 7x - 3}{9x^2 - 1} = \frac{8x^3 + 27}{27x^3 + 1} \cdot \frac{9x^2 - 1}{6x^2 + 7x - 3}$$
$$= \frac{(2x+3)(4x^2 - 6x + 9)}{(3x+1)(9x^2 - 3x + 1)} \cdot \frac{(3x+1)(3x-1)}{(2x+3)(3x-1)}$$
$$= \frac{4x^2 - 6x + 9}{9x^2 - 3x + 1}$$

111. Simplifying: $5x - 4x + 6x = 7x$

113. Simplifying: $35xy^2 - 8xy^2 = 27xy^2$

115. Simplifying: $\dfrac{1}{2}x + \dfrac{1}{3}x = \dfrac{3}{6}x + \dfrac{2}{6}x = \dfrac{5}{6}x$

117. Simplifying: $\sqrt{18} = \sqrt{9 \cdot 2} = 3\sqrt{2}$

119. Simplifying: $\sqrt{75xy^3} = \sqrt{25y^2 \cdot 3xy} = 5y\sqrt{3xy}$

121. Simplifying: $\sqrt[3]{8a^4b^2} = \sqrt[3]{8a^3 \cdot ab^2} = 2a\sqrt[3]{ab^2}$

123. The prime factorization of 8,640 is: $8640 = 2^6 \cdot 3^3 \cdot 5$. Therefore: $\sqrt[3]{8640} = \sqrt[3]{2^6 \cdot 3^3 \cdot 5} = 2^2 \cdot 3\sqrt[3]{5} = 12\sqrt[3]{5}$

125. The prime factorization of 10,584 is: $10584 = 2^3 \cdot 3^3 \cdot 7^2$. Therefore: $\sqrt[3]{10584} = \sqrt[3]{2^3 \cdot 3^3 \cdot 7^2} = 2 \cdot 3\sqrt[3]{7^2} = 6\sqrt[3]{49}$

127. Rationalizing the denominator: $\dfrac{1}{\sqrt[10]{a^3}} = \dfrac{1}{\sqrt[10]{a^3}} \cdot \dfrac{\sqrt[10]{a^7}}{\sqrt[10]{a^7}} = \dfrac{\sqrt[10]{a^7}}{a}$

129. Rationalizing the denominator: $\dfrac{1}{\sqrt[20]{a^{11}}} = \dfrac{1}{\sqrt[20]{a^{11}}} \cdot \dfrac{\sqrt[20]{a^9}}{\sqrt[20]{a^9}} = \dfrac{\sqrt[20]{a^9}}{a}$

131.   Graphing each function:

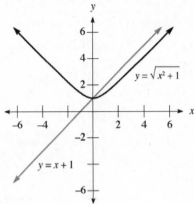

Note that the two graphs are not the same.

133.   When $x = 2$, the distance apart is: $(2+1) - \sqrt{2^2 + 1} = 3 - \sqrt{5} \approx 0.76$

135.   The two expressions are equal when $x = 0$.

7.4 Addition and Subtraction of Radical Expressions

1.   Combining radicals: $3\sqrt{5} + 4\sqrt{5} = 7\sqrt{5}$

3.   Combining radicals: $3x\sqrt{7} - 4x\sqrt{7} = -x\sqrt{7}$

5.   Combining radicals: $5\sqrt[3]{10} - 4\sqrt[3]{10} = \sqrt[3]{10}$

7.   Combining radicals: $8\sqrt[5]{6} - 2\sqrt[5]{6} + 3\sqrt[5]{6} = 9\sqrt[5]{6}$

9.   Combining radicals: $3x\sqrt{2} - 4x\sqrt{2} + x\sqrt{2} = 0$

11.   Combining radicals: $\sqrt{20} - \sqrt{80} + \sqrt{45} = 2\sqrt{5} - 4\sqrt{5} + 3\sqrt{5} = \sqrt{5}$

13.   Combining radicals: $4\sqrt{8} - 2\sqrt{50} - 5\sqrt{72} = 8\sqrt{2} - 10\sqrt{2} - 30\sqrt{2} = -32\sqrt{2}$

15.   Combining radicals: $5x\sqrt{8} + 3\sqrt{32x^2} - 5\sqrt{50x^2} = 10x\sqrt{2} + 12x\sqrt{2} - 25x\sqrt{2} = -3x\sqrt{2}$

17.   Combining radicals: $5\sqrt[3]{16} - 4\sqrt[3]{54} = 10\sqrt[3]{2} - 12\sqrt[3]{2} = -2\sqrt[3]{2}$

19.   Combining radicals: $\sqrt[3]{x^4 y^2} + 7x\sqrt[3]{xy^2} = x\sqrt[3]{xy^2} + 7x\sqrt[3]{xy^2} = 8x\sqrt[3]{xy^2}$

21.   Combining radicals: $5a^2\sqrt{27ab^3} - 6b\sqrt{12a^5 b} = 15a^2 b\sqrt{3ab} - 12a^2 b\sqrt{3ab} = 3a^2 b\sqrt{3ab}$

23.   Combining radicals: $b\sqrt[3]{24a^5 b} + 3a\sqrt[3]{81a^2 b^4} = 2ab\sqrt[3]{3a^2 b} + 9ab\sqrt[3]{3a^2 b} = 11ab\sqrt[3]{3a^2 b}$

25.   Combining radicals: $5x\sqrt[4]{3y^5} + y\sqrt[4]{243x^4 y} + \sqrt[4]{48x^4 y^5} = 5xy\sqrt[4]{3y} + 3xy\sqrt[4]{3y} + 2xy\sqrt[4]{3y} = 10xy\sqrt[4]{3y}$

27.   Combining radicals: $\dfrac{\sqrt{2}}{2} + \dfrac{1}{\sqrt{2}} = \dfrac{\sqrt{2}}{2} + \dfrac{1}{\sqrt{2}} \cdot \dfrac{\sqrt{2}}{\sqrt{2}} = \dfrac{\sqrt{2}}{2} + \dfrac{\sqrt{2}}{2} = \sqrt{2}$

29.   Combining radicals: $\dfrac{\sqrt{5}}{3} + \dfrac{1}{\sqrt{5}} = \dfrac{\sqrt{5}}{3} + \dfrac{1}{\sqrt{5}} \cdot \dfrac{\sqrt{5}}{\sqrt{5}} = \dfrac{\sqrt{5}}{3} + \dfrac{\sqrt{5}}{5} = \dfrac{5\sqrt{5}}{15} + \dfrac{3\sqrt{5}}{15} = \dfrac{8\sqrt{5}}{15}$

31.   Combining radicals: $\sqrt{x} - \dfrac{1}{\sqrt{x}} = \sqrt{x} - \dfrac{1}{\sqrt{x}} \cdot \dfrac{\sqrt{x}}{\sqrt{x}} = \sqrt{x} - \dfrac{\sqrt{x}}{x} = \dfrac{x\sqrt{x}}{x} - \dfrac{\sqrt{x}}{x} = \dfrac{(x-1)\sqrt{x}}{x}$

33.   Combining radicals: $\dfrac{\sqrt{18}}{6} + \sqrt{\dfrac{1}{2}} + \dfrac{\sqrt{2}}{2} = \dfrac{3\sqrt{2}}{6} + \dfrac{1}{\sqrt{2}} \cdot \dfrac{\sqrt{2}}{\sqrt{2}} + \dfrac{\sqrt{2}}{2} = \dfrac{\sqrt{2}}{2} + \dfrac{\sqrt{2}}{2} + \dfrac{\sqrt{2}}{2} = \dfrac{3\sqrt{2}}{2}$

35.   Combining radicals: $\sqrt{6} - \sqrt{\dfrac{2}{3}} + \sqrt{\dfrac{1}{6}} = \sqrt{6} - \dfrac{\sqrt{2}}{\sqrt{3}} \cdot \dfrac{\sqrt{3}}{\sqrt{3}} + \dfrac{1}{\sqrt{6}} \cdot \dfrac{\sqrt{6}}{\sqrt{6}} = \sqrt{6} - \dfrac{\sqrt{6}}{3} + \dfrac{\sqrt{6}}{6} = \dfrac{6\sqrt{6}}{6} - \dfrac{2\sqrt{6}}{6} + \dfrac{\sqrt{6}}{6} = \dfrac{5\sqrt{6}}{6}$

37.   Combining radicals: $\sqrt[3]{25} + \dfrac{3}{\sqrt[3]{5}} = \sqrt[3]{25} + \dfrac{3}{\sqrt[3]{5}} \cdot \dfrac{\sqrt[3]{25}}{\sqrt[3]{25}} = \sqrt[3]{25} + \dfrac{3\sqrt[3]{25}}{5} = \dfrac{5\sqrt[3]{25}}{5} + \dfrac{3\sqrt[3]{25}}{5} = \dfrac{8\sqrt[3]{25}}{5}$

39. **a.** Finding the function: $f(x)+g(x)=\sqrt{8x}+\sqrt{72x}=2\sqrt{2x}+6\sqrt{2x}=8\sqrt{2x}$

 b. Finding the function: $f(x)-g(x)=\sqrt{8x}-\sqrt{72x}=2\sqrt{2x}-6\sqrt{2x}=-4\sqrt{2x}$

41. **a.** Finding the function: $f(x)+g(x)=3\sqrt{2x}+\sqrt{2x}=4\sqrt{2x}$

 b. Finding the function: $f(x)-g(x)=3\sqrt{2x}-\sqrt{2x}=2\sqrt{2x}$

43. **a.** Finding the function: $f(x)+g(x)=x\sqrt{2}+2x\sqrt{2}=3x\sqrt{2}$

 b. Finding the function: $f(x)-g(x)=x\sqrt{2}-2x\sqrt{2}=-x\sqrt{2}$

45. **a.** Finding the function: $f(x)+g(x)=\left(\sqrt{2x}-2\right)+\left(2\sqrt{2x}+5\right)=3\sqrt{2x}+3$

 b. Finding the function: $f(x)-g(x)=\left(\sqrt{2x}-2\right)-\left(2\sqrt{2x}+5\right)=-\sqrt{2x}-7$

47. Using a calculator:
$$\sqrt{12}\approx 3.464 \qquad 2\sqrt{3}\approx 3.464$$

49. It is equal to the decimal approximation for $\sqrt{50}$:
$$\sqrt{8}+\sqrt{18}\approx 7.071\approx \sqrt{50} \qquad \sqrt{26}\approx 5.099$$

51. Answers will vary.

53. Answers will vary.

55. First use the Pythagorean theorem to find the height h:
$$x^2+h^2=(2x)^2$$
$$x^2+h^2=4x^2$$
$$h^2=3x^2$$
$$h=x\sqrt{3}$$

Therefore the ratio is: $\dfrac{x\sqrt{3}}{2x}=\dfrac{\sqrt{3}}{2}$

57. Combining the fractions: $\dfrac{2a-4}{a+2}-\dfrac{a-6}{a+2}=\dfrac{2a-4-a+6}{a+2}=\dfrac{a+2}{a+2}=1$

59. Combining the fractions: $3+\dfrac{4}{3-t}=3\bullet\dfrac{3-t}{3-t}+\dfrac{4}{3-t}=\dfrac{9-3t+4}{3-t}=\dfrac{13-3t}{3-t}$

61. Combining the fractions:
$$\frac{3}{2x-5}-\frac{39}{8x^2-14x-15}=\frac{3}{2x-5}\bullet\frac{4x+3}{4x+3}-\frac{39}{(2x-5)(4x+3)}$$
$$=\frac{12x+9-39}{(2x-5)(4x+3)}$$
$$=\frac{12x-30}{(2x-5)(4x+3)}$$
$$=\frac{6(2x-5)}{(2x-5)(4x+3)}$$
$$=\frac{6}{4x+3}$$

63. Combining the fractions:

$$\frac{1}{x-y} - \frac{3xy}{x^3-y^3} = \frac{1}{x-y} \cdot \frac{x^2+xy+y^2}{x^2+xy+y^2} - \frac{3xy}{(x-y)(x^2+xy+y^2)}$$

$$= \frac{x^2+xy+y^2-3xy}{(x-y)(x^2+xy+y^2)}$$

$$= \frac{x^2-2xy+y^2}{(x-y)(x^2+xy+y^2)}$$

$$= \frac{(x-y)^2}{(x-y)(x^2+xy+y^2)}$$

$$= \frac{x-y}{x^2+xy+y^2}$$

65. Simplifying: $3 \cdot 2 = 6$

67. Simplifying: $(x+y)(4x-y) = 4x^2 - xy + 4xy - y^2 = 4x^2 + 3xy - y^2$

69. Simplifying: $(x+3)^2 = x^2 + 2(3x) + 3^2 = x^2 + 6x + 9$

71. Simplifying: $(x-2)(x+2) = x^2 - 2^2 = x^2 - 4$

73. Simplifying: $2\sqrt{18} = 2\sqrt{9 \cdot 2} = 2 \cdot 3\sqrt{2} = 6\sqrt{2}$ **75.** Simplifying: $\left(\sqrt{6}\right)^2 = 6$

77. Simplifying: $\left(3\sqrt{x}\right)^2 = 9x$ **79.** Rationalizing the denominator: $\dfrac{\sqrt{3}}{\sqrt{2}} = \dfrac{\sqrt{3}}{\sqrt{2}} \cdot \dfrac{\sqrt{2}}{\sqrt{2}} = \dfrac{\sqrt{6}}{2}$

81. Simplifying: $\sqrt[5]{32x^5y^5} - y\sqrt[3]{27x^3} = 2xy - 3xy = -xy$

83. Simplifying: $3\sqrt[9]{x^9y^{18}z^{27}} - 4\sqrt[6]{x^6y^{12}z^{18}} = 3xy^2z^3 - 4xy^2z^3 = -xy^2z^3$

85. Simplifying:

$$3c\sqrt[8]{4a^6b^{18}} + b\sqrt[4]{32a^3b^5c^4} = 3c\left(2^2\,a^6b^{18}\right)^{1/8} + b \cdot 2bc\sqrt[4]{2a^3b}$$

$$= 3c\left(2a^3b^9\right)^{1/4} + 2b^2c\sqrt[4]{2b}$$

$$= 3c\left(2a^3b^9\right)^{1/4} + 2b^2c\sqrt[4]{2a^3b}$$

$$= 3b^2c\sqrt[4]{2a^3b} + 2b^2c\sqrt[4]{2a^3b}$$

$$= 5b^2c\sqrt[4]{2a^3b}$$

87. Simplifying:

$$3\sqrt[9]{8a^{12}b^9} + b\sqrt[3]{16a^4} - 8\sqrt[6]{4a^8b^6} = 3\left(2^3\,a^{12}b^9\right)^{1/9} + b \cdot 2a\sqrt[3]{2a} - 8\left(2^2\,a^8b^6\right)^{1/6}$$

$$= 3b\left(2a^4\right)^{1/3} + 2ab\sqrt[3]{2a} - 8b\left(2a^4\right)^{1/3}$$

$$= 3ab\sqrt[3]{2a} + 2ab\sqrt[3]{2a} - 8ab\sqrt[3]{2a}$$

$$= -3ab\sqrt[3]{2a}$$

7.5 Multiplication and Division of Radical Expressions

1. Multiplying: $\sqrt{6}\sqrt{3} = \sqrt{18} = 3\sqrt{2}$

3. Multiplying: $\left(2\sqrt{3}\right)\left(5\sqrt{7}\right) = 10\sqrt{21}$

5. Multiplying: $\left(4\sqrt{6}\right)\left(2\sqrt{15}\right)\left(3\sqrt{10}\right) = 24\sqrt{900} = 24 \cdot 30 = 720$

7. Multiplying: $\left(3\sqrt[3]{3}\right)\left(6\sqrt[3]{9}\right) = 18\sqrt[3]{27} = 18 \cdot 3 = 54$

9. Multiplying: $\sqrt{3}\left(\sqrt{2} - 3\sqrt{3}\right) = \sqrt{6} - 3\sqrt{9} = \sqrt{6} - 9$

11. Multiplying: $6\sqrt[3]{4}\left(2\sqrt[3]{2} + 1\right) = 12\sqrt[3]{8} + 6\sqrt[3]{4} = 24 + 6\sqrt[3]{4}$

13. Multiplying: $\sqrt[3]{4}\left(\sqrt[3]{2} + \sqrt[3]{6}\right) = \sqrt[3]{8} + \sqrt[3]{24} = 2 + 2\sqrt[3]{2}$

15. Multiplying: $\sqrt[3]{x}\left(\sqrt[3]{x^2 y^4} + \sqrt[3]{x^5 y}\right) = \sqrt[3]{x^3 y^4} + \sqrt[3]{x^6 y} = xy\sqrt[3]{y} + x^2\sqrt[3]{y}$

17. Multiplying: $\sqrt[4]{2x^3}\left(\sqrt[4]{8x^6} + \sqrt[4]{16x^9}\right) = \sqrt[4]{16x^9} + \sqrt[4]{32x^{12}} = 2x^2\sqrt[4]{x} + 2x^3\sqrt[4]{2}$

19. Multiplying: $\left(\sqrt{3} + \sqrt{2}\right)\left(3\sqrt{3} - \sqrt{2}\right) = 3\sqrt{9} - \sqrt{6} + 3\sqrt{6} - \sqrt{4} = 9 + 2\sqrt{6} - 2 = 7 + 2\sqrt{6}$

21. Multiplying: $\left(\sqrt{x} + 5\right)\left(\sqrt{x} - 3\right) = x - 3\sqrt{x} + 5\sqrt{x} - 15 = x + 2\sqrt{x} - 15$

23. Multiplying: $\left(3\sqrt{6} + 4\sqrt{2}\right)\left(\sqrt{6} + 2\sqrt{2}\right) = 3\sqrt{36} + 4\sqrt{12} + 6\sqrt{12} + 8\sqrt{4} = 18 + 8\sqrt{3} + 12\sqrt{3} + 16 = 34 + 20\sqrt{3}$

25. Multiplying: $\left(\sqrt{3} + 4\right)^2 = \left(\sqrt{3} + 4\right)\left(\sqrt{3} + 4\right) = \sqrt{9} + 4\sqrt{3} + 4\sqrt{3} + 16 = 19 + 8\sqrt{3}$

27. Multiplying: $\left(\sqrt{x} - 3\right)^2 = \left(\sqrt{x} - 3\right)\left(\sqrt{x} - 3\right) = x - 3\sqrt{x} - 3\sqrt{x} + 9 = x - 6\sqrt{x} + 9$

29. Multiplying: $\left(2\sqrt{a} - 3\sqrt{b}\right)^2 = \left(2\sqrt{a} - 3\sqrt{b}\right)\left(2\sqrt{a} - 3\sqrt{b}\right) = 4a - 6\sqrt{ab} - 6\sqrt{ab} + 9b = 4a - 12\sqrt{ab} + 9b$

31. Multiplying: $\left(\sqrt{x-4} + 2\right)^2 = \left(\sqrt{x-4} + 2\right)\left(\sqrt{x-4} + 2\right) = x - 4 + 2\sqrt{x-4} + 2\sqrt{x-4} + 4 = x + 4\sqrt{x-4}$

33. Multiplying: $\left(\sqrt{x-5} - 3\right)^2 = \left(\sqrt{x-5} - 3\right)\left(\sqrt{x-5} - 3\right) = x - 5 - 3\sqrt{x-5} - 3\sqrt{x-5} + 9 = x + 4 - 6\sqrt{x-5}$

35. Multiplying: $\left(\sqrt{3} - \sqrt{2}\right)\left(\sqrt{3} + \sqrt{2}\right) = \left(\sqrt{3}\right)^2 - \left(\sqrt{2}\right)^2 = 3 - 2 = 1$

37. Multiplying: $\left(\sqrt{a} + 7\right)\left(\sqrt{a} - 7\right) = \left(\sqrt{a}\right)^2 - \left(7\right)^2 = a - 49$

39. Multiplying: $\left(5 - \sqrt{x}\right)\left(5 + \sqrt{x}\right) = \left(5\right)^2 - \left(\sqrt{x}\right)^2 = 25 - x$

41. Multiplying: $\left(\sqrt{x-4} + 2\right)\left(\sqrt{x-4} - 2\right) = \left(\sqrt{x-4}\right)^2 - \left(2\right)^2 = x - 4 - 4 = x - 8$

43. Multiplying: $\left(\sqrt{3} + 1\right)^3 = \left(\sqrt{3} + 1\right)\left(3 + 2\sqrt{3} + 1\right) = \left(\sqrt{3} + 1\right)\left(4 + 2\sqrt{3}\right) = 4\sqrt{3} + 4 + 6 + 2\sqrt{3} = 10 + 6\sqrt{3}$

45. Multiplying: $\left(\sqrt[3]{3} + \sqrt[3]{2}\right)\left(\sqrt[3]{9} + \sqrt[3]{4}\right) = \sqrt[3]{27} + \sqrt[3]{12} + \sqrt[3]{18} + \sqrt[3]{8} = 3 + \sqrt[3]{12} + \sqrt[3]{18} + 2 = 5 + \sqrt[3]{12} + \sqrt[3]{18}$

47. Multiplying: $\left(\sqrt[3]{x^5} + \sqrt[3]{y}\right)\left(\sqrt[3]{x} + \sqrt[3]{y^2}\right) = \sqrt[3]{x^6} + \sqrt[3]{x^5 y^2} + \sqrt[3]{xy} + \sqrt[3]{y^3} = x^2 + x\sqrt[3]{x^2 y^2} + \sqrt[3]{xy} + y$

49. Rationalizing the denominator: $\dfrac{1}{\sqrt{2}} = \dfrac{1}{\sqrt{2}} \cdot \dfrac{\sqrt{2}}{\sqrt{2}} = \dfrac{\sqrt{2}}{2}$

51. Rationalizing the denominator: $\dfrac{1}{\sqrt{x}} = \dfrac{1}{\sqrt{x}} \cdot \dfrac{\sqrt{x}}{\sqrt{x}} = \dfrac{\sqrt{x}}{x}$

53. Rationalizing the denominator: $\dfrac{4}{\sqrt{3}} = \dfrac{4}{\sqrt{3}} \cdot \dfrac{\sqrt{3}}{\sqrt{3}} = \dfrac{4\sqrt{3}}{3}$

55.　Rationalizing the denominator: $\dfrac{2x}{\sqrt{6}} = \dfrac{2x}{\sqrt{6}} \cdot \dfrac{\sqrt{6}}{\sqrt{6}} = \dfrac{2x\sqrt{6}}{6} = \dfrac{x\sqrt{6}}{3}$

57.　Rationalizing the denominator: $\dfrac{4}{\sqrt{10x}} = \dfrac{4}{\sqrt{10x}} \cdot \dfrac{\sqrt{10x}}{\sqrt{10x}} = \dfrac{4\sqrt{10x}}{10x} = \dfrac{2\sqrt{10x}}{5x}$

59.　Rationalizing the denominator: $\dfrac{2}{\sqrt{8}} = \dfrac{2}{\sqrt{8}} \cdot \dfrac{\sqrt{2}}{\sqrt{2}} = \dfrac{2\sqrt{2}}{4} = \dfrac{\sqrt{2}}{2}$

61.　Rationalizing the denominator: $\sqrt{\dfrac{32x^3 y}{4xy^2}} = \sqrt{\dfrac{8x^2}{y}} = \dfrac{2x\sqrt{2}}{\sqrt{y}} \cdot \dfrac{\sqrt{y}}{\sqrt{y}} = \dfrac{2x\sqrt{2y}}{y}$

63.　Rationalizing the denominator: $\sqrt{\dfrac{12a^3 b^3 c}{8ab^5 c^2}} = \sqrt{\dfrac{3a^2}{2b^2 c}} = \dfrac{a\sqrt{3}}{\sqrt{2b^2 c}} \cdot \dfrac{\sqrt{2c}}{\sqrt{2c}} = \dfrac{a\sqrt{6c}}{2bc}$

65.　Rationalizing the denominator: $\sqrt[3]{\dfrac{16a^4 b^2 c^4}{2ab^3 c}} = \sqrt[3]{\dfrac{8a^3 c^3}{b}} = \dfrac{2ac}{\sqrt[3]{b}} \cdot \dfrac{\sqrt[3]{b^2}}{\sqrt[3]{b^2}} = \dfrac{2ac\sqrt[3]{b^2}}{b}$

67.　Rationalizing the denominator: $\dfrac{\sqrt{2}}{\sqrt{6}-\sqrt{2}} = \dfrac{\sqrt{2}}{\sqrt{6}-\sqrt{2}} \cdot \dfrac{\sqrt{6}+\sqrt{2}}{\sqrt{6}+\sqrt{2}} = \dfrac{\sqrt{12}+2}{6-2} = \dfrac{2\sqrt{3}+2}{4} = \dfrac{1+\sqrt{3}}{2}$

69.　Rationalizing the denominator: $\dfrac{\sqrt{5}}{\sqrt{5}+1} = \dfrac{\sqrt{5}}{\sqrt{5}+1} \cdot \dfrac{\sqrt{5}-1}{\sqrt{5}-1} = \dfrac{5-\sqrt{5}}{5-1} = \dfrac{5-\sqrt{5}}{4}$

71.　Rationalizing the denominator: $\dfrac{\sqrt{x}}{\sqrt{x}-3} = \dfrac{\sqrt{x}}{\sqrt{x}-3} \cdot \dfrac{\sqrt{x}+3}{\sqrt{x}+3} = \dfrac{x+3\sqrt{x}}{x-9}$

73.　Rationalizing the denominator: $\dfrac{\sqrt{5}}{2\sqrt{5}-3} = \dfrac{\sqrt{5}}{2\sqrt{5}-3} \cdot \dfrac{2\sqrt{5}+3}{2\sqrt{5}+3} = \dfrac{2\sqrt{25}+3\sqrt{5}}{20-9} = \dfrac{10+3\sqrt{5}}{11}$

75.　Rationalizing the denominator: $\dfrac{3}{\sqrt{x}-\sqrt{y}} = \dfrac{3}{\sqrt{x}-\sqrt{y}} \cdot \dfrac{\sqrt{x}+\sqrt{y}}{\sqrt{x}+\sqrt{y}} = \dfrac{3\sqrt{x}+3\sqrt{y}}{x-y}$

77.　Rationalizing the denominator: $\dfrac{\sqrt{6}+\sqrt{2}}{\sqrt{6}-\sqrt{2}} = \dfrac{\sqrt{6}+\sqrt{2}}{\sqrt{6}-\sqrt{2}} \cdot \dfrac{\sqrt{6}+\sqrt{2}}{\sqrt{6}+\sqrt{2}} = \dfrac{6+2\sqrt{12}+2}{6-2} = \dfrac{8+4\sqrt{3}}{4} = 2+\sqrt{3}$

79.　Rationalizing the denominator: $\dfrac{\sqrt{7}-2}{\sqrt{7}+2} = \dfrac{\sqrt{7}-2}{\sqrt{7}+2} \cdot \dfrac{\sqrt{7}-2}{\sqrt{7}-2} = \dfrac{7-4\sqrt{7}+4}{7-4} = \dfrac{11-4\sqrt{7}}{3}$

81.　Rationalizing the denominator: $\dfrac{\sqrt{a}+\sqrt{b}}{\sqrt{a}-\sqrt{b}} = \dfrac{\sqrt{a}+\sqrt{b}}{\sqrt{a}-\sqrt{b}} \cdot \dfrac{\sqrt{a}+\sqrt{b}}{\sqrt{a}+\sqrt{b}} = \dfrac{a+2\sqrt{ab}+b}{a-b}$

83.　Rationalizing the denominator: $\dfrac{\sqrt{x}+2}{\sqrt{x}-2} = \dfrac{\sqrt{x}+2}{\sqrt{x}-2} \cdot \dfrac{\sqrt{x}+2}{\sqrt{x}+2} = \dfrac{x+4\sqrt{x}+4}{x-4}$

85.　Rationalizing the denominator:

$\dfrac{2\sqrt{3}-\sqrt{7}}{3\sqrt{3}+\sqrt{7}} = \dfrac{2\sqrt{3}-\sqrt{7}}{3\sqrt{3}+\sqrt{7}} \cdot \dfrac{3\sqrt{3}-\sqrt{7}}{3\sqrt{3}-\sqrt{7}} = \dfrac{18-3\sqrt{21}-2\sqrt{21}+7}{27-7} = \dfrac{25-5\sqrt{21}}{20} = \dfrac{5-\sqrt{21}}{4}$

87.　Rationalizing the denominator: $\dfrac{3\sqrt{x}+2}{1+\sqrt{x}} = \dfrac{3\sqrt{x}+2}{1+\sqrt{x}} \cdot \dfrac{1-\sqrt{x}}{1-\sqrt{x}} = \dfrac{3\sqrt{x}+2-3x-2\sqrt{x}}{1-x} = \dfrac{\sqrt{x}-3x+2}{1-x}$

89.　Simplifying: $\dfrac{2}{\sqrt{3}} + \sqrt{12} = \dfrac{2}{\sqrt{3}} \cdot \dfrac{\sqrt{3}}{\sqrt{3}} + 2\sqrt{3} = \dfrac{2\sqrt{3}}{3} + \dfrac{2\sqrt{3}}{1} \cdot \dfrac{3}{3} = \dfrac{2\sqrt{3}}{3} + \dfrac{6\sqrt{3}}{3} = \dfrac{8\sqrt{3}}{3}$

91.　Simplifying: $\dfrac{1}{\sqrt{5}} + \sqrt{20} = \dfrac{1}{\sqrt{5}} \cdot \dfrac{\sqrt{5}}{\sqrt{5}} + 2\sqrt{5} = \dfrac{\sqrt{5}}{5} + \dfrac{2\sqrt{5}}{1} \cdot \dfrac{5}{5} = \dfrac{\sqrt{5}}{5} + \dfrac{10\sqrt{5}}{5} = \dfrac{11\sqrt{5}}{5}$

93.　Simplifying: $\dfrac{6}{\sqrt{12}} - \sqrt{75} = \dfrac{6}{2\sqrt{3}} \cdot \dfrac{\sqrt{3}}{\sqrt{3}} - 5\sqrt{3} = \dfrac{6\sqrt{3}}{6} - 5\sqrt{3} = \sqrt{3} - 5\sqrt{3} = -4\sqrt{3}$

95. Simplifying:

$$\frac{1}{\sqrt{3}} + \sqrt{48} + \frac{4}{\sqrt{12}} = \frac{1}{\sqrt{3}} \cdot \frac{\sqrt{3}}{\sqrt{3}} + 4\sqrt{3} + \frac{4}{2\sqrt{3}} \cdot \frac{\sqrt{3}}{\sqrt{3}}$$

$$= \frac{\sqrt{3}}{3} + \frac{4\sqrt{3}}{1} \cdot \frac{3}{3} + \frac{4\sqrt{3}}{6}$$

$$= \frac{\sqrt{3}}{3} + \frac{12\sqrt{3}}{3} + \frac{2\sqrt{3}}{3}$$

$$= \frac{15\sqrt{3}}{3}$$

$$= 5\sqrt{3}$$

97. Simplifying the product: $\left(\sqrt[3]{2} + \sqrt[3]{3}\right)\left(\sqrt[3]{4} - \sqrt[3]{6} + \sqrt[3]{9}\right) = \sqrt[3]{8} - \sqrt[3]{12} + \sqrt[3]{18} + \sqrt[3]{12} - \sqrt[3]{18} + \sqrt[3]{27} = 2 + 3 = 5$

99. The correct statement is: $5\left(2\sqrt{3}\right) = 10\sqrt{3}$

101. The correct statement is: $\left(\sqrt{x} + 3\right)^2 = \left(\sqrt{x} + 3\right)\left(\sqrt{x} + 3\right) = x + 6\sqrt{x} + 9$

103. The correct statement is: $\left(5\sqrt{3}\right)^2 = \left(5\sqrt{3}\right)\left(5\sqrt{3}\right) = 25 \cdot 3 = 75$

105. Substituting $h = 50$: $t = \frac{\sqrt{100 - 50}}{4} = \frac{\sqrt{50}}{4} = \frac{5\sqrt{2}}{4}$ second

Substituting $h = 0$: $t = \frac{\sqrt{100 - 0}}{4} = \frac{\sqrt{100}}{4} = \frac{10}{4} = \frac{5}{2}$ second

107. Since the large rectangle is a golden rectangle and $AC = 6$, then $CE = 6\left(\frac{1 + \sqrt{5}}{2}\right) = 3 + 3\sqrt{5}$. Since $CD = 6$, then

$DE = 3 + 3\sqrt{5} - 6 = 3\sqrt{5} - 3$. Now computing the ratio:

$$\frac{EF}{DE} = \frac{6}{3\sqrt{5} - 3} \cdot \frac{3\sqrt{5} + 3}{3\sqrt{5} + 3} = \frac{18\left(\sqrt{5} + 1\right)}{45 - 9} = \frac{18\left(\sqrt{5} + 1\right)}{36} = \frac{1 + \sqrt{5}}{2}$$

Therefore the smaller rectangle $BDEF$ is also a golden rectangle.

109. Since the large rectangle is a golden rectangle and $AC = 2x$, then $CE = 2x\left(\frac{1 + \sqrt{5}}{2}\right) = x\left(1 + \sqrt{5}\right)$. Since $CD = 2x$, then

$DE = x\left(1 + \sqrt{5}\right) - 2x = x\left(-1 + \sqrt{5}\right)$. Now computing the ratio:

$$\frac{EF}{DE} = \frac{2x}{x\left(-1 + \sqrt{5}\right)} = \frac{2}{-1 + \sqrt{5}} \cdot \frac{-1 - \sqrt{5}}{-1 - \sqrt{5}} = \frac{-2\left(\sqrt{5} + 1\right)}{1 - 5} = \frac{-2\left(\sqrt{5} + 1\right)}{-4} = \frac{1 + \sqrt{5}}{2}$$

Therefore the smaller rectangle $BDEF$ is also a golden rectangle.

111. Simplifying the complex fraction: $\dfrac{\frac{1}{4} - \frac{1}{3}}{\frac{1}{2} + \frac{1}{6}} = \dfrac{\left(\frac{1}{4} - \frac{1}{3}\right) \cdot 12}{\left(\frac{1}{2} + \frac{1}{6}\right) \cdot 12} = \dfrac{3 - 4}{6 + 2} = -\dfrac{1}{8}$

113. Simplifying the complex fraction: $\dfrac{1 - \dfrac{2}{y}}{1 + \dfrac{2}{y}} = \dfrac{\left(1 - \dfrac{2}{y}\right) \cdot y}{\left(1 + \dfrac{2}{y}\right) \cdot y} = \dfrac{y - 2}{y + 2}$

115. Simplifying the complex fraction: $\dfrac{4 + \dfrac{4}{x} + \dfrac{1}{x^2}}{4 - \dfrac{1}{x^2}} = \dfrac{\left(4 + \dfrac{4}{x} + \dfrac{1}{x^2}\right) \cdot x^2}{\left(4 - \dfrac{1}{x^2}\right) \cdot x^2} = \dfrac{4x^2 + 4x + 1}{4x^2 - 1} = \dfrac{\left(2x + 1\right)^2}{\left(2x + 1\right)\left(2x - 1\right)} = \dfrac{2x + 1}{2x - 1}$

117. Simplifying: $(t+5)^2 = t^2 + 2(5t) + 5^2 = t^2 + 10t + 25$

119. Simplifying: $\sqrt{x} \bullet \sqrt{x} = \sqrt{x^2} = x$

121. Solving the equation:

$$3x + 4 = 5^2$$
$$3x + 4 = 25$$
$$3x = 21$$
$$x = 7$$

123. Solving the equation:

$$t^2 + 7t + 12 = 0$$
$$(t+4)(t+3) = 0$$
$$t = -4, -3$$

125. Solving the equation:

$$t^2 + 10t + 25 = t + 7$$
$$t^2 + 9t + 18 = 0$$
$$(t+6)(t+3) = 0$$
$$t = -6, -3$$

127. Solving the equation:

$$(x+4)^2 = x + 6$$
$$x^2 + 8x + 16 = x + 6$$
$$x^2 + 7x + 10 = 0$$
$$(x+5)(x+2) = 0$$
$$x = -5, -2$$

129. Rationalizing the denominator: $\dfrac{x}{\sqrt{x-2}+4} = \dfrac{x}{\sqrt{x-2}+4} \bullet \dfrac{\sqrt{x-2}-4}{\sqrt{x-2}-4} = \dfrac{x\left(\sqrt{x-2}-4\right)}{x-2-16} = \dfrac{x\left(\sqrt{x-2}-4\right)}{x-18}$

131. Rationalizing the denominator: $\dfrac{x}{\sqrt{x+5}-5} = \dfrac{x}{\sqrt{x+5}-5} \bullet \dfrac{\sqrt{x+5}+5}{\sqrt{x+5}+5} = \dfrac{x\left(\sqrt{x+5}+5\right)}{x+5-25} = \dfrac{x\left(\sqrt{x+5}+5\right)}{x-20}$

133. Rationalizing the denominator: $\dfrac{3x}{\sqrt{5x}+x} = \dfrac{3x}{\sqrt{5x}+x} \bullet \dfrac{\sqrt{5x}-x}{\sqrt{5x}-x} = \dfrac{3x\left(\sqrt{5x}-x\right)}{5x-x^2} = \dfrac{3x\left(\sqrt{5x}-x\right)}{x(5-x)} = \dfrac{3\left(\sqrt{5x}-x\right)}{5-x}$

7.6 Equations with Radicals

1. Solving the equation:

$$\sqrt{2x+1} = 3$$
$$\left(\sqrt{2x+1}\right)^2 = 3^2$$
$$2x + 1 = 9$$
$$2x = 8$$
$$x = 4$$

3. Solving the equation:

$$\sqrt{4x+1} = -5$$
$$\left(\sqrt{4x+1}\right)^2 = (-5)^2$$
$$4x + 1 = 25$$
$$4x = 24$$
$$x = 6$$

Since this value does not check, the solution set is $\varnothing$.

5. Solving the equation:

$$\sqrt{2y-1} = 3$$
$$\left(\sqrt{2y-1}\right)^2 = 3^2$$
$$2y - 1 = 9$$
$$2y = 10$$
$$y = 5$$

7. Solving the equation:

$$\sqrt{5x-7} = -1$$
$$\left(\sqrt{5x-7}\right)^2 = (-1)^2$$
$$5x - 7 = 1$$
$$5x = 8$$
$$x = \frac{8}{5}$$

Since this value does not check, the solution set is $\varnothing$.

9. Solving the equation:

$$\sqrt{2x-3} - 2 = 4$$
$$\sqrt{2x-3} = 6$$
$$\left(\sqrt{2x-3}\right)^2 = 6^2$$
$$2x - 3 = 36$$
$$2x = 39$$
$$x = \frac{39}{2}$$

11. Solving the equation:

$$\sqrt{4a+1} + 3 = 2$$
$$\sqrt{4a+1} = -1$$
$$\left(\sqrt{4a+1}\right)^2 = (-1)^2$$
$$4a + 1 = 1$$
$$4a = 0$$
$$a = 0$$

Since this value does not check, the solution set is $\varnothing$.

13. Solving the equation:

$$\sqrt[4]{3x+1}=2$$
$$\left(\sqrt[4]{3x+1}\right)^4=2^4$$
$$3x+1=16$$
$$3x=15$$
$$x=5$$

15. Solving the equation:

$$\sqrt[3]{2x-5}=1$$
$$\left(\sqrt[3]{2x-5}\right)^3=1^3$$
$$2x-5=1$$
$$2x=6$$
$$x=3$$

17. Solving the equation:

$$\sqrt[3]{3a+5}=-3$$
$$\left(\sqrt[3]{3a+5}\right)^3=(-3)^3$$
$$3a+5=-27$$
$$3a=-32$$
$$a=-\frac{32}{3}$$

19. Solving the equation:

$$\sqrt{y-3}=y-3$$
$$\left(\sqrt{y-3}\right)^2=(y-3)^2$$
$$y-3=y^2-6y+9$$
$$0=y^2-7y+12$$
$$0=(y-3)(y-4)$$
$$y=3,4$$

21. Solving the equation:

$$\sqrt{a+2}=a+2$$
$$\left(\sqrt{a+2}\right)^2=(a+2)^2$$
$$a+2=a^2+4a+4$$
$$0=a^2+3a+2$$
$$0=(a+2)(a+1)$$
$$a=-2,-1$$

23. Solving the equation:

$$\sqrt{2x+3}=\frac{2x-7}{3}$$
$$\left(\sqrt{2x+3}\right)^2=\left(\frac{2x-7}{3}\right)^2$$
$$2x+3=\frac{4x^2-28x+49}{9}$$
$$18x+27=4x^2-28x+49$$
$$0=4x^2-46x+22$$
$$0=2x^2-23x+11$$
$$0=(2x-1)(x-11)$$
$$x=\tfrac{1}{2},11$$

The solution is 11 (1/2 does not check).

25. Solving the equation:

$$\sqrt{4x-3}=\frac{x+3}{2}$$
$$\left(\sqrt{4x-3}\right)^2=\left(\frac{x+3}{2}\right)^2$$
$$4x-3=\frac{x^2+6x+9}{4}$$
$$16x-12=x^2+6x+9$$
$$0=x^2-10x+21$$
$$0=(x-3)(x-7)$$
$$x=3,7$$

27. Solving the equation:

$$\sqrt{7x+2}=\frac{2x+2}{3}$$
$$\left(\sqrt{7x+2}\right)^2=\left(\frac{2x+2}{3}\right)^2$$
$$7x+2=\frac{4x^2+8x+4}{9}$$
$$63x+18=4x^2+8x+4$$
$$0=4x^2-55x-14$$
$$0=(4x+1)(x-14)$$
$$x=-\tfrac{1}{4},14$$

29. Solving the equation:

$$\sqrt{2x+4}=\sqrt{1-x}$$
$$\left(\sqrt{2x+4}\right)^2=\left(\sqrt{1-x}\right)^2$$
$$2x+4=1-x$$
$$3x=-3$$
$$x=-1$$

31. Solving the equation:

$$\sqrt{4a+7}=-\sqrt{a+2}$$
$$\left(\sqrt{4a+7}\right)^2=\left(-\sqrt{a+2}\right)^2$$
$$4a+7=a+2$$
$$3a=-5$$
$$a=-\frac{5}{3}$$

Since this value does not check, the solution set is $\varnothing$.

33.   Solving the equation:

$$\sqrt[4]{5x-8} = \sqrt[4]{4x-1}$$
$$\left(\sqrt[4]{5x-8}\right)^4 = \left(\sqrt[4]{4x-1}\right)^4$$
$$5x-8 = 4x-1$$
$$x = 7$$

35.   Solving the equation:

$$x+1 = \sqrt{5x+1}$$
$$(x+1)^2 = \left(\sqrt{5x+1}\right)^2$$
$$x^2 + 2x + 1 = 5x + 1$$
$$x^2 - 3x = 0$$
$$x(x-3) = 0$$
$$x = 0, 3$$

37.   Solving the equation:

$$t+5 = \sqrt{2t+9}$$
$$(t+5)^2 = \left(\sqrt{2t+9}\right)^2$$
$$t^2 + 10t + 25 = 2t + 9$$
$$t^2 + 8t + 16 = 0$$
$$(t+4)^2 = 0$$
$$t = -4$$

39.   Solving the equation:

$$\sqrt{y-8} = \sqrt{8-y}$$
$$\left(\sqrt{y-8}\right)^2 = \left(\sqrt{8-y}\right)^2$$
$$y-8 = 8-y$$
$$2y = 16$$
$$y = 8$$

41.   Solving the equation:

$$\sqrt[3]{3x+5} = \sqrt[3]{5-2x}$$
$$\left(\sqrt[3]{3x+5}\right)^3 = \left(\sqrt[3]{5-2x}\right)^3$$
$$3x+5 = 5-2x$$
$$5x = 0$$
$$x = 0$$

43.   Solving the equation:

$$\sqrt{x-8} = \sqrt{x} - 2$$
$$\left(\sqrt{x-8}\right)^2 = \left(\sqrt{x}-2\right)^2$$
$$x-8 = x - 4\sqrt{x} + 4$$
$$-12 = -4\sqrt{x}$$
$$\sqrt{x} = 3$$
$$x = 9$$

45.   Solving the equation:

$$\sqrt{x+1} = \sqrt{x} + 1$$
$$\left(\sqrt{x+1}\right)^2 = \left(\sqrt{x}+1\right)^2$$
$$x+1 = x + 2\sqrt{x} + 1$$
$$0 = 2\sqrt{x}$$
$$\sqrt{x} = 0$$
$$x = 0$$

47.   Solving the equation:

$$\sqrt{x+8} = \sqrt{x-4} + 2$$
$$\left(\sqrt{x+8}\right)^2 = \left(\sqrt{x-4}+2\right)^2$$
$$x+8 = x - 4 + 4\sqrt{x-4} + 4$$
$$8 = 4\sqrt{x-4}$$
$$\sqrt{x-4} = 2$$
$$x-4 = 4$$
$$x = 8$$

49.   Solving the equation:

$$\sqrt{x-5} - 3 = \sqrt{x-8}$$
$$\left(\sqrt{x-5}-3\right)^2 = \left(\sqrt{x-8}\right)^2$$
$$x-5 - 6\sqrt{x-5} + 9 = x-8$$
$$-6\sqrt{x-5} = -12$$
$$\sqrt{x-5} = 2$$
$$x-5 = 4$$
$$x = 9$$

Since this value does not check, the solution set is $\varnothing$.

51.   Solving the equation:

$$\sqrt{x+4} = 2 - \sqrt{2x}$$
$$\left(\sqrt{x+4}\right)^2 = \left(2-\sqrt{2x}\right)^2$$
$$x+4 = 4 - 4\sqrt{2x} + 2x$$
$$-x = -4\sqrt{2x}$$
$$(-x)^2 = \left(-4\sqrt{2x}\right)^2$$
$$x^2 = 32x$$
$$x^2 - 32x = 0$$
$$x(x-32) = 0$$
$$x = 0, 32$$

The solution is 0 (32 does not check).

53. Solving the equation:

$$\sqrt{2x+4} = \sqrt{x+3}+1$$
$$\left(\sqrt{2x+4}\right)^2 = \left(\sqrt{x+3}+1\right)^2$$
$$2x+4 = x+3+2\sqrt{x+3}+1$$
$$x = 2\sqrt{x+3}$$
$$x^2 = \left(2\sqrt{x+3}\right)^2$$
$$x^2 = 4x+12$$
$$x^2 - 4x - 12 = 0$$
$$(x-6)(x+2) = 0$$
$$x = -2, 6$$

The solution is 6 (−2 does not check).

57. Solving the equation:

$$f(x) = 2x-1$$
$$\sqrt{2x-1} = 2x-1$$
$$\left(\sqrt{2x-1}\right)^2 = (2x-1)^2$$
$$2x-1 = 4x^2 - 4x + 1$$
$$0 = 4x^2 - 6x + 2$$
$$0 = 2x^2 - 3x + 1$$
$$0 = (2x-1)(x-1)$$
$$x = \tfrac{1}{2}, 1$$

61. Solving the equation:

$$g(x) = 0$$
$$\sqrt{2x+3} = 0$$
$$\left(\sqrt{2x+3}\right)^2 = (0)^2$$
$$2x+3 = 0$$
$$2x = -3$$
$$x = -\tfrac{3}{2}$$

65. Solving the equation:

$$f(x) = g(x)$$
$$\sqrt{2x} - 1 = 0$$
$$\sqrt{2x} = 1$$
$$\left(\sqrt{2x}\right)^2 = (1)^2$$
$$2x = 1$$
$$x = \tfrac{1}{2}$$

55. Solving the equation:

$$f(x) = 0$$
$$\sqrt{2x-1} = 0$$
$$\left(\sqrt{2x-1}\right)^2 = (0)^2$$
$$2x-1 = 0$$
$$2x = 1$$
$$x = \tfrac{1}{2}$$

59. Solving the equation:

$$f(x) = \sqrt{x-4}+2$$
$$\sqrt{2x-1} = \sqrt{x-4}+2$$
$$\left(\sqrt{2x-1}\right)^2 = \left(\sqrt{x-4}+2\right)^2$$
$$2x-1 = x-4+4\sqrt{x-4}+4$$
$$x-1 = 4\sqrt{x-4}$$
$$(x-1)^2 = \left(4\sqrt{x-4}\right)^2$$
$$x^2 - 2x + 1 = 16(x-4)$$
$$x^2 - 2x + 1 = 16x - 64$$
$$x^2 - 18x + 65 = 0$$
$$(x-5)(x-13) = 0$$
$$x = 5, 13$$

63. Solving the equation:

$$g(x) = -\sqrt{5x}$$
$$\sqrt{2x+3} = -\sqrt{5x}$$
$$\left(\sqrt{2x+3}\right)^2 = \left(-\sqrt{5x}\right)^2$$
$$2x+3 = 5x$$
$$3 = 3x$$
$$x = 1$$

Since this value does not check, the solution set is $\varnothing$.

67. Solving the equation:

$$f(x) = g(x)$$
$$\sqrt{2x} - 1 = \sqrt{2x+5}$$
$$\left(\sqrt{2x} - 1\right)^2 = \left(\sqrt{2x+5}\right)^2$$
$$2x - 2\sqrt{2x} + 1 = 2x + 5$$
$$-2\sqrt{2x} = 4$$
$$\sqrt{2x} = -2$$
$$\left(\sqrt{2x}\right)^2 = (-2)^2$$
$$2x = 4$$
$$x = 2$$

Since this value does not check, the solution set is $\varnothing$.

69. Solving the equation:
$$h(x) = f(x)$$
$$\sqrt[3]{3x+5} = 2$$
$$\left(\sqrt[3]{3x+5}\right)^3 = (2)^3$$
$$3x+5 = 8$$
$$3x = 3$$
$$x = 1$$

71. Solving the equation:
$$h(x) = f(x)$$
$$\sqrt[3]{5-2x} = 3$$
$$\left(\sqrt[3]{5-2x}\right)^3 = (3)^3$$
$$5-2x = 27$$
$$-2x = 22$$
$$x = -11$$

73. Graphing the equation:

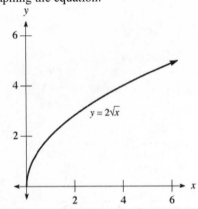

75. Graphing the equation:

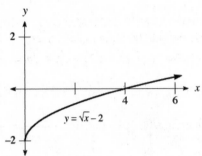

77. Graphing the equation:

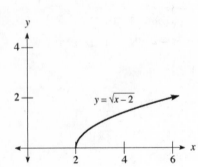

79. Graphing the equation:

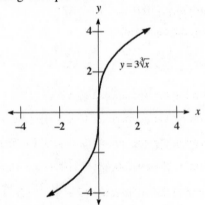

81. Graphing the equation:

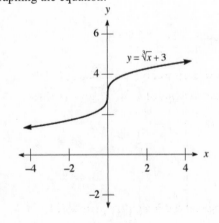

83. Graphing the equation:

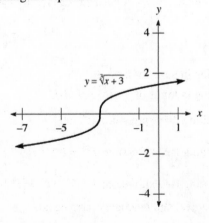

85. Solving for h:

$$t = \frac{\sqrt{100 - h}}{4}$$
$$4t = \sqrt{100 - h}$$
$$16t^2 = 100 - h$$
$$h = 100 - 16t^2$$

87. Solving for L:

$$2 = 2\left(\frac{22}{7}\right)\sqrt{\frac{L}{32}}$$
$$\frac{7}{22} = \sqrt{\frac{L}{32}}$$
$$\left(\frac{7}{22}\right)^2 = \frac{L}{32}$$
$$L = 32\left(\frac{7}{22}\right)^2 \approx 3.24 \text{ feet}$$

89. The width is $\sqrt{25} = 5$ meters.

91. Solving the equation:

$$\sqrt{x} = 50$$
$$x = 2500$$

The plume is 2,500 meters down river.

93. Multiplying: $\sqrt{2}\left(\sqrt{3} - \sqrt{2}\right) = \sqrt{6} - \sqrt{4} = \sqrt{6} - 2$

95. Multiplying: $\left(\sqrt{x} + 5\right)^2 = \left(\sqrt{x} + 5\right)\left(\sqrt{x} + 5\right) = x + 5\sqrt{x} + 5\sqrt{x} + 25 = x + 10\sqrt{x} + 25$

97. Rationalizing the denominator: $\dfrac{\sqrt{x}}{\sqrt{x} + 3} = \dfrac{\sqrt{x}}{\sqrt{x} + 3} \bullet \dfrac{\sqrt{x} - 3}{\sqrt{x} - 3} = \dfrac{x - 3\sqrt{x}}{x - 9}$

99. Simplifying: $\sqrt{25} = 5$

101. Simplifying: $\sqrt{12} = \sqrt{4 \bullet 3} = 2\sqrt{3}$

103. Simplifying: $(-1)^{15} = -1$

105. Simplifying: $(-1)^{50} = 1$

107. Solving the equation:

$$3x = 12$$
$$x = 4$$

109. Solving the equation:

$$4x - 3 = 5$$
$$4x = 8$$
$$x = 2$$

111. Performing the operations: $(3 + 4x) + (7 - 6x) = 10 - 2x$

113. Performing the operations: $(7 + 3x) - (5 + 6x) = 7 + 3x - 5 - 6x = 2 - 3x$

115. Performing the operations: $(3 - 4x)(2 + 5x) = 6 + 15x - 8x - 20x^2 = 6 + 7x - 20x^2$

117. Performing the operations: $2x(4 - 6x) = 8x - 12x^2$

119. Performing the operations: $(2 + 3x)^2 = 2^2 + 2(2)(3x) + (3x)^2 = 4 + 12x + 9x^2$

121. Performing the operations: $(2 - 3x)(2 + 3x) = 2^2 - (3x)^2 = 4 - 9x^2$

7.7 Complex Numbers

1. Writing in terms of i: $\sqrt{-36} = 6i$

3. Writing in terms of i: $-\sqrt{-25} = -5i$

5. Writing in terms of i: $\sqrt{-72} = 6i\sqrt{2}$

7. Writing in terms of i: $-\sqrt{-12} = -2i\sqrt{3}$

9. Rewriting the expression: $i^{28} = \left(i^4\right)^7 = (1)^7 = 1$

11. Rewriting the expression: $i^{26} = i^{24}i^2 = \left(i^4\right)^6 i^2 = (1)^6(-1) = -1$

13. Rewriting the expression: $i^{75} = i^{72}i^3 = \left(i^4\right)^{18} i^2 i = (1)^{18}(-1)i = -i$

15. Setting real and imaginary parts equal:

$$2x = 6 \qquad 3y = -3$$
$$x = 3 \qquad y = -1$$

17. Setting real and imaginary parts equal:

$$-x = 2 \qquad 10y = -5$$
$$x = -2 \qquad y = -\frac{1}{2}$$

19. Setting real and imaginary parts equal:

$$2x = -16 \qquad -2y = 10$$
$$x = -8 \qquad y = -5$$

21. Setting real and imaginary parts equal:

$$2x - 4 = 10 \qquad -6y = -3$$
$$2x = 14 \qquad y = \frac{1}{2}$$
$$x = 7$$

23. Setting real and imaginary parts equal:

$$7x - 1 = 2 \qquad 5y + 2 = 4$$
$$7x = 3 \qquad 5y = 2$$
$$x = \frac{3}{7} \qquad y = \frac{2}{5}$$

25. Combining the numbers: $(2+3i)+(3+6i) = 5+9i$

27. Combining the numbers: $(3-5i)+(2+4i) = 5-i$

29. Combining the numbers: $(5+2i)-(3+6i) = 5+2i-3-6i = 2-4i$

31. Combining the numbers: $(3-5i)-(2+i) = 3-5i-2-i = 1-6i$

33. Combining the numbers: $[(3+2i)-(6+i)]+(5+i) = 3+2i-6-i+5+i = 2+2i$

35. Combining the numbers: $[(7-i)-(2+4i)]-(6+2i) = 7-i-2-4i-6-2i = -1-7i$

37. Combining the numbers:

$$(3+2i)-[(3-4i)-(6+2i)] = (3+2i)-(3-4i-6-2i) = (3+2i)-(-3-6i) = 3+2i+3+6i = 6+8i$$

39. Combining the numbers: $(4-9i)+[(2-7i)-(4+8i)] = (4-9i)+(2-7i-4-8i) = (4-9i)+(-2-15i) = 2-24i$

41. Finding the product: $3i(4+5i) = 12i+15i^2 = -15+12i$

43. Finding the product: $6i(4-3i) = 24i-18i^2 = 18+24i$

45. Finding the product: $(3+2i)(4+i) = 12+8i+3i+2i^2 = 12+11i-2 = 10+11i$

47. Finding the product: $(4+9i)(3-i) = 12+27i-4i-9i^2 = 12+23i+9 = 21+23i$

49. Finding the product: $(1+i)^3 = (1+i)(1+i)^2 = (1+i)(1+2i-1) = (1+i)(2i) = -2+2i$

51. Finding the product: $(2-i)^3 = (2-i)(2-i)^2 = (2-i)(4-4i-1) = (2-i)(3-4i) = 6-11i-4 = 2-11i$

53. Finding the product: $(2+5i)^2 = (2+5i)(2+5i) = 4+10i+10i-25 = -21+20i$

55. Finding the product: $(1-i)^2 = (1-i)(1-i) = 1-i-i-1 = -2i$

57. Finding the product: $(3-4i)^2 = (3-4i)(3-4i) = 9-12i-12i-16 = -7-24i$

59. Finding the product: $(2+i)(2-i) = 4-i^2 = 4+1 = 5$

61. Finding the product: $(6-2i)(6+2i) = 36-4i^2 = 36+4 = 40$

63. Finding the product: $(2+3i)(2-3i) = 4-9i^2 = 4+9 = 13$

65. Finding the product: $(10+8i)(10-8i) = 100-64i^2 = 100+64 = 164$

67. Finding the quotient: $\dfrac{2-3i}{i} = \dfrac{2-3i}{i} \cdot \dfrac{i}{i} = \dfrac{2i+3}{-1} = -3-2i$

69. Finding the quotient: $\dfrac{5+2i}{-i} = \dfrac{5+2i}{-i} \cdot \dfrac{i}{i} = \dfrac{5i-2}{1} = -2+5i$

71. Finding the quotient: $\dfrac{4}{2-3i} = \dfrac{4}{2-3i} \cdot \dfrac{2+3i}{2+3i} = \dfrac{8+12i}{4+9} = \dfrac{8+12i}{13} = \dfrac{8}{13}+\dfrac{12}{13}i$

73. Finding the quotient: $\dfrac{6}{-3+2i} = \dfrac{6}{-3+2i} \cdot \dfrac{-3-2i}{-3-2i} = \dfrac{-18-12i}{9+4} = \dfrac{-18-12i}{13} = -\dfrac{18}{13}-\dfrac{12}{13}i$

75. Finding the quotient: $\dfrac{2+3i}{2-3i} = \dfrac{2+3i}{2-3i} \cdot \dfrac{2+3i}{2+3i} = \dfrac{4+12i-9}{4+9} = \dfrac{-5+12i}{13} = -\dfrac{5}{13}+\dfrac{12}{13}i$

77. Finding the quotient: $\dfrac{5+4i}{3+6i} = \dfrac{5+4i}{3+6i} \cdot \dfrac{3-6i}{3-6i} = \dfrac{15-18i+24}{9+36} = \dfrac{39-18i}{45} = \dfrac{13}{15} - \dfrac{2}{5}i$

79. Dividing to find R: $R = \dfrac{80+20i}{-6+2i} = \dfrac{80+20i}{-6+2i} \cdot \dfrac{-6-2i}{-6-2i} = \dfrac{-480-280i+40}{36+4} = \dfrac{-440-280i}{40} = (-11-7i)$ ohms

81. Solving the equation:

$$\dfrac{t}{3} - \dfrac{1}{2} = -1$$
$$6\left(\dfrac{t}{3} - \dfrac{1}{2}\right) = 6(-1)$$
$$2t - 3 = -6$$
$$2t = -3$$
$$t = -\dfrac{3}{2}$$

83. Solving the equation:

$$2 + \dfrac{5}{y} = \dfrac{3}{y^2}$$
$$y^2\left(2 + \dfrac{5}{y}\right) = y^2\left(\dfrac{3}{y^2}\right)$$
$$2y^2 + 5y = 3$$
$$2y^2 + 5y - 3 = 0$$
$$(2y-1)(y+3) = 0$$
$$y = -3, \dfrac{1}{2}$$

85. Let x represent the number. The equation is:

$$x + \dfrac{1}{x} = \dfrac{41}{20}$$
$$20x\left(x + \dfrac{1}{x}\right) = 20x\left(\dfrac{41}{20}\right)$$
$$20x^2 + 20 = 41x$$
$$20x^2 - 41x + 20 = 0$$
$$(5x-4)(4x-5) = 0$$
$$x = \dfrac{4}{5}, \dfrac{5}{4}$$

The number is either $\dfrac{5}{4}$ or $\dfrac{4}{5}$.

87. Simplifying: $\dfrac{1}{i} = \dfrac{1}{i} \cdot \dfrac{i}{i} = \dfrac{i}{i^2} = \dfrac{i}{-1} = -i$

89. Substituting into the equation: $x^2 - 2x + 2 = (1+i)^2 - 2(1+i) + 2 = 1 + 2i - 1 - 2 - 2i + 2 = 0$

Thus $x = 1+i$ is a solution to the equation.

91. Substituting into the equation:

$$\begin{aligned} x^3 - 11x + 20 &= (2+i)^3 - 11(2+i) + 20 \\ &= (2+i)(4+4i-1) - 22 - 11i + 20 \\ &= (2+i)(3+4i) - 2 - 11i \\ &= 6 + 11i - 4 - 2 - 11i \\ &= 0 \end{aligned}$$

Thus $x = 2+i$ is a solution to the equation.

Chapter 7 Review/Test

1. Simplifying: $\sqrt{49} = 7$

2. Simplifying: $(-27)^{1/3} = -3$

3. Simplifying: $16^{1/4} = 2$

4. Simplifying: $9^{3/2} = \left(9^{1/2}\right)^3 = 3^3 = 27$

5. Simplifying: $\sqrt[5]{32x^{15}y^{10}} = 2x^3y^2$

6. Simplifying: $8^{-4/3} = \left(8^{1/3}\right)^{-4} = 2^{-4} = \dfrac{1}{2^4} = \dfrac{1}{16}$

7. Simplifying: $x^{2/3} \cdot x^{4/3} = x^{2/3+4/3} = x^2$

8. Simplifying: $\left(a^{2/3}b^{4/3}\right)^3 = a^{3\cdot 2/3}b^{3\cdot 4/3} = a^2b^4$

9. Simplifying: $\dfrac{a^{3/5}}{a^{1/4}} = a^{3/5-1/4} = a^{12/20-5/20} = a^{7/20}$

10. Simplifying: $\dfrac{a^{2/3}b^3}{a^{1/4}b^{1/3}} = a^{2/3-1/4}b^{3-1/3} = a^{8/12-3/12}b^{9/3-1/3} = a^{5/12}b^{8/3}$

11. Multiplying: $\left(3x^{1/2} + 5y^{1/2}\right)\left(4x^{1/2} - 3y^{1/2}\right) = 12x - 9x^{1/2}y^{1/2} + 20x^{1/2}y^{1/2} - 15y = 12x + 11x^{1/2}y^{1/2} - 15y$

12. Multiplying: $\left(a^{1/3} - 5\right)^2 = \left(a^{1/3} - 5\right)\left(a^{1/3} - 5\right) = a^{2/3} - 5a^{1/3} - 5a^{1/3} + 25 = a^{2/3} - 10a^{1/3} + 25$

13. Dividing: $\dfrac{28x^{5/6} + 14x^{7/6}}{7x^{1/3}} = \dfrac{28x^{5/6}}{7x^{1/3}} + \dfrac{14x^{7/6}}{7x^{1/3}} = 4x^{5/6-1/3} + 2x^{7/6-1/3} = 4x^{1/2} + 2x^{5/6}$

14. Factoring: $8(x-3)^{5/4} - 2(x-3)^{1/4} = 2(x-3)^{1/4}\left[4(x-3) - 1\right] = 2(x-3)^{1/4}\left(4x - 12 - 1\right) = 2(x-3)^{1/4}\left(4x - 13\right)$

15. Simplifying: $x^{3/4} + \dfrac{5}{x^{1/4}} = x^{3/4} \cdot \dfrac{x^{1/4}}{x^{1/4}} + \dfrac{5}{x^{1/4}} = \dfrac{x+5}{x^{1/4}}$

16. Simplifying: $\sqrt{12} = \sqrt{4 \cdot 3} = 2\sqrt{3}$ 17. Simplifying: $\sqrt{50} = \sqrt{25 \cdot 2} = 5\sqrt{2}$

18. Simplifying: $\sqrt[3]{16} = \sqrt[3]{8 \cdot 2} = 2\sqrt[3]{2}$ 19. Simplifying: $\sqrt{18x^2} = \sqrt{9x^2 \cdot 2} = 3x\sqrt{2}$

20. Simplifying: $\sqrt{80a^3b^4c^2} = \sqrt{16a^2b^4c^2 \cdot 5a} = 4ab^2c\sqrt{5a}$

21. Simplifying: $\sqrt[4]{32a^4b^5c^6} = \sqrt[4]{16a^4b^4c^4 \cdot 2bc^2} = 2abc\sqrt[4]{2bc^2}$

22. Rationalizing the denominator: $\dfrac{3}{\sqrt{2}} = \dfrac{3}{\sqrt{2}} \cdot \dfrac{\sqrt{2}}{\sqrt{2}} = \dfrac{3\sqrt{2}}{2}$

23. Rationalizing the denominator: $\dfrac{6}{\sqrt[3]{2}} = \dfrac{6}{\sqrt[3]{2}} \cdot \dfrac{\sqrt[3]{4}}{\sqrt[3]{4}} = \dfrac{6\sqrt[3]{4}}{2} = 3\sqrt[3]{4}$

24. Simplifying: $\sqrt{\dfrac{48x^3}{7y}} = \dfrac{4x\sqrt{3x}}{\sqrt{7y}} \cdot \dfrac{\sqrt{7y}}{\sqrt{7y}} = \dfrac{4x\sqrt{21xy}}{7y}$

25. Simplifying: $\sqrt[3]{\dfrac{40x^2y^3}{3z}} = \dfrac{2y\sqrt[3]{5x^2}}{\sqrt[3]{3z}} \cdot \dfrac{\sqrt[3]{9z^2}}{\sqrt[3]{9z^2}} = \dfrac{2y\sqrt[3]{45x^2z^2}}{3z}$

26. Combining the expressions: $5x\sqrt{6} + 2x\sqrt{6} - 9x\sqrt{6} = -2x\sqrt{6}$

27. Combining the expressions: $\sqrt{12} + \sqrt{3} = 2\sqrt{3} + \sqrt{3} = 3\sqrt{3}$

28. Combining the expressions: $\dfrac{3}{\sqrt{5}} + \sqrt{5} = \dfrac{3}{\sqrt{5}} \cdot \dfrac{\sqrt{5}}{\sqrt{5}} + \sqrt{5} = \dfrac{3\sqrt{5}}{5} + \dfrac{5\sqrt{5}}{5} = \dfrac{8\sqrt{5}}{5}$

29. Combining the expressions: $3\sqrt{8} - 4\sqrt{72} + 5\sqrt{50} = 6\sqrt{2} - 24\sqrt{2} + 25\sqrt{2} = 7\sqrt{2}$

30. Combining the expressions: $3b\sqrt{27a^5b} + 2a\sqrt{3a^3b^3} = 9a^2b\sqrt{3ab} + 2a^2b\sqrt{3ab} = 11a^2b\sqrt{3ab}$

31. Combining the expressions: $2x\sqrt[3]{xy^3z^2} - 6y\sqrt[3]{x^4z^2} = 2xy\sqrt[3]{xz^2} - 6xy\sqrt[3]{xz^2} = -4xy\sqrt[3]{xz^2}$

32. Multiplying: $\sqrt{2}\left(\sqrt{3} - 2\sqrt{2}\right) = \sqrt{6} - 2\sqrt{4} = \sqrt{6} - 4$

33. Multiplying: $\left(\sqrt{x} - 2\right)\left(\sqrt{x} - 3\right) = x - 2\sqrt{x} - 3\sqrt{x} + 6 = x - 5\sqrt{x} + 6$

34. Rationalizing the denominator: $\dfrac{3}{\sqrt{5} - 2} = \dfrac{3}{\sqrt{5} - 2} \cdot \dfrac{\sqrt{5} + 2}{\sqrt{5} + 2} = \dfrac{3\left(\sqrt{5} + 2\right)}{5 - 4} = 3\sqrt{5} + 6$

35. Rationalizing the denominator: $\dfrac{\sqrt{7} + \sqrt{5}}{\sqrt{7} - \sqrt{5}} = \dfrac{\sqrt{7} + \sqrt{5}}{\sqrt{7} - \sqrt{5}} \cdot \dfrac{\sqrt{7} + \sqrt{5}}{\sqrt{7} + \sqrt{5}} = \dfrac{7 + 2\sqrt{35} + 5}{7 - 5} = \dfrac{12 + 2\sqrt{35}}{2} = 6 + \sqrt{35}$

36. Rationalizing the denominator: $\dfrac{3\sqrt{7}}{3\sqrt{7} - 4} = \dfrac{3\sqrt{7}}{3\sqrt{7} - 4} \cdot \dfrac{3\sqrt{7} + 4}{3\sqrt{7} + 4} = \dfrac{9 \cdot 7 + 12\sqrt{7}}{63 - 16} = \dfrac{63 + 12\sqrt{7}}{47}$

37. Solving the equation:
$$\sqrt{4a+1} = 1$$
$$\left(\sqrt{4a+1}\right)^2 = (1)^2$$
$$4a+1 = 1$$
$$4a = 0$$
$$a = 0$$

38. Solving the equation:
$$\sqrt[3]{3x-8} = 1$$
$$\left(\sqrt[3]{3x-8}\right)^3 = (1)^3$$
$$3x-8 = 1$$
$$3x = 9$$
$$x = 3$$

39. Solving the equation:
$$\sqrt{3x+1} - 3 = 1$$
$$\sqrt{3x+1} = 4$$
$$\left(\sqrt{3x+1}\right)^2 = (4)^2$$
$$3x+1 = 16$$
$$3x = 15$$
$$x = 5$$

40. Solving the equation:
$$\sqrt{x+4} = \sqrt{x} - 2$$
$$\left(\sqrt{x+4}\right)^2 = \left(\sqrt{x} - 2\right)^2$$
$$x+4 = x - 2\sqrt{x} + 4$$
$$0 = -2\sqrt{x}$$
$$x = 0$$

There is no solution (0 does not check).

41. Graphing the equation:

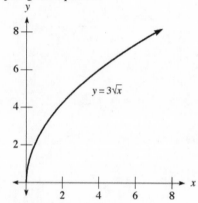

$$y = 3\sqrt{x}$$

42. Graphing the equation:

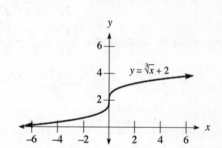

$$y = \sqrt[3]{x} + 2$$

43. Writing in terms of i: $i^{24} = \left(i^4\right)^6 = (1)^6 = 1$

44. Writing in terms of i: $i^{27} = i^{24} \cdot i^2 \cdot i = \left(i^4\right)^6 \cdot (-1) \cdot i = (1)^6 (-i) = -i$

45. Setting real and imaginary parts equal:
$$-2x = 3 \qquad 8y = -4$$
$$x = -\frac{3}{2} \qquad y = -\frac{1}{2}$$

46. Setting real and imaginary parts equal:
$$3x + 2 = -4$$
$$3x = -6 \qquad 2y = -8$$
$$x = -2 \qquad y = -4$$

47. Combining the numbers: $(3+5i) + (6-2i) = 9 + 3i$

48. Combining the numbers: $(2+5i) - \left[(3+2i) + (6-i)\right] = (2+5i) - (9+i) = 2 + 5i - 9 - i = -7 + 4i$

49. Multiplying: $3i(4+2i) = 12i + 6i^2 = -6 + 12i$

50. Multiplying: $(2+3i)(4+i) = 8 + 12i + 2i + 3i^2 = 8 + 14i - 3 = 5 + 14i$

51. Multiplying: $(4+2i)^2 = (4+2i)(4+2i) = 16 + 8i + 8i + 4i^2 = 16 + 16i - 4 = 12 + 16i$

52. Multiplying: $(4+3i)(4-3i) = 16 - 9i^2 = 16 + 9 = 25$

53. Dividing: $\dfrac{3+i}{i} = \dfrac{3+i}{i} \cdot \dfrac{i}{i} = \dfrac{3i-1}{-1} = 1 - 3i$

54. Dividing: $\dfrac{-3}{2+i} = \dfrac{-3}{2+i} \cdot \dfrac{2-i}{2-i} = \dfrac{-6+3i}{4+1} = -\dfrac{6}{5} + \dfrac{3}{5}i$

55.   Let l represent the desired length. Using the Pythagorean theorem:

$$18^2 + 13.5^2 = l^2$$
$$l^2 = 506.25$$
$$l = \sqrt{506.25} = 22.5$$

The length of one side of the roof is 22.5 feet.

56.   Let d represent the distance across the pond. Using the Pythagorean theorem:

$$25^2 + 60^2 = d^2$$
$$d^2 = 4225$$
$$d = \sqrt{4225} = 65$$

The distance across the pond is 65 yards.

Chapter 8
Quadratic Functions

8.1 Completing the Square $1-4\text{?}$ odd

1. Solving the equation:

$$x^2 = 25$$
$$x = \pm\sqrt{25} = \pm 5$$

3. Solving the equation:

$$y^2 = \frac{3}{4}$$
$$y = \pm\sqrt{\frac{3}{4}} = \pm\frac{\sqrt{3}}{2}$$

5. Solving the equation:

$$x^2 + 12 = 0$$
$$x^2 = -12$$
$$x = \pm\sqrt{-12} = \pm 2i\sqrt{3}$$

7. Solving the equation:

$$4a^2 - 45 = 0$$
$$4a^2 = 45$$
$$a^2 = \frac{45}{4}$$
$$a = \pm\sqrt{\frac{45}{4}} = \pm\frac{3\sqrt{5}}{2}$$

9. Solving the equation:

$$(2y-1)^2 = 25$$
$$2y - 1 = \pm\sqrt{25} = \pm 5$$
$$2y - 1 = -5, 5$$
$$2y = -4, 6$$
$$y = -2, 3$$

11. Solving the equation:

$$(2a+3)^2 = -9$$
$$2a + 3 = \pm\sqrt{-9} = \pm 3i$$
$$2a = -3 \pm 3i$$
$$a = \frac{-3 \pm 3i}{2}$$

13. Solving the equation:

$$x^2 + 8x + 16 = -27$$
$$(x+4)^2 = -27$$
$$x + 4 = \pm\sqrt{-27} = \pm 3i\sqrt{3}$$
$$x = -4 \pm 3i\sqrt{3}$$

15. Solving the equation:

$$4a^2 - 12a + 9 = -4$$
$$(2a-3)^2 = -4$$
$$2a - 3 = \pm\sqrt{-4} = \pm 2i$$
$$2a = 3 \pm 2i$$
$$a = \frac{3 \pm 2i}{2}$$

17. Completing the square: $x^2 + 12x + 36 = (x+6)^2$

19. Completing the square: $x^2 - 4x + 4 = (x-2)^2$

21. Completing the square: $a^2 - 10a + 25 = (a-5)^2$

23. Completing the square: $x^2 + 5x + \frac{25}{4} = \left(x + \frac{5}{2}\right)^2$

25. Completing the square: $y^2 - 7y + \frac{49}{4} = \left(y - \frac{7}{2}\right)^2$

27. Completing the square: $x^2 + \frac{1}{2}x + \frac{1}{16} = \left(x + \frac{1}{4}\right)^2$

29. Completing the square: $x^2 + \frac{2}{3}x + \frac{1}{9} = \left(x + \frac{1}{3}\right)^2$

31. Solving the equation:
$$x^2 + 12x = -27$$
$$x^2 + 12x + 36 = -27 + 36$$
$$(x+6)^2 = 9$$
$$x + 6 = \pm\sqrt{9} = \pm 3$$
$$x + 6 = -3, 3$$
$$x = -9, -3$$

33. Solving the equation:
$$a^2 - 2a + 5 = 0$$
$$a^2 - 2a + 1 = -5 + 1$$
$$(a-1)^2 = -4$$
$$a - 1 = \pm\sqrt{-4} = \pm 2i$$
$$a = 1 \pm 2i$$

35. Solving the equation:
$$y^2 - 8y + 1 = 0$$
$$y^2 - 8y + 16 = -1 + 16$$
$$(y-4)^2 = 15$$
$$y - 4 = \pm\sqrt{15}$$
$$y = 4 \pm \sqrt{15}$$

37. Solving the equation:
$$x^2 - 5x - 3 = 0$$
$$x^2 - 5x + \frac{25}{4} = 3 + \frac{25}{4}$$
$$\left(x - \frac{5}{2}\right)^2 = \frac{37}{4}$$
$$x - \frac{5}{2} = \pm\frac{\sqrt{37}}{2}$$
$$x = \frac{5 \pm \sqrt{37}}{2}$$

39. Solving the equation:
$$2x^2 - 4x - 8 = 0$$
$$x^2 - 2x - 4 = 0$$
$$x^2 - 2x + 1 = 4 + 1$$
$$(x-1)^2 = 5$$
$$x - 1 = \pm\sqrt{5}$$
$$x = 1 \pm \sqrt{5}$$

41. Solving the equation:
$$3t^2 - 8t + 1 = 0$$
$$t^2 - \frac{8}{3}t + \frac{1}{3} = 0$$
$$t^2 - \frac{8}{3}t + \frac{16}{9} = -\frac{1}{3} + \frac{16}{9}$$
$$\left(t - \frac{4}{3}\right)^2 = \frac{13}{9}$$
$$t - \frac{4}{3} = \pm\sqrt{\frac{13}{9}} = \pm\frac{\sqrt{13}}{3}$$
$$t = \frac{4 \pm \sqrt{13}}{3}$$

43. Solving the equation:
$$4x^2 - 3x + 5 = 0$$
$$x^2 - \frac{3}{4}x + \frac{5}{4} = 0$$
$$x^2 - \frac{3}{4}x + \frac{9}{64} = -\frac{5}{4} + \frac{9}{64}$$
$$\left(x - \frac{3}{8}\right)^2 = -\frac{71}{64}$$
$$x - \frac{3}{8} = \pm\sqrt{-\frac{71}{64}} = \pm\frac{i\sqrt{71}}{8}$$
$$x = \frac{3 \pm i\sqrt{71}}{8}$$

45. **a.** No, it cannot be solved by factoring.
   b. Solving the equation:
$$x^2 = -9$$
$$x = \pm\sqrt{-9}$$
$$x = \pm 3i$$

47. **a.** Solving by factoring:

$$x^2 - 6x = 0$$
$$x(x-6) = 0$$
$$x = 0, 6$$

b. Solving by completing the square:

$$x^2 - 6x = 0$$
$$x^2 - 6x + 9 = 0 + 9$$
$$(x-3)^2 = 9$$
$$x - 3 = \pm\sqrt{9}$$
$$x - 3 = -3, 3$$
$$x = 0, 6$$

49. **a.** Solving by factoring:

$$x^2 + 2x = 35$$
$$x^2 + 2x - 35 = 0$$
$$(x+7)(x-5) = 0$$
$$x = -7, 5$$

b. Solving by completing the square:

$$x^2 + 2x = 35$$
$$x^2 + 2x + 1 = 35 + 1$$
$$(x+1)^2 = 36$$
$$x + 1 = \pm\sqrt{36}$$
$$x + 1 = -6, 6$$
$$x = -7, 5$$

51. Substituting: $x^2 - 6x - 7 = \left(-3+\sqrt{2}\right)^2 - 6\left(-3+\sqrt{2}\right) - 7 = 9 - 6\sqrt{2} + 2 + 18 - 6\sqrt{2} - 7 = 22 - 12\sqrt{2}$

No, $x = -3 + \sqrt{2}$ is not a solution to the equation.

53. **a.** Solving the equation:

$$5x - 7 = 0$$
$$5x = 7$$
$$x = \frac{7}{5}$$

b. Solving the equation:

$$5x - 7 = 8$$
$$5x = 15$$
$$x = 3$$

c. Solving the equation:

$$(5x-7)^2 = 8$$
$$5x - 7 = \pm\sqrt{8}$$
$$5x - 7 = \pm 2\sqrt{2}$$
$$5x = 7 \pm 2\sqrt{2}$$
$$x = \frac{7 \pm 2\sqrt{2}}{5}$$

d. Solving the equation:

$$\sqrt{5x-7} = 8$$
$$\left(\sqrt{5x-7}\right)^2 = (8)^2$$
$$5x - 7 = 64$$
$$5x = 71$$
$$x = \frac{71}{5}$$

e. Solving the equation:

$$\frac{5}{2} - \frac{7}{2x} = \frac{4}{x}$$
$$2x\left(\frac{5}{2} - \frac{7}{2x}\right) = 2x\left(\frac{4}{x}\right)$$
$$5x - 7 = 8$$
$$5x = 15$$
$$x = 3$$

55. Solving the equation:

$$(x+5)^2 + (x-5)^2 = 52$$
$$x^2 + 10x + 25 + x^2 - 10x + 25 = 52$$
$$2x^2 + 50 = 52$$
$$2x^2 = 2$$
$$x^2 = 1$$
$$x = \pm\sqrt{1} = \pm 1$$

57. Solving the equation:

$$(2x+3)^2 + (2x-3)^2 = 26$$
$$4x^2 + 12x + 9 + 4x^2 - 12x + 9 = 26$$
$$8x^2 + 18 = 26$$
$$8x^2 = 8$$
$$x^2 = 1$$
$$x = \pm\sqrt{1} = \pm 1$$

59. Solving the equation:

$$(3x+4)(3x-4)-(x+2)(x-2)=-4$$
$$9x^2-16-x^2+4=-4$$
$$8x^2-12=-4$$
$$8x^2=8$$
$$x^2=1$$
$$x=\pm\sqrt{1}=\pm1$$

61. Completing the table:

x	$f(x)$	$g(x)$	$h(x)$
–2	49	49	25
–1	25	25	13
0	9	9	9
1	1	1	13
2	1	1	25

63. Solving the equation:

$$f(x)=0$$
$$(x-3)^2=0$$
$$x-3=0$$
$$x=3$$

65. **a.** Finding the x-intercepts:

$$x^2-5x-6=0$$
$$(x+1)(x-6)=0$$
$$x=-1,6$$

b. Setting $f(x)=0$:

$$f(x)=0$$
$$x^2-5x-6=0$$
$$(x+1)(x-6)=0$$
$$x=-1,6$$

c. Finding the value: $f(0)=(0)^2-5(0)-6=-6$

d. Finding the value: $f(1)=(1)^2-5(1)-6=-10$

67. The other two sides are $\dfrac{\sqrt{3}}{2}$ inch, 1 inch .

69. The hypotenuse is $\sqrt{2}$ inches.

71. Let x represent the horizontal distance. Using the Pythagorean theorem:

$$x^2+120^2=790^2$$
$$x^2+14400=624100$$
$$x^2=609700$$
$$x=\sqrt{609700}\approx781 \text{ feet}$$

73. Solving for r:

$$3456=3000(1+r)^2$$
$$(1+r)^2=1.152$$
$$1+r=\sqrt{1.152}$$
$$r=\sqrt{1.152}-1\approx0.073$$

The annual interest rate is 7.3%.

75. Its length is $20\sqrt{2}\approx28$ feet.

77. Simplifying: $\sqrt{45}=\sqrt{9\cdot5}=3\sqrt{5}$

79. Simplifying: $\sqrt{27y^5}=\sqrt{9y^4\cdot3y}=3y^2\sqrt{3y}$

81. Simplifying: $\sqrt[3]{54x^6y^5}=\sqrt[3]{27x^6y^3\cdot2y^2}=3x^2y\sqrt[3]{2y^2}$

83. Rationalizing the denominator: $\dfrac{3}{\sqrt{2}}=\dfrac{3}{\sqrt{2}}\cdot\dfrac{\sqrt{2}}{\sqrt{2}}=\dfrac{3\sqrt{2}}{2}$

85. Rationalizing the denominator: $\dfrac{2}{\sqrt[3]{4}}=\dfrac{2}{\sqrt[3]{4}}\cdot\dfrac{\sqrt[3]{2}}{\sqrt[3]{2}}=\dfrac{2\sqrt[3]{2}}{2}=\sqrt[3]{2}$

87. Simplifying: $\sqrt{49-4(6)(-5)}=\sqrt{49+120}=\sqrt{169}=13$

89. Simplifying: $\sqrt{(-27)^2-4(0.1)(1,700)}=\sqrt{729-680}=\sqrt{49}=7$

91. Simplifying: $\dfrac{-7+\sqrt{169}}{12}=\dfrac{-7+13}{12}=\dfrac{6}{12}=\dfrac{1}{2}$

93. Substituting values: $\sqrt{b^2-4ac}=\sqrt{7^2-4(6)(-5)}=\sqrt{49+120}=\sqrt{169}=13$

95. Factoring: $27t^3-8=(3t-2)(9t^2+6t+4)$

97. Solving for x:

$$(x+a)^2+(x-a)^2=10a^2$$
$$x^2+2ax+a^2+x^2-2ax+a^2=10a^2$$
$$2x^2+2a^2=10a^2$$
$$2x^2=8a^2$$
$$x^2=4a^2$$
$$x=\pm2a$$

99. Solving for x:

$$x^2+px+q=0$$
$$x^2+px+\frac{p^2}{4}=-q+\frac{p^2}{4}$$
$$\left(x+\frac{p}{2}\right)^2=\frac{p^2-4q}{4}$$
$$x+\frac{p}{2}=\pm\sqrt{\frac{p^2-4q}{4}}=\pm\frac{\sqrt{p^2-4q}}{2}$$
$$x=\frac{-p\pm\sqrt{p^2-4q}}{2}$$

101. Solving for x:

$$3x^2+px+q=0$$
$$x^2+\frac{p}{3}x+\frac{q}{3}=0$$
$$x^2+\frac{p}{3}x+\frac{p^2}{36}=-\frac{q}{3}+\frac{p^2}{36}$$
$$\left(x+\frac{p}{6}\right)^2=\frac{p^2-12q}{36}$$
$$x+\frac{p}{6}=\pm\sqrt{\frac{p^2-12q}{36}}=\pm\frac{\sqrt{p^2-12q}}{6}$$
$$x=\frac{-p\pm\sqrt{p^2-12q}}{6}$$

103. Completing the square:

$$x^2-10x+y^2-6y=-30$$
$$\left(x^2-10x+25\right)+\left(y^2-6y+9\right)=-30+25+9$$
$$(x-5)^2+(y-3)^2=4$$
$$(x-5)^2+(y-3)^2=2^2$$

8.2 The Quadratic Formula

1. **a.** Using the quadratic formula:

$$x=\frac{-4\pm\sqrt{4^2-4(3)(-2)}}{2(3)}=\frac{-4\pm\sqrt{16+24}}{6}=\frac{-4\pm\sqrt{40}}{6}=\frac{-4\pm2\sqrt{10}}{6}=\frac{-2\pm\sqrt{10}}{3}$$

b. Using the quadratic formula: $x=\dfrac{4\pm\sqrt{(-4)^2-4(3)(-2)}}{2(3)}=\dfrac{4\pm\sqrt{16+24}}{6}=\dfrac{4\pm\sqrt{40}}{6}=\dfrac{4\pm2\sqrt{10}}{6}=\dfrac{2\pm\sqrt{10}}{3}$

c. Using the quadratic formula: $x=\dfrac{-4\pm\sqrt{4^2-4(3)(2)}}{2(3)}=\dfrac{-4\pm\sqrt{16-24}}{6}=\dfrac{-4\pm\sqrt{-8}}{6}=\dfrac{-4\pm2i\sqrt{2}}{6}=\dfrac{-2\pm i\sqrt{2}}{3}$

d. Using the quadratic formula:

$$x=\frac{-4\pm\sqrt{4^2-4(2)(-3)}}{2(2)}=\frac{-4\pm\sqrt{16+24}}{4}=\frac{-4\pm\sqrt{40}}{4}=\frac{-4\pm2\sqrt{10}}{4}=\frac{-2\pm\sqrt{10}}{2}$$

e. Using the quadratic formula: $x=\dfrac{4\pm\sqrt{(-4)^2-4(2)(3)}}{2(2)}=\dfrac{4\pm\sqrt{16-24}}{4}=\dfrac{4\pm\sqrt{-8}}{4}=\dfrac{4\pm2i\sqrt{2}}{4}=\dfrac{2\pm i\sqrt{2}}{2}$

3.   **a.**   Using the quadratic formula: $x = \dfrac{2 \pm \sqrt{(-2)^2 - 4(1)(2)}}{2(1)} = \dfrac{2 \pm \sqrt{4-8}}{2} = \dfrac{2 \pm \sqrt{-4}}{2} = \dfrac{2 \pm 2i}{2} = 1 \pm i$

    b.   Using the quadratic formula: $x = \dfrac{2 \pm \sqrt{(-2)^2 - 4(1)(5)}}{2(1)} = \dfrac{2 \pm \sqrt{4-20}}{2} = \dfrac{2 \pm \sqrt{-16}}{2} = \dfrac{2 \pm 4i}{2} = 1 \pm 2i$

    c.   Using the quadratic formula: $x = \dfrac{-2 \pm \sqrt{2^2 - 4(1)(2)}}{2(1)} = \dfrac{-2 \pm \sqrt{4-8}}{2} = \dfrac{-2 \pm \sqrt{-4}}{2} = \dfrac{-2 \pm 2i}{2} = -1 \pm i$

5.   Solving the equation:
$$\tfrac{1}{6}x^2 - \tfrac{1}{2}x + \tfrac{1}{3} = 0$$
$$x^2 - 3x + 2 = 0$$
$$(x-1)(x-2) = 0$$
$$x = 1, 2$$

7.   Solving the equation:
$$\frac{x^2}{2} + 1 = \frac{2x}{3}$$
$$3x^2 + 6 = 4x$$
$$3x^2 - 4x + 6 = 0$$
$$x = \frac{4 \pm \sqrt{16-72}}{6} = \frac{4 \pm \sqrt{-56}}{6} = \frac{4 \pm 2i\sqrt{14}}{6} = \frac{2 \pm i\sqrt{14}}{3}$$

9.   Solving the equation: 11.   Solving the equation:
$$y^2 - 5y = 0 \qquad\qquad\qquad\qquad\qquad 30x^2 + 40x = 0$$
$$y(y-5) = 0 \qquad\qquad\qquad\qquad\qquad 10x(3x+4) = 0$$
$$y = 0, 5 \qquad\qquad\qquad\qquad\qquad\qquad x = -\tfrac{4}{3}, 0$$

13.   Solving the equation:
$$\frac{2t^2}{3} - t = -\frac{1}{6}$$
$$4t^2 - 6t = -1$$
$$4t^2 - 6t + 1 = 0$$
$$t = \frac{6 \pm \sqrt{36-16}}{8} = \frac{6 \pm \sqrt{20}}{8} = \frac{6 \pm 2\sqrt{5}}{8} = \frac{3 \pm \sqrt{5}}{4}$$

15.   Solving the equation:
$$0.01x^2 + 0.06x - 0.08 = 0$$
$$x^2 + 6x - 8 = 0$$
$$x = \frac{-6 \pm \sqrt{36+32}}{2} = \frac{-6 \pm \sqrt{68}}{2} = \frac{-6 \pm 2\sqrt{17}}{2} = -3 \pm \sqrt{17}$$

17.   Solving the equation:
$$2x + 3 = -2x^2$$
$$2x^2 + 2x + 3 = 0$$
$$x = \frac{-2 \pm \sqrt{4-24}}{4} = \frac{-2 \pm \sqrt{-20}}{4} = \frac{-2 \pm 2i\sqrt{5}}{4} = \frac{-1 \pm i\sqrt{5}}{2}$$

19.   Solving the equation:
$$100x^2 - 200x + 100 = 0$$
$$100\left(x^2 - 2x + 1\right) = 0$$
$$100(x-1)^2 = 0$$
$$x = 1$$

21.   Solving the equation:

$$\tfrac{1}{2}r^2 = \tfrac{1}{6}r - \tfrac{2}{3}$$
$$3r^2 = r - 4$$
$$3r^2 - r + 4 = 0$$
$$r = \frac{1 \pm \sqrt{1 - 48}}{6} = \frac{1 \pm \sqrt{-47}}{6} = \frac{1 \pm i\sqrt{47}}{6}$$

23.   Solving the equation:

$$(x - 3)(x - 5) = 1$$
$$x^2 - 8x + 15 = 1$$
$$x^2 - 8x + 14 = 0$$
$$x = \frac{8 \pm \sqrt{64 - 56}}{2} = \frac{8 \pm \sqrt{8}}{2} = \frac{8 \pm 2\sqrt{2}}{2} = 4 \pm \sqrt{2}$$

25.   Solving the equation:

$$(x + 3)^2 + (x - 8)(x - 1) = 16$$
$$x^2 + 6x + 9 + x^2 - 9x + 8 = 16$$
$$2x^2 - 3x + 1 = 0$$
$$(2x - 1)(x - 1) = 0$$
$$x = \tfrac{1}{2}, 1$$

27.   Solving the equation:

$$\frac{x^2}{3} - \frac{5x}{6} = \tfrac{1}{2}$$
$$2x^2 - 5x = 3$$
$$2x^2 - 5x - 3 = 0$$
$$(2x + 1)(x - 3) = 0$$
$$x = -\tfrac{1}{2}, 3$$

29.   Solving the equation:

$$\frac{1}{x + 1} - \frac{1}{x} = \tfrac{1}{2}$$
$$2x(x + 1)\left(\frac{1}{x + 1} - \frac{1}{x}\right) = 2x(x + 1) \cdot \tfrac{1}{2}$$
$$2x - (2x + 2) = x^2 + x$$
$$2x - 2x - 2 = x^2 + x$$
$$x^2 + x + 2 = 0$$
$$x = \frac{-1 \pm \sqrt{1 - 8}}{2} = \frac{-1 \pm \sqrt{-7}}{2} = \frac{-1 \pm i\sqrt{7}}{2}$$

31.   Solving the equation:

$$\frac{1}{y - 1} + \frac{1}{y + 1} = 1$$
$$(y + 1)(y - 1)\left(\frac{1}{y - 1} + \frac{1}{y + 1}\right) = (y + 1)(y - 1) \cdot 1$$
$$y + 1 + y - 1 = y^2 - 1$$
$$2y = y^2 - 1$$
$$y^2 - 2y - 1 = 0$$
$$y = \frac{2 \pm \sqrt{4 + 4}}{2} = \frac{2 \pm \sqrt{8}}{2} = \frac{2 \pm 2\sqrt{2}}{2} = 1 \pm \sqrt{2}$$

33.   Solving the equation:

$$\frac{1}{x+2}+\frac{1}{x+3}=1$$

$$(x+2)(x+3)\left(\frac{1}{x+2}+\frac{1}{x+3}\right)=(x+2)(x+3)\bullet 1$$

$$x+3+x+2=x^2+5x+6$$

$$2x+5=x^2+5x+6$$

$$x^2+3x+1=0$$

$$x=\frac{-3\pm\sqrt{9-4}}{2}=\frac{-3\pm\sqrt{5}}{2}$$

35.   Solving the equation:

$$\frac{6}{r^2-1}-\frac{1}{2}=\frac{1}{r+1}$$

$$2(r+1)(r-1)\left(\frac{6}{(r+1)(r-1)}-\frac{1}{2}\right)=2(r+1)(r-1)\bullet\frac{1}{r+1}$$

$$12-\left(r^2-1\right)=2r-2$$

$$12-r^2+1=2r-2$$

$$r^2+2r-15=0$$

$$(r+5)(r-3)=0$$

$$r=-5,3$$

37.   Solving the equation:

$$x^3-8=0$$

$$(x-2)\left(x^2+2x+4\right)=0$$

$$x=2 \quad\text{or}\quad x=\frac{-2\pm\sqrt{4-16}}{2}=\frac{-2\pm\sqrt{-12}}{2}=\frac{-2\pm 2i\sqrt{3}}{2}=-1\pm i\sqrt{3}$$

$$x=2,-1\pm i\sqrt{3}$$

39.   Solving the equation:

$$8a^3+27=0$$

$$(2a+3)\left(4a^2-6a+9\right)=0$$

$$a=-\frac{3}{2} \quad\text{or}\quad a=\frac{6\pm\sqrt{36-144}}{8}=\frac{6\pm\sqrt{-108}}{8}=\frac{6\pm 6i\sqrt{3}}{8}=\frac{3\pm 3i\sqrt{3}}{4}$$

$$a=-\frac{3}{2},\frac{3\pm 3i\sqrt{3}}{4}$$

41.   Solving the equation:

$$125t^3-1=0$$

$$(5t-1)\left(25t^2+5t+1\right)=0$$

$$t=\frac{1}{5} \quad\text{or}\quad t=\frac{-5\pm\sqrt{25-100}}{50}=\frac{-5\pm\sqrt{-75}}{50}=\frac{-5\pm 5i\sqrt{3}}{50}=\frac{-1\pm i\sqrt{3}}{10}$$

$$t=\frac{1}{5},\frac{-1\pm i\sqrt{3}}{10}$$

43. Solving the equation:

$$2x^3 + 2x^2 + 3x = 0$$
$$x\left(2x^2 + 2x + 3\right) = 0$$

$x = 0$ or $x = \dfrac{-2 \pm \sqrt{4 - 24}}{4} = \dfrac{-2 \pm \sqrt{-20}}{4} = \dfrac{-2 \pm 2i\sqrt{5}}{4} = \dfrac{-1 \pm i\sqrt{5}}{2}$

$$x = 0, \dfrac{-1 \pm i\sqrt{5}}{2}$$

45. Solving the equation:

$$3y^4 = 6y^3 - 6y^2$$
$$3y^4 - 6y^3 + 6y^2 = 0$$
$$3y^2\left(y^2 - 2y + 2\right) = 0$$

$y = 0$ or $y = \dfrac{2 \pm \sqrt{4 - 8}}{2} = \dfrac{2 \pm \sqrt{-4}}{2} = \dfrac{2 \pm 2i}{2} = 1 \pm i$

$$y = 0, 1 \pm i$$

47. Solving the equation:

$$6t^5 + 4t^4 = -2t^3$$
$$6t^5 + 4t^4 + 2t^3 = 0$$
$$2t^3\left(3t^2 + 2t + 1\right) = 0$$

$t = 0$ or $t = \dfrac{-2 \pm \sqrt{4 - 12}}{6} = \dfrac{-2 \pm \sqrt{-8}}{6} = \dfrac{-2 \pm 2i\sqrt{2}}{6} = \dfrac{-1 \pm i\sqrt{2}}{3}$

$$t = 0, \dfrac{-1 \pm i\sqrt{2}}{3}$$

49. The expressions from **a** and **b** are equivalent, since: $\dfrac{6 + 2\sqrt{3}}{4} = \dfrac{2\left(3 + \sqrt{3}\right)}{4} = \dfrac{3 + \sqrt{3}}{2}$

51. **a.** Solving by factoring:

$$3x^2 - 5x = 0$$
$$x\left(3x - 5\right) = 0$$
$$x = 0, \dfrac{5}{3}$$

   b. Using the quadratic formula: $x = \dfrac{5 \pm \sqrt{(-5)^2 - 4(3)(0)}}{2(3)} = \dfrac{5 \pm \sqrt{25 - 0}}{6} = \dfrac{5 \pm 5}{6} = 0, \dfrac{5}{3}$

53. No, it cannot be solved by factoring. Using the quadratic formula:

$$x = \dfrac{4 \pm \sqrt{(-4)^2 - 4(1)(7)}}{2(1)} = \dfrac{4 \pm \sqrt{16 - 28}}{2} = \dfrac{4 \pm \sqrt{-12}}{2} = \dfrac{4 \pm 2i\sqrt{3}}{2} = 2 \pm i\sqrt{3}$$

55. Substituting: $x^2 + 2x = (-1 + i)^2 + 2(-1 + i) = 1 - 2i + i^2 - 2 + 2i = 1 - 2i - 1 - 2 + 2i = -2$

Yes, $x = -1 + i$ is a solution to the equation.

57. **a.** Solving the equation: **b.** Solving the equation:

$$f(x) = 0$$
$$x^2 - 2x - 3 = 0$$
$$(x + 1)(x - 3) = 0$$
$$x = -1, 3$$

$$f(x) = -11$$
$$x^2 - 2x - 3 = -11$$
$$x^2 - 2x = -8$$
$$x^2 - 2x + 1 = -8 + 1$$
$$(x - 1)^2 = -7$$
$$x - 1 = \pm i\sqrt{7}$$
$$x = 1 \pm i\sqrt{7}$$

c. Solving the equation:

$$f(x) = -2x + 1$$
$$x^2 - 2x - 3 = -2x + 1$$
$$x^2 - 4 = 0$$
$$(x+2)(x-2) = 0$$
$$x = -2, 2$$

d. Solving the equation:

$$f(x) = 2x + 1$$
$$x^2 - 2x - 3 = 2x + 1$$
$$x^2 - 4x = 4$$
$$x^2 - 4x + 4 = 4 + 4$$
$$(x-2)^2 = 8$$
$$x - 2 = \pm 2\sqrt{2}$$
$$x = 2 \pm 2\sqrt{2}$$

59. **a.** Solving the equation:

$$f(x) = g(x)$$
$$\frac{10}{x^2} = 3 + \frac{1}{x}$$
$$x^2 \left(\frac{10}{x^2} \right) = x^2 \left(3 + \frac{1}{x} \right)$$
$$10 = 3x^2 + x$$
$$0 = 3x^2 + x - 10$$
$$0 = (3x - 5)(x + 2)$$
$$x = -2, \frac{5}{3}$$

b. Solving the equation:

$$f(x) = g(x)$$
$$\frac{10}{x^2} = 8x - \frac{17}{x^2}$$
$$x^2 \left(\frac{10}{x^2} \right) = x^2 \left(8x - \frac{17}{x^2} \right)$$
$$10 = 8x^3 - 17$$
$$0 = 8x^3 - 27$$
$$0 = (2x - 3)\left(4x^2 + 6x + 9 \right)$$
$$x = \frac{3}{2}, x = \frac{-6 \pm \sqrt{6^2 - 4(4)(9)}}{2(4)} = \frac{-6 \pm \sqrt{-108}}{8} = \frac{-6 \pm 6i\sqrt{3}}{8} = \frac{-3 \pm 3i\sqrt{3}}{4}$$

c. Solving the equation:

$$f(x) = g(x)$$
$$\frac{10}{x^2} = 0$$
$$10 = 0$$
(impossible)

There is no solution ($\varnothing$).

d. Solving the equation:

$$f(x) = g(x)$$
$$\frac{10}{x^2} = 10$$
$$x^2 \left(\frac{10}{x^2} \right) = x^2 (10)$$
$$10 = 10x^2$$
$$x^2 = 1$$
$$x = \pm 1$$

61. Substituting $s = 74$:

$$5t + 16t^2 = 74$$
$$16t^2 + 5t - 74 = 0$$
$$(16t + 37)(t - 2) = 0$$
$$t = 2 \quad \left(t = -\frac{37}{16} \text{ is impossible} \right)$$

It will take 2 seconds for the object to fall 74 feet.

63. Since profit is revenue minus the cost, the equation is:

$$100x - 0.5x^2 - (60x + 300) = 300$$
$$100x - 0.5x^2 - 60x - 300 = 300$$
$$-0.5x^2 + 40x - 600 = 0$$
$$x^2 - 80x + 1,200 = 0$$
$$(x - 20)(x - 60) = 0$$
$$x = 20, 60$$

The weekly profit is $300 if 20 items or 60 items are sold.

65. Let x represent the width of strip being cut off. After removing the strip, the overall area is 80% of its original area. The equation is:

$$(10.5 - 2x)(8.2 - 2x) = 0.80(10.5 \times 8.2)$$
$$86.1 - 37.4x + 4x^2 = 68.88$$
$$4x^2 - 37.4x + 17.22 = 0$$

$$x = \frac{37.4 \pm \sqrt{(-37.4)^2 - 4(4)(17.22)}}{8} = \frac{37.4 \pm \sqrt{1123.24}}{8} \approx \frac{37.4 \pm 33.5}{8} \approx 0.49, 8.86$$

The width of strip is 0.49 centimeter (8.86 cm is impossible).

67. **a.** The two equations are: $l + w = 10, lw = 15$

b. Since $l = 10 - w$, the equation is:

$$w(10 - w) = 15$$
$$10w - w^2 = 15$$
$$w^2 - 10w + 15 = 0$$

$$w = \frac{10 \pm \sqrt{100 - 60}}{2} = \frac{10 \pm \sqrt{40}}{2} = \frac{10 \pm 2\sqrt{10}}{2} = 5 \pm \sqrt{10} \approx 1.84, 8.16$$

The length and width are 8.16 yards and 1.84 yards.

c. Two answers are possible because either dimension (long or short) may be considered the length.

69. Dividing using long division:

$$\begin{array}{r} 4y + 1 \\ 2y - 7 \overline{\smash{\big)} 8y^2 - 26y - 9} \\ \underline{8y^2 - 28y } \\ 2y - 9 \\ \underline{2y - 7} \\ -2 \end{array}$$

The quotient is $4y + 1 - \dfrac{2}{2y - 7}$.

71. Dividing using long division:

$$\begin{array}{r} x^2 + 7x + 12 \\ x + 2 \overline{\smash{\big)} x^3 + 9x^2 + 26x + 24} \\ \underline{x^3 + 2x^2 } \\ 7x^2 + 26x \\ \underline{7x^2 + 14x } \\ 12x + 24 \\ \underline{12x + 24} \\ 0 \end{array}$$

The quotient is $x^2 + 7x + 12$.

73. Simplifying: $25^{1/2} = \sqrt{25} = 5$

75. Simplifying: $\left(\frac{9}{25}\right)^{3/2} = \left(\frac{3}{5}\right)^3 = \frac{27}{125}$

77. Simplifying: $8^{-2/3} = \left(8^{-1/3}\right)^2 = \left(\frac{1}{2}\right)^2 = \frac{1}{4}$

79. Simplifying: $\dfrac{\left(49x^8 y^{-4}\right)^{1/2}}{\left(27x^{-3} y^9\right)^{-1/3}} = \dfrac{7x^4 y^{-2}}{\frac{1}{3} xy^{-3}} = 21x^{4-1} y^{-2+3} = 21x^3 y$

81. Evaluating $b^2 - 4ac$: $b^2 - 4ac = (-3)^2 - 4(1)(-40) = 9 + 160 = 169$

83. Evaluating $b^2 - 4ac$: $b^2 - 4ac = 12^2 - 4(4)(9) = 144 - 144 = 0$

85. Solving the equation:
$$k^2 - 144 = 0$$
$$(k+12)(k-12) = 0$$
$$k = -12, 12$$

87. Multiplying: $(x-3)(x+2) = x^2 + 2x - 3x - 6 = x^2 - x - 6$

89. Multiplying:
$$(x-3)(x-3)(x+2) = \left(x^2 - 6x + 9\right)(x+2)$$
$$= x^3 - 6x^2 + 9x + 2x^2 - 12x + 18$$
$$= x^3 - 4x^2 - 3x + 18$$

91. Solving the equation:
$$x^2 + \sqrt{3}x - 6 = 0$$
$$\left(x + 2\sqrt{3}\right)\left(x - \sqrt{3}\right) = 0$$
$$x = -2\sqrt{3}, \sqrt{3}$$

93. Solving the equation:
$$\sqrt{2}x^2 + 2x - \sqrt{2} = 0$$
$$x = \frac{-2 \pm \sqrt{4 + 4(2)}}{2\sqrt{2}} = \frac{-2 \pm \sqrt{12}}{2\sqrt{2}} = \frac{-2 \pm 2\sqrt{3}}{2\sqrt{2}} \cdot \frac{\sqrt{2}}{\sqrt{2}} = \frac{-2\sqrt{2} \pm 2\sqrt{6}}{4} = \frac{-\sqrt{2} \pm \sqrt{6}}{2}$$

95. Solving the equation:
$$x^2 + ix + 2 = 0$$
$$x = \frac{-i \pm \sqrt{-1-8}}{2} = \frac{-i \pm 3i}{2} = -2i, i$$

8.3 Additional Items Involving Solutions to Equations

1. Computing the discriminant: $D = (-6)^2 - 4(1)(5) = 36 - 20 = 16$. The equation will have two rational solutions.

3. First write the equation as $4x^2 - 4x + 1 = 0$. Computing the discriminant: $D = (-4)^2 - 4(4)(1) = 16 - 16 = 0$
The equation will have one rational solution.

5. Computing the discriminant: $D = 1^2 - 4(1)(-1) = 1 + 4 = 5$. The equation will have two irrational solutions.

7. First write the equation as $2y^2 - 3y - 1 = 0$. Computing the discriminant: $D = (-3)^2 - 4(2)(-1) = 9 + 8 = 17$
The equation will have two irrational solutions.

9. Computing the discriminant: $D = 0^2 - 4(1)(-9) = 36$. The equation will have two rational solutions.

11. First write the equation as $5a^2 - 4a - 5 = 0$. Computing the discriminant: $D = (-4)^2 - 4(5)(-5) = 16 + 100 = 116$
The equation will have two irrational solutions.

13. Setting the discriminant equal to 0:
$$(-k)^2 - 4(1)(25) = 0$$
$$k^2 - 100 = 0$$
$$k^2 = 100$$
$$k = \pm 10$$

15. First write the equation as $x^2 - kx + 36 = 0$. Setting the discriminant equal to 0:
$$(-k)^2 - 4(1)(36) = 0$$
$$k^2 - 144 = 0$$
$$k^2 = 144$$
$$k = \pm 12$$

17. Setting the discriminant equal to 0:
$$(-12)^2 - 4(4)(k) = 0$$
$$144 - 16k = 0$$
$$16k = 144$$
$$k = 9$$

19. First write the equation as $kx^2 - 40x - 25 = 0$. Setting the discriminant equal to 0:
$$(-40)^2 - 4(k)(-25) = 0$$
$$1600 + 100k = 0$$
$$100k = -1600$$
$$k = -16$$

21. Setting the discriminant equal to 0:
$$(-k)^2 - 4(3)(2) = 0$$
$$k^2 - 24 = 0$$
$$k^2 = 24$$
$$k = \pm\sqrt{24} = \pm 2\sqrt{6}$$

23. Writing the equation:
$$(x-5)(x-2) = 0$$
$$x^2 - 7x + 10 = 0$$

25. Writing the equation:
$$(t+3)(t-6) = 0$$
$$t^2 - 3t - 18 = 0$$

27. Writing the equation:
$$(y-2)(y+2)(y-4) = 0$$
$$\left(y^2 - 4\right)(y-4) = 0$$
$$y^3 - 4y^2 - 4y + 16 = 0$$

29. Writing the equation:
$$(2x-1)(x-3) = 0$$
$$2x^2 - 7x + 3 = 0$$

31. Writing the equation:
$$(4t+3)(t-3) = 0$$
$$4t^2 - 9t - 9 = 0$$

33. Writing the equation:
$$(x-3)(x+3)(6x-5) = 0$$
$$\left(x^2 - 9\right)(6x-5) = 0$$
$$6x^3 - 5x^2 - 54x + 45 = 0$$

35. Writing the equation:
$$(2a+1)(5a-3) = 0$$
$$10a^2 - a - 3 = 0$$

37. Writing the equation:
$$(3x+2)(3x-2)(x-1) = 0$$
$$\left(9x^2 - 4\right)(x-1) = 0$$
$$9x^3 - 9x^2 - 4x + 4 = 0$$

39. Writing the equation:
$$(x-2)(x+2)(x-3)(x+3) = 0$$
$$\left(x^2 - 4\right)\left(x^2 - 9\right) = 0$$
$$x^4 - 13x^2 + 36 = 0$$

41. Writing the equation:
$$(x-1)(x+2) = 0$$
$$x^2 + x - 2 = 0$$

So $f(x) = x^2 + x - 2$.

43. Starting with the solutions:
$$x = 3 \pm \sqrt{2}$$
$$x - 3 = \pm\sqrt{2}$$
$$(x-3)^2 = \left(\pm\sqrt{2}\right)^2$$
$$x^2 - 6x + 9 = 2$$
$$x^2 - 6x + 7 = 0$$
So $f(x) = x^2 - 6x + 7$.

45. Starting with the solutions:
$$x = 2 \pm i$$
$$x - 2 = \pm i$$
$$(x-2)^2 = (\pm i)^2$$
$$x^2 - 4x + 4 = -1$$
$$x^2 - 4x + 5 = 0$$
So $f(x) = x^2 - 4x + 5$.

47.   Starting with the solutions:

$$x = \frac{5 \pm \sqrt{7}}{2}$$
$$2x = 5 \pm \sqrt{7}$$
$$(2x - 5)^2 = (\pm\sqrt{7})^2$$
$$4x^2 - 20x + 25 = 7$$
$$4x^2 - 20x + 18 = 0$$
$$2x^2 - 10x + 9 = 0$$

So $f(x) = 2x^2 - 10x + 9$.

49.   Starting with the solutions:

$$x = \frac{3 \pm i\sqrt{5}}{2}$$
$$2x = 3 \pm i\sqrt{5}$$
$$(2x - 3)^2 = (\pm i\sqrt{5})^2$$
$$4x^2 - 12x + 9 = -5$$
$$4x^2 - 12x + 14 = 0$$
$$2x^2 - 6x + 7 = 0$$

So $f(x) = 2x^2 - 6x + 7$.

51.   Starting with the solutions:

$$x = 2 \pm i\sqrt{3}$$
$$(x - 2)^2 = (\pm i\sqrt{3})^2$$
$$x^2 - 4x + 4 = -3$$
$$x^2 - 4x + 7 = 0$$

So $f(x) = x(x^2 - 4x + 7) = x^3 - 4x^2 + 7x$.

53.   **a.**   This graph has one x-intercept, which is $y = g(x)$.

b.   This graph has no x-intercepts, which is $y = h(x)$.

c.   This graph has two x-intercepts, which is $y = f(x)$.

55.   Multiplying: $(a^{1/2} - 5)(a^{1/2} + 3) = a - 2a^{1/2} - 15$

57.   Multiplying: $(x^{1/2} - 8)(x^{1/2} + 8) = (x^{1/2})^2 - (8)^2 = x - 64$

59.   Dividing: $\dfrac{45x^{5/3}y^{7/3} - 36x^{8/3}y^{4/3}}{9x^{2/3}y^{1/3}} = \dfrac{45x^{5/3}y^{7/3}}{9x^{2/3}y^{1/3}} - \dfrac{36x^{8/3}y^{4/3}}{9x^{2/3}y^{1/3}} = 5xy^2 - 4x^2y$

61.   Factoring: $8(x+1)^{4/3} - 2(x+1)^{1/3} = 2(x+1)^{1/3}[4(x+1) - 1] = 2(x+1)^{1/3}(4x + 3)$

63.   Factoring: $9x^{2/3} + 12x^{1/3} + 4 = (3x^{1/3} + 2)^2$

65.   Simplifying: $(x - 2)^2 - 3(x - 2) - 10 = x^2 - 4x + 4 - 3x + 6 - 10 = x^2 - 7x$

67.   Simplifying: $(3a - 2)^2 + 2(3a - 2) - 3 = 9a^2 - 12a + 4 + 6a - 4 - 3 = 9a^2 - 6a - 3$

69.   Simplifying:

$$6(2a+4)^2 - (2a+4) - 2 = 6(4a^2 + 16a + 16) - (2a + 4) - 2$$
$$= 24a^2 + 96a + 96 - 2a - 4 - 2$$
$$= 24a^2 + 94a + 90$$

71.   Solving the equation:

$$x^2 = -2$$
$$x = \pm\sqrt{-2} = \pm i\sqrt{2}$$

73.   Solving the equation:

$$\sqrt{x} = 2$$
$$x = 2^2 = 4$$

75.   Solving the equation:

$$x + 3 = -2$$
$$x = -5$$

77. Solving the equation:
$$y^2 + y - 6 = 0$$
$$(y+3)(y-2) = 0$$
$$y = -3, 2$$

79. Solving the equation:
$$6x^2 - 13x - 5 = 0$$
$$(3x+1)(2x-5) = 0$$
$$x = -\frac{1}{3}, \frac{5}{2}$$

81. Solving the equation:
$$4x^2 = 8x + 5$$
$$4x^2 - 8x - 5 = 0$$
$$(2x-5)(2x+1) = 0$$
$$x = -\frac{1}{2}, \frac{5}{2}$$

83. Solving the equation:
$$4x^2 - 1 = 4x$$
$$4x^2 - 4x - 1 = 0$$
$$x = \frac{4 \pm \sqrt{16+16}}{8} = \frac{4 \pm \sqrt{32}}{8} = \frac{4 \pm 4\sqrt{2}}{8} = \frac{1 \pm \sqrt{2}}{2} \approx -0.21, 1.21$$

85. Solving the equation:
$$3x^3 - 2x^2 - 3x + 2 = 0$$
$$x^2(3x-2) - 1(3x-2) = 0$$
$$(3x-2)(x^2-1) = 0$$
$$(3x-2)(x+1)(x-1) = 0$$
$$x = -1, \frac{2}{3}, 1$$

87. Solving the equation:
$$3x^3 - 9x = 2x^2 - 6$$
$$3x^3 - 2x^2 - 9x + 6 = 0$$
$$x^2(3x-2) - 3(3x-2) = 0$$
$$(3x-2)(x^2-3) = 0$$
$$x = \frac{2}{3}, \pm\sqrt{3}$$
$$x \approx -1.73, 0.67, 1.73$$

8.4 Equations Quadratic in Form

1. Solving the equation:
$$(x-3)^2 + 3(x-3) + 2 = 0$$
$$(x-3+2)(x-3+1) = 0$$
$$(x-1)(x-2) = 0$$
$$x = 1, 2$$

3. Solving the equation:
$$2(x+4)^2 + 5(x+4) - 12 = 0$$
$$[2(x+4)-3][(x+4)+4] = 0$$
$$(2x+5)(x+8) = 0$$
$$x = -8, -\frac{5}{2}$$

5. Solving the equation:
$$x^4 - 10x^2 + 9 = 0$$
$$(x^2-1)(x^2-9) = 0$$
$$x^2 = 1, 9$$
$$x = \pm 1, \pm 3$$

7. Solving the equation:
$$x^4 - 7x^2 + 12 = 0$$
$$(x^2-4)(x^2-3) = 0$$
$$x^2 = 4, 3$$
$$x = \pm 2, \pm\sqrt{3}$$

9. Solving the equation:
$$x^4 - 6x^2 - 27 = 0$$
$$(x^2-9)(x^2+3) = 0$$
$$x^2 = 9, -3$$
$$x = \pm 3, \pm i\sqrt{3}$$

11. Solving the equation:
$$x^4 + 9x^2 = -20$$
$$x^4 + 9x^2 + 20 = 0$$
$$(x^2+4)(x^2+5) = 0$$
$$x^2 = -4, -5$$
$$x = \pm 2i, \pm i\sqrt{5}$$

13. Solving the equation:
$$(2a-3)^2 - 9(2a-3) = -20$$
$$(2a-3)^2 - 9(2a-3) + 20 = 0$$
$$(2a-3-4)(2a-3-5) = 0$$
$$(2a-7)(2a-8) = 0$$
$$a = \frac{7}{2}, 4$$

15. Solving the equation:
$$2(4a+2)^2 = 3(4a+2) + 20$$
$$2(4a+2)^2 - 3(4a+2) - 20 = 0$$
$$[2(4a+2)+5][(4a+2)-4] = 0$$
$$(8a+9)(4a-2) = 0$$
$$a = -\frac{9}{8}, \frac{1}{2}$$

17. Solving the equation:
$$6t^4 = -t^2 + 5$$
$$6t^4 + t^2 - 5 = 0$$
$$(6t^2 - 5)(t^2 + 1) = 0$$
$$t^2 = \frac{5}{6}, -1$$
$$t = \pm\sqrt{\frac{5}{6}} = \pm\frac{\sqrt{30}}{6}, \pm i$$

19. Solving the equation:
$$9x^4 - 49 = 0$$
$$(3x^2 - 7)(3x^2 + 7) = 0$$
$$x^2 = \frac{7}{3}, -\frac{7}{3}$$
$$x = \pm\sqrt{\frac{7}{3}}, \pm\sqrt{-\frac{7}{3}}$$
$$t = \pm\frac{\sqrt{21}}{3}, \pm\frac{i\sqrt{21}}{3}$$

21. Solving the equation:
$$x - 7\sqrt{x} + 10 = 0$$
$$(\sqrt{x} - 5)(\sqrt{x} - 2) = 0$$
$$\sqrt{x} = 2, 5$$
$$x = 4, 25$$
Both values check in the original equation.

23. Solving the equation:
$$t - 2\sqrt{t} - 15 = 0$$
$$(\sqrt{t} - 5)(\sqrt{t} + 3) = 0$$
$$\sqrt{t} = -3, 5$$
$$t = 9, 25$$
Only $t = 25$ checks in the original equation.

25. Solving the equation:
$$6x + 11\sqrt{x} = 35$$
$$6x + 11\sqrt{x} - 35 = 0$$
$$(3\sqrt{x} - 5)(2\sqrt{x} + 7) = 0$$
$$\sqrt{x} = \frac{5}{3}, -\frac{7}{2}$$
$$x = \frac{25}{9}, \frac{49}{4}$$

Only $x = \frac{25}{9}$ checks in the original equation.

27. Solving the equation:
$$x - 2\sqrt{x} - 8 = 0$$
$$(\sqrt{x} - 4)(\sqrt{x} + 2) = 0$$
$$\sqrt{x} = -2, 4$$
$$x = 4, 16$$

Only $x = 16$ checks in the original equation.

29. Solving the equation:
$$x + 3\sqrt{x} - 18 = 0$$
$$(\sqrt{x} - 3)(\sqrt{x} + 6) = 0$$
$$\sqrt{x} = -6, 3$$
$$x = 36, 9$$

Only $x = 9$ checks in the original equation.

31. Solving the equation:
$$2x + 9\sqrt{x} - 5 = 0$$
$$(2\sqrt{x} - 1)(\sqrt{x} + 5) = 0$$
$$\sqrt{x} = -5, \frac{1}{2}$$
$$x = 25, \frac{1}{4}$$

Only $x = \frac{1}{4}$ checks in the original equation.

33. Solving the equation:
$$(a-2) - 11\sqrt{a-2} + 30 = 0$$
$$(\sqrt{a-2} - 6)(\sqrt{a-2} - 5) = 0$$
$$\sqrt{a-2} = 5, 6$$
$$a - 2 = 25, 36$$
$$a = 27, 38$$

35. Solving the equation:
$$(2x+1) - 8\sqrt{2x+1} + 15 = 0$$
$$(\sqrt{2x+1} - 3)(\sqrt{2x+1} - 5) = 0$$
$$\sqrt{2x+1} = 3, 5$$
$$2x + 1 = 9, 25$$
$$2x = 8, 24$$
$$x = 4, 12$$

37. Solving the equation:
$$\left(x^2+1\right)^2-2\left(x^2+1\right)-15=0$$
$$\left(x^2+1-5\right)\left(x^2+1+3\right)=0$$
$$\left(x^2-4\right)\left(x^2+4\right)=0$$
$$x^2=4,-4$$
$$x=\pm2,\pm2i$$

39. Solving the equation:
$$\left(x^2+5\right)^2-6\left(x^2+5\right)-27=0$$
$$\left(x^2+5-9\right)\left(x^2+5+3\right)=0$$
$$\left(x^2-4\right)\left(x^2+8\right)=0$$
$$x^2=4,-8$$
$$x=\pm2,\pm2i\sqrt{2}$$

41. Solving the equation:
$$x^{-2}-3x^{-1}+2=0$$
$$\left(x^{-1}-2\right)\left(x^{-1}-1\right)=0$$
$$x^{-1}=2,1$$
$$x=\tfrac{1}{2},1$$

43. Solving the equation:
$$y^{-4}-5y^{-2}=0$$
$$y^{-2}\left(y^{-2}-5\right)=0$$
$$y^{-2}=0,5$$
$$y^2=\frac{1}{5}$$
$$y=\pm\sqrt{\frac{1}{5}}=\pm\frac{\sqrt{5}}{5}$$

45. Solving the equation:
$$x^{2/3}+4x^{1/3}-12=0$$
$$\left(x^{1/3}+6\right)\left(x^{1/3}-2\right)=0$$
$$x^{1/3}=-6,2$$
$$x=(-6)^3,2^3$$
$$x=-216,8$$

47. Solving the equation:
$$x^{4/3}+6x^{2/3}+9=0$$
$$\left(x^{2/3}+3\right)^2=0$$
$$x^{2/3}=-3$$
$$x^2=(-3)^3=-27$$
$$x=\pm\sqrt{-27}=\pm3i\sqrt{3}$$

49. Solving for t:
$$16t^2-vt-h=0$$
$$t=\frac{v\pm\sqrt{v^2-4(16)(-h)}}{32}=\frac{v\pm\sqrt{v^2+64h}}{32}$$

51. Solving for x:
$$kx^2+8x+4=0$$
$$x=\frac{-8\pm\sqrt{64-16k}}{2k}=\frac{-8\pm4\sqrt{4-k}}{2k}=\frac{-4\pm2\sqrt{4-k}}{k}$$

53. Solving for x:
$$x^2+2xy+y^2=0$$
$$x=\frac{-2y\pm\sqrt{4y^2-4y^2}}{2}=\frac{-2y}{2}=-y$$

55. Solving for t (note that $t>0$):
$$16t^2-8t-h=0$$
$$t=\frac{8+\sqrt{64+64h}}{32}=\frac{8+8\sqrt{1+h}}{32}=\frac{1+\sqrt{1+h}}{4}$$

57. **a.** Sketching the graph:

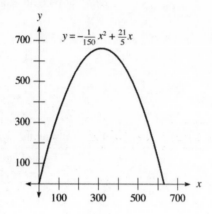

b. Finding the x-intercepts:

$$-\frac{1}{150}x^2 + \frac{21}{5}x = 0$$

$$-\frac{1}{150}x(x - 630) = 0$$

$$x = 0, 630$$

The width is 630 feet.

59. Let $x = BC$. Solving the proportion:

$$\frac{AB}{BC} = \frac{BC}{AC}$$

$$\frac{4}{x} = \frac{x}{4 + x}$$

$$16 + 4x = x^2$$

$$0 = x^2 - 4x - 16$$

$$x = \frac{4 \pm \sqrt{16 + 64}}{2} = \frac{4 \pm \sqrt{80}}{2} = \frac{4 \pm 4\sqrt{5}}{2} = 2 \pm 2\sqrt{5}$$

Thus $BC = 2 + 2\sqrt{5} = 4\left(\dfrac{1 + \sqrt{5}}{2}\right)$.

61. Combining: $5\sqrt{7} - 2\sqrt{7} = 3\sqrt{7}$

63. Combining: $\sqrt{18} - \sqrt{8} + \sqrt{32} = 3\sqrt{2} - 2\sqrt{2} + 4\sqrt{2} = 5\sqrt{2}$

65. Combining: $9x\sqrt{20x^3y^2} + 7y\sqrt{45x^5} = 9x \bullet 2xy\sqrt{5x} + 7y \bullet 3x^2\sqrt{5x} = 18x^2y\sqrt{5x} + 21x^2y\sqrt{5x} = 39x^2y\sqrt{5x}$

67. Multiplying: $\left(\sqrt{5} - 2\right)\left(\sqrt{5} + 8\right) = 5 - 2\sqrt{5} + 8\sqrt{5} - 16 = -11 + 6\sqrt{5}$

69. Multiplying: $\left(\sqrt{x} + 2\right)^2 = \left(\sqrt{x}\right)^2 + 4\sqrt{x} + 4 = x + 4\sqrt{x} + 4$

71. Rationalizing the denominator: $\dfrac{\sqrt{7}}{\sqrt{7} - 2} \bullet \dfrac{\sqrt{7} + 2}{\sqrt{7} + 2} = \dfrac{7 + 2\sqrt{7}}{7 - 4} = \dfrac{7 + 2\sqrt{7}}{3}$

73. Evaluating when $x = 1$: $y = 3(1)^2 - 6(1) + 1 = 3 - 6 + 1 = -2$

75. Evaluating: $P(135) = -0.1(135)^2 + 27(135) - 500 = -1,822.5 + 3,645 - 500 = 1,322.5$

77. Solving the equation:

$$0 = a(80)^2 + 70$$

$$0 = 6400a + 70$$

$$6400a = -70$$

$$a = -\frac{7}{640}$$

79. Solving the equation:

$$x^2 - 6x + 5 = 0$$

$$(x - 1)(x - 5) = 0$$

$$x = 1, 5$$

81. Solving the equation:
$$-x^2 - 2x + 3 = 0$$
$$x^2 + 2x - 3 = 0$$
$$(x+3)(x-1) = 0$$
$$x = -3, 1$$

83. Solving the equation:
$$2x^2 - 6x + 5 = 0$$
$$x = \frac{6 \pm \sqrt{(-6)^2 - 4(2)(5)}}{2(2)} = \frac{6 \pm \sqrt{36-40}}{4} = \frac{6 \pm \sqrt{-4}}{4} = \frac{6 \pm 2i}{4} = \frac{3 \pm i}{2} = \frac{3}{2} \pm \frac{1}{2}i$$

85. Completing the square: $x^2 - 6x + 9 = (x-3)^2$ 87. Completing the square: $y^2 + 2y + 1 = (y+1)^2$

89. To find the x-intercepts, set $y = 0$:
$$x^3 - 4x = 0$$
$$x(x^2 - 4) = 0$$
$$x(x+2)(x-2) = 0$$
$$x = -2, 0, 2$$
To find the y-intercept, set $x = 0$: $y = 0^3 - 4(0) = 0$. The x-intercepts are –2,0,2 and the y-intercept is 0.

91. To find the x-intercepts, set $y = 0$:
$$3x^3 + x^2 - 27x - 9 = 0$$
$$x^2(3x+1) - 9(3x+1) = 0$$
$$(3x+1)(x^2-9) = 0$$
$$(3x+1)(x+3)(x-3) = 0$$
$$x = -3, -\frac{1}{3}, 3$$

To find the y-intercept, set $x = 0$: $y = 3(0)^3 + (0)^2 - 27(0) - 9 = -9$

The x-intercepts are $-3, -\frac{1}{3}, 3$ and the y-intercept is –9.

93. Using long division:
$$\begin{array}{r} 2x^2 + x - 1 \\ x-4\overline{)2x^3 - 7x^2 - 5x + 4} \\ \underline{2x^3 - 8x^2} \\ x^2 - 5x \\ \underline{x^2 - 4x} \\ -x + 4 \\ \underline{-x + 4} \\ 0 \end{array}$$

Now solve the equation:
$$2x^2 + x - 1 = 0$$
$$(2x-1)(x+1) = 0$$
$$x = -1, \frac{1}{2}$$
It also crosses the x-axis at $\frac{1}{2}, -1$.

8.5 Graphing Parabolas

1. First complete the square: $y = x^2 + 2x - 3 = \left(x^2 + 2x + 1\right) - 1 - 3 = \left(x + 1\right)^2 - 4$

    The x-intercepts are $-3, 1$ and the vertex is $(-1, -4)$. Graphing the parabola:

    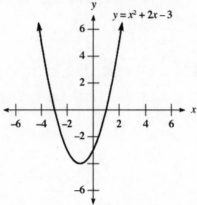

3. First complete the square: $y = -x^2 - 4x + 5 = -\left(x^2 + 4x + 4\right) + 4 + 5 = -\left(x + 2\right)^2 + 9$

    The x-intercepts are $-5, 1$ and the vertex is $(-2, 9)$. Graphing the parabola:

    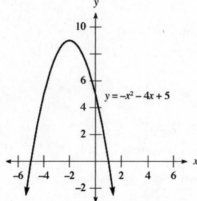

5. The x-intercepts are $-1, 1$ and the vertex is $(0, -1)$. Graphing the parabola:

    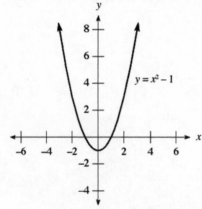

7.   The x-intercepts are $-3,3$ and the vertex is $(0,9)$. Graphing the parabola:

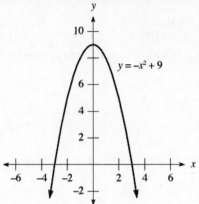

9.   First complete the square: $y = 2x^2 - 4x - 6 = 2\left(x^2 - 2x + 1\right) - 2 - 6 = 2(x-1)^2 - 8$

The x-intercepts are $-1,3$ and the vertex is $(1,-8)$. Graphing the parabola:

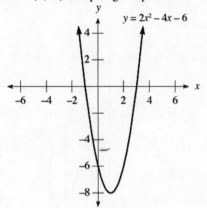

11.   First complete the square: $y = x^2 - 2x - 4 = \left(x^2 - 2x + 1\right) - 1 - 4 = (x-1)^2 - 5$

The x-intercepts are $1 \pm \sqrt{5}$ and the vertex is $(1,-5)$. Graphing the parabola:

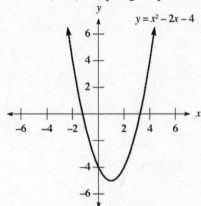

13. First complete the square: $y = x^2 - 4x - 4 = \left(x^2 - 4x + 4\right) - 4 - 4 = (x - 2)^2 - 8$

The vertex is (2,–8). Graphing the parabola:

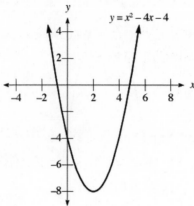

15. First complete the square: $y = -x^2 + 2x - 5 = -\left(x^2 - 2x + 1\right) + 1 - 5 = -(x - 1)^2 - 4$

The vertex is (1,–4). Graphing the parabola:

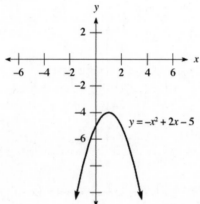

17. The vertex is (0,1). Graphing the parabola: 19. The vertex is (0,–3). Graphing the parabola:

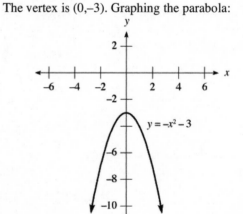

21. Completing the square: $y = x^2 - 6x + 5 = \left(x^2 - 6x + 9\right) - 9 + 5 = (x - 3)^2 - 4$

The vertex is (3,–4), which is the lowest point on the graph.

23. Completing the square: $y = -x^2 + 2x + 8 = -\left(x^2 - 2x + 1\right) + 1 + 8 = -(x - 1)^2 + 9$

The vertex is (1,9), which is the highest point on the graph.

25. Completing the square: $y = -x^2 + 4x + 12 = -\left(x^2 - 4x + 4\right) + 4 + 12 = -(x - 2)^2 + 16$

The vertex is (2,16), which is the highest point on the graph.

27.   Completing the square: $y = -x^2 - 8x = -\left(x^2 + 8x + 16\right) + 16 = -(x+4)^2 + 16$

The vertex is (–4,16), which is the highest point on the graph.

29.   First complete the square:

$$P(x) = -0.002x^2 + 3.5x - 800 = -0.002\left(x^2 - 1750x + 765,625\right) + 1,531.25 - 800 = -0.002(x - 875)^2 + 731.25$$

It must sell 875 patterns to obtain a maximum profit of \$731.25.

31.   The ball is in her hand at times 0 sec and 2 sec.

Completing the square: $h(t) = -16t^2 + 32t = -16\left(t^2 - 2t + 1\right) + 16 = -16(t-1)^2 + 16$

The maximum height of the ball is 16 feet.

33.   Completing the square: $R = xp = 1200p - 100p^2 = -100\left(p^2 - 12p + 36\right) + 3600 = -100(p-6)^2 + 3600$

The price is \$6.00 and the maximum revenue is \$3,600. Sketching the graph:

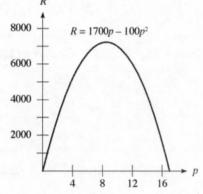

35.   Completing the square: $R = xp = 1700p - 100p^2 = -100\left(p^2 - 17p + 72.25\right) + 7225 = -100(p - 8.5)^2 + 7225$

The price is \$8.50 and the maximum revenue is \$7,225. Sketching the graph:

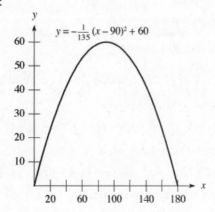

37.   The equation is given on the graph:

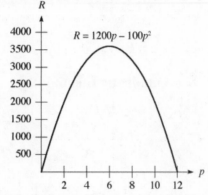

39. Performing the operations: $(3-5i)-(2-4i) = 3-5i-2+4i = 1-i$

41. Performing the operations: $(3+2i)(7-3i) = 21+5i-6i^2 = 27+5i$

43. Performing the operations: $\dfrac{i}{3+i} = \dfrac{i}{3+i} \cdot \dfrac{3-i}{3-i} = \dfrac{3i-i^2}{9-i^2} = \dfrac{3i+1}{9+1} = \dfrac{1}{10} + \dfrac{3}{10}i$

45. Solving the equation:

$$x^2 - 2x - 8 = 0$$
$$(x-4)(x+2) = 0$$
$$x = -2, 4$$

47. Solving the equation:

$$6x^2 - x = 2$$
$$6x^2 - x - 2 = 0$$
$$(2x+1)(3x-2) = 0$$
$$x = -\tfrac{1}{2}, \tfrac{2}{3}$$

49. Solving the equation:

$$x^2 - 6x + 9 = 0$$
$$(x-3)^2 = 0$$
$$x = 3$$

51. The equation is $y = (x-2)^2 - 4$.

8.6 Quadratic Inequalities

1. Factoring the inequality:
$$x^2 + x - 6 > 0$$
$$(x+3)(x-2) > 0$$

Forming the sign chart:

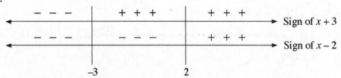

The solution set is $x < -3$ or $x > 2$. Graphing the solution set:

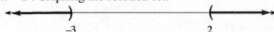

3. Factoring the inequality:
$$x^2 - x - 12 \le 0$$
$$(x+3)(x-4) \le 0$$

Forming the sign chart:

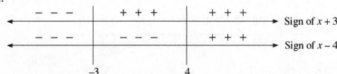

The solution set is $-3 \le x \le 4$. Graphing the solution set:

5.   Factoring the inequality:
$$x^2 + 5x \geq -6$$
$$x^2 + 5x + 6 \geq 0$$
$$(x+2)(x+3) \geq 0$$
Forming the sign chart:

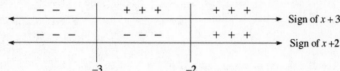

The solution set is $x \leq -3$ or $x \geq -2$. Graphing the solution set:

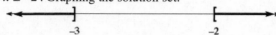

7.   Factoring the inequality:
$$6x^2 < 5x - 1$$
$$6x^2 - 5x + 1 < 0$$
$$(3x-1)(2x-1) < 0$$
Forming the sign chart:

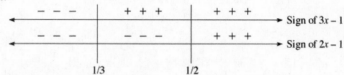

The solution set is $\frac{1}{3} < x < \frac{1}{2}$. Graphing the solution set:

9.   Factoring the inequality:
$$x^2 - 9 < 0$$
$$(x+3)(x-3) < 0$$
Forming the sign chart:

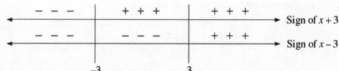

The solution set is $-3 < x < 3$. Graphing the solution set:

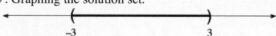

11.   Factoring the inequality:
$$4x^2 - 9 \geq 0$$
$$(2x+3)(2x-3) \geq 0$$
Forming the sign chart:

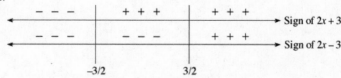

The solution set is $x \leq -\frac{3}{2}$ or $x \geq \frac{3}{2}$. Graphing the solution set:

13. Factoring the inequality:
$$2x^2 - x - 3 < 0$$
$$(2x - 3)(x + 1) < 0$$

Forming the sign chart:

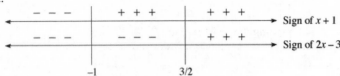

The solution set is $-1 < x < \frac{3}{2}$. Graphing the solution set:

15. Factoring the inequality:
$$x^2 - 4x + 4 \geq 0$$
$$(x - 2)^2 \geq 0$$

Since this inequality is always true, the solution set is all real numbers. Graphing the solution set:

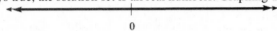

17. Factoring the inequality:
$$x^2 - 10x + 25 < 0$$
$$(x - 5)^2 < 0$$

Since this inequality is never true, there is no solution.

19. Forming the sign chart:

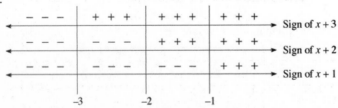

The solution set is $2 < x < 3$ or $x > 4$. Graphing the solution set:

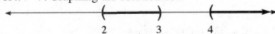

21. Forming the sign chart:

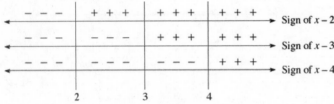

The solution set is $x \leq -3$ or $-2 \leq x \leq -1$. Graphing the solution set:

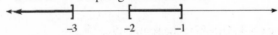

23.   Forming the sign chart:

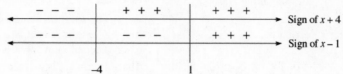

-4   1

The solution set is $-4 < x \le 1$. Graphing the solution set:

-4   1

25.   Forming the sign chart:

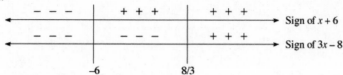

-6   8/3

The solution set is $-6 < x < \frac{8}{3}$. Graphing the solution set:

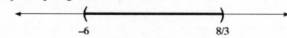

-6   8/3

27.   Forming the sign chart:

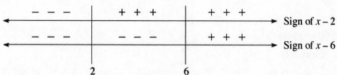

2   6

The solution set is $x < 2$ or $x > 6$. Graphing the solution set:

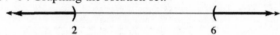

2   6

29.   Forming the sign chart:

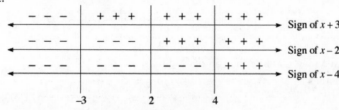

-3   2   4

The solution set is $x < -3$ or $2 < x < 4$. Graphing the solution set:

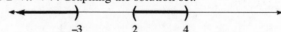

-3   2   4

31.   Forming the sign chart:

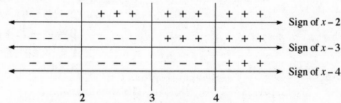

2   3   4

The solution set is $2 < x < 3$ or $x > 4$. Graphing the solution set:

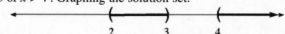

2   3   4

33.   **a.**   Writing the statement: $x - 1 > 0$ **b.**   Writing the statement: $x - 1 \ge 0$
       c.   Writing the statement: $x - 1 \ge 0$

35.   This statement is never true, so there is no solution ($\varnothing$).

37.   This statement is only true if $x = 1$.   **39.**   This statement is true for all real numbers.

41.   This statement is true if $x \ne 1$, or $x < 1$ or $x > 1$.

43.   Since $x^2 - 6x + 9 = (x - 3)^2$, this statement is never true, so there is no solution ($\varnothing$).

45. Since $x^2 - 6x + 9 = (x-3)^2$, this statement is true if $x \neq 3$, or $x < 3$ or $x > 3$.

47. **a.** The solution set is $-2 < x < 2$.
    b. The solution set is $x < -2$ or $x > 2$.
    c. The solution set is $x = -2, 2$.

49. **a.** The solution set is $-2 < x < 5$.
    b. The solution set is $x < -2$ or $x > 5$.
    c. The solution set is $x = -2, 5$.

51. **a.** The solution set is $x < -1$ or $1 < x < 3$.

    b. The solution set is $-1 < x < 1$ or $x > 3$.

    c. The solution set is $x = -1, 1, 3$.

53. Let w represent the width and $2w + 3$ represent the length. Using the area formula:

$$w(2w + 3) \geq 44$$
$$2w^2 + 3w \geq 44$$
$$2w^2 + 3w - 44 \geq 0$$
$$(2w + 11)(w - 4) \geq 0$$

Forming the sign chart:

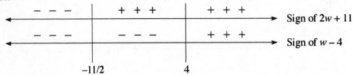

The width is at least 4 inches.

55. Solving the inequality:

$$1300p - 100p^2 \geq 4000$$
$$-100p^2 + 1300p - 4000 \geq 0$$
$$p^2 - 13p + 40 \leq 0$$
$$(p - 8)(p - 5) \leq 0$$

Forming the sign chart:

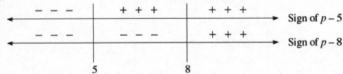

She should charge at least \$5 but no more than \$8 per radio.

57. Using a calculator: $\dfrac{50,000}{32,000} = 1.5625$

59. Using a calculator: $\dfrac{1}{2}\left(\dfrac{4.5926}{1.3876} - 2\right) \approx 0.6549$

61. Solving the equation:

$$\sqrt{3t - 1} = 2$$
$$\left(\sqrt{3t - 1}\right)^2 = (2)^2$$
$$3t - 1 = 4$$
$$3t = 5$$
$$t = \tfrac{5}{3}$$

The solution is $\tfrac{5}{3}$.

63. Solving the equation:

$$\sqrt{x + 3} = x - 3$$
$$\left(\sqrt{x + 3}\right)^2 = (x - 3)^2$$
$$x + 3 = x^2 - 6x + 9$$
$$0 = x^2 - 7x + 6$$
$$0 = (x - 6)(x - 1)$$
$$x = 1, 6 \qquad (x = 1 \text{ does not check})$$

The solution is 6.

65. Graphing the equation:

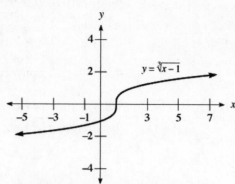

67. Using the quadratic formula: $x = \dfrac{2 \pm \sqrt{4 - 4(-1)}}{2(1)} = \dfrac{2 \pm \sqrt{8}}{2} = \dfrac{2 \pm 2\sqrt{2}}{2} = 1 \pm \sqrt{2}$

The inequality is satisfied when $1 - \sqrt{2} < x < 1 + \sqrt{2}$.

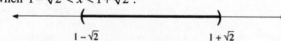

69. Using the quadratic formula: $x = \dfrac{8 \pm \sqrt{64 - 4(13)}}{2(1)} = \dfrac{8 \pm \sqrt{12}}{2} = \dfrac{8 \pm 2\sqrt{3}}{2} = 4 \pm \sqrt{3}$

The inequality is satisfied when $x < 4 - \sqrt{3}$ or $x > 4 + \sqrt{3}$.

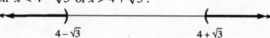

Chapter 8 Review/Test

1. Solving the equation:
$$(2t - 5)^2 = 25$$
$$2t - 5 = \pm 5$$
$$2t - 5 = -5, 5$$
$$2t = 0, 10$$
$$t = 0, 5$$

2. Solving the equation:
$$(3t - 2)^2 = 4$$
$$3t - 2 = \pm 2$$
$$3t - 2 = -2, 2$$
$$3t = 0, 4$$
$$t = 0, \tfrac{4}{3}$$

3. Solving the equation:
$$(3y - 4)^2 = -49$$
$$3y - 4 = \pm\sqrt{-49}$$
$$3y - 4 = \pm 7i$$
$$3y = 4 \pm 7i$$
$$y = \dfrac{4 \pm 7i}{3}$$

4. Solving the equation:
$$(2x + 6)^2 = 12$$
$$2x + 6 = \pm\sqrt{12}$$
$$2x + 6 = \pm 2\sqrt{3}$$
$$2x = -6 \pm 2\sqrt{3}$$
$$x = -3 \pm \sqrt{3}$$

5. Solving by completing the square:
$$2x^2 + 6x - 20 = 0$$
$$x^2 + 3x = 10$$
$$x^2 + 3x + \tfrac{9}{4} = 10 + \tfrac{9}{4}$$
$$\left(x + \tfrac{3}{2}\right)^2 = \tfrac{49}{4}$$
$$x + \tfrac{3}{2} = -\tfrac{7}{2}, \tfrac{7}{2}$$
$$x = -5, 2$$

6. Solving by completing the square:
$$3x^2 + 15x = -18$$
$$x^2 + 5x = -6$$
$$x^2 + 5x + \tfrac{25}{4} = -6 + \tfrac{25}{4}$$
$$\left(x + \tfrac{5}{2}\right)^2 = \tfrac{1}{4}$$
$$x + \tfrac{5}{2} = -\tfrac{1}{2}, \tfrac{1}{2}$$
$$x = -3, -2$$

7.   Solving by completing the square:
$$a^2 + 9 = 6a$$
$$a^2 - 6a = -9$$
$$a^2 - 6a + 9 = -9 + 9$$
$$(a - 3)^2 = 0$$
$$a - 3 = 0$$
$$a = 3$$

8.   Solving by completing the square:
$$a^2 + 4 = 4a$$
$$a^2 - 4a = -4$$
$$a^2 - 4a + 4 = -4 + 4$$
$$(a - 2)^2 = 0$$
$$a - 2 = 0$$
$$a = 2$$

9.   Solving by completing the square:
$$2y^2 + 6y = -3$$
$$y^2 + 3y = -\frac{3}{2}$$
$$y^2 + 3y + \frac{9}{4} = -\frac{3}{2} + \frac{9}{4}$$
$$\left(y + \frac{3}{2}\right)^2 = \frac{3}{4}$$
$$y + \frac{3}{2} = \pm\frac{\sqrt{3}}{2}$$
$$y = \frac{-3 \pm \sqrt{3}}{2}$$

10.   Solving by completing the square:
$$3y^2 + 3 = 9y$$
$$3y^2 - 9y = -3$$
$$y^2 - 3y = -1$$
$$y^2 - 3y + \frac{9}{4} = -1 + \frac{9}{4}$$
$$\left(y - \frac{3}{2}\right)^2 = \frac{5}{4}$$
$$y - \frac{3}{2} = \pm\frac{\sqrt{5}}{2}$$
$$y = \frac{3 \pm \sqrt{5}}{2}$$

11.   Solving the equation:
$$\frac{1}{6}x^2 + \frac{1}{2}x - \frac{5}{3} = 0$$
$$x^2 + 3x - 10 = 0$$
$$(x + 5)(x - 2) = 0$$
$$x = -5, 2$$

12.   Solving the equation:
$$8x^2 - 18x = 0$$
$$2x(4x - 9) = 0$$
$$x = 0, \frac{9}{4}$$

13.   Solving the equation:
$$4t^2 - 8t + 19 = 0$$
$$t = \frac{8 \pm \sqrt{64 - 304}}{8} = \frac{8 \pm \sqrt{-240}}{8} = \frac{8 \pm 4i\sqrt{15}}{8} = \frac{2 \pm i\sqrt{15}}{2}$$

14.   Solving the equation:
$$100x^2 - 200x = 100$$
$$x^2 - 2x - 1 = 0$$
$$x = \frac{2 \pm \sqrt{4 + 4}}{2} = \frac{2 \pm \sqrt{8}}{2} = \frac{2 \pm 2\sqrt{2}}{2} = 1 \pm \sqrt{2}$$

15.   Solving the equation:
$$0.06a^2 + 0.05a = 0.04$$
$$0.06a^2 + 0.05a - 0.04 = 0$$
$$6a^2 + 5a - 4 = 0$$
$$(2a - 1)(3a + 4) = 0$$
$$a = -\frac{4}{3}, \frac{1}{2}$$

16.   Solving the equation:
$$9 - 6x = -x^2$$
$$x^2 - 6x + 9 = 0$$
$$(x - 3)^2 = 0$$
$$x = 3$$

17.   Solving the equation:
$$(2x + 1)(x - 5) - (x + 3)(x - 2) = -17$$
$$2x^2 - 9x - 5 - x^2 - x + 6 = -17$$
$$x^2 - 10x + 1 = -17$$
$$x^2 - 10x + 18 = 0$$
$$x = \frac{10 \pm \sqrt{100 - 72}}{2} = \frac{10 \pm \sqrt{28}}{2} = \frac{10 \pm 2\sqrt{7}}{2} = 5 \pm \sqrt{7}$$

18. Solving the equation:
$$2y^3 + 2y = 10y^2$$
$$2y^3 - 10y^2 + 2y = 0$$
$$2y(y^2 - 5y + 1) = 0$$
$$y = 0, \frac{5 \pm \sqrt{25 - 4}}{2} = 0, \frac{5 \pm \sqrt{21}}{2}$$

19. Solving the equation:
$$5x^2 = -2x + 3$$
$$5x^2 + 2x - 3 = 0$$
$$(x + 1)(5x - 3) = 0$$
$$x = -1, \frac{3}{5}$$

20. Solving the equation:
$$x^3 - 27 = 0$$
$$(x - 3)(x^2 + 3x + 9) = 0$$
$$x = 3, \frac{-3 \pm \sqrt{9 - 36}}{2} = \frac{-3 \pm \sqrt{-27}}{2} = \frac{-3 \pm 3i\sqrt{3}}{2}$$

21. Solving the equation:
$$3 - \frac{2}{x} + \frac{1}{x^2} = 0$$
$$3x^2 - 2x + 1 = 0$$
$$x = \frac{2 \pm \sqrt{4 - 12}}{6} = \frac{2 \pm \sqrt{-8}}{6} = \frac{2 \pm 2i\sqrt{2}}{6} = \frac{1 \pm i\sqrt{2}}{3}$$

22. Solving the equation:
$$\frac{1}{x - 3} + \frac{1}{x + 2} = 1$$
$$x + 2 + x - 3 = (x - 3)(x + 2)$$
$$2x - 1 = x^2 - x - 6$$
$$0 = x^2 - 3x - 5$$
$$x = \frac{3 \pm \sqrt{9 + 20}}{2} = \frac{3 \pm \sqrt{29}}{2}$$

23. The profit equation is given by:
$$34x - 0.1x^2 - 7x - 400 = 1300$$
$$-0.1x^2 + 27x - 1700 = 0$$
$$x^2 - 270x + 17000 = 0$$
$$(x - 100)(x - 170) = 0$$
$$x = 100, 170$$
The company must sell either 100 or 170 items for its weekly profit to be $1,300.

24. The profit equation is given by:
$$110x - 0.5x^2 - 70x - 300 = 300$$
$$-0.5x^2 + 40x - 600 = 0$$
$$x^2 - 80x + 1200 = 0$$
$$(x - 20)(x - 60) = 0$$
$$x = 20, 60$$
The company must sell either 20 or 60 items for its weekly profit to be $300.

25. First write the equation as $2x^2 - 8x + 8 = 0$. Computing the discriminant: $D = (-8)^2 - 4(2)(8) = 64 - 64 = 0$
The equation will have one rational solution.

26. First write the equation as $4x^2 - 8x + 4 = 0$. Computing the discriminant: $D = (-8)^2 - 4(4)(4) = 64 - 64 = 0$
    The equation will have one rational solution.

27. Computing the discriminant: $D = (1)^2 - 4(2)(-3) = 1 + 24 = 25$. The equation will have two rational solutions.

28. First write the equation as $5x^2 + 11x - 12 = 0$. Computing the discriminant:
    $$D = (11)^2 - 4(5)(-12) = 121 + 240 = 361$$
    The equation will have two rational solutions.

29. First write the equation as $x^2 - x - 1 = 0$. Computing the discriminant: $D = (-1)^2 - 4(1)(-1) = 1 + 4 = 5$
    The equation will have two irrational solutions.

30. First write the equation as $x^2 - 5x + 5 = 0$. Computing the discriminant: $D = (-5)^2 - 4(1)(5) = 25 - 20 = 5$
    The equation will have two irrational solutions.

31. First write the equation as $3x^2 + 5x + 4 = 0$. Computing the discriminant: $D = (5)^2 - 4(3)(4) = 25 - 48 = -23$
    The equation will have two complex solutions.

32. First write the equation as $4x^2 - 3x + 6 = 0$. Computing the discriminant: $D = (-3)^2 - 4(4)(6) = 9 - 96 = -87$
    The equation will have two complex solutions.

33. Setting the discriminant equal to 0:
    $$(-k)^2 - 4(25)(4) = 0$$
    $$k^2 - 400 = 0$$
    $$k^2 = 400$$
    $$k = \pm 20$$

34. Setting the discriminant equal to 0:
    $$k^2 - 4(4)(25) = 0$$
    $$k^2 - 400 = 0$$
    $$k^2 = 400$$
    $$k = \pm 20$$

35. Setting the discriminant equal to 0:
    $$(12)^2 - 4(k)(9) = 0$$
    $$144 - 36k = 0$$
    $$36k = 144$$
    $$k = 4$$

36. Setting the discriminant equal to 0:
    $$(-16)^2 - 4(k)(16) = 0$$
    $$256 - 64k = 0$$
    $$64k = 256$$
    $$k = 4$$

37. Setting the discriminant equal to 0:
    $$30^2 - 4(9)(k) = 0$$
    $$900 - 36k = 0$$
    $$36k = 900$$
    $$k = 25$$

38. Setting the discriminant equal to 0:
    $$28^2 - 4(4)(k) = 0$$
    $$784 - 16k = 0$$
    $$16k = 784$$
    $$k = 49$$

39. Writing the equation:
    $$(x - 3)(x - 5) = 0$$
    $$x^2 - 8x + 15 = 0$$

40. Writing the equation:
    $$(x + 2)(x - 4) = 0$$
    $$x^2 - 2x - 8 = 0$$

41. Writing the equation:
    $$(2y - 1)(y + 4) = 0$$
    $$2y^2 + 7y - 4 = 0$$

42. Writing the equation:
    $$(t - 3)(t + 3)(t - 5) = 0$$
    $$(t^2 - 9)(t - 5) = 0$$
    $$t^3 - 5t^2 - 9t + 45 = 0$$

43. Solving the equation:
    $$(x - 2)^2 - 4(x - 2) - 60 = 0$$
    $$(x - 2 - 10)(x - 2 + 6) = 0$$
    $$(x - 12)(x + 4) = 0$$
    $$x = -4, 12$$

44. Solving the equation:
    $$6(2y + 1)^2 - (2y + 1) - 2 = 0$$
    $$[3(2y + 1) - 2][2(2y + 1) + 1] = 0$$
    $$(6y + 1)(4y + 3) = 0$$
    $$y = -\frac{3}{4}, -\frac{1}{6}$$

45. Solving the equation:
$$x^4 - x^2 = 12$$
$$x^4 - x^2 - 12 = 0$$
$$\left(x^2 - 4\right)\left(x^2 + 3\right) = 0$$
$$x^2 = 4, -3$$
$$x = \pm 2, \pm i\sqrt{3}$$

46. Solving the equation:
$$x - \sqrt{x} - 2 = 0$$
$$\left(\sqrt{x} - 2\right)\left(\sqrt{x} + 1\right) = 0$$
$$\sqrt{x} = 2, -1$$
$$x = 4, 1 \qquad \left(x = 1 \text{ does not check}\right)$$

47. Solving the equation:
$$2x - 11\sqrt{x} = -12$$
$$2x - 11\sqrt{x} + 12 = 0$$
$$\left(2\sqrt{x} - 3\right)\left(\sqrt{x} - 4\right) = 0$$
$$\sqrt{x} = \tfrac{3}{2}, 4$$
$$x = \tfrac{9}{4}, 16$$

48. Solving the equation:
$$\sqrt{x+5} = \sqrt{x} + 1$$
$$\left(\sqrt{x+5}\right)^2 = \left(\sqrt{x} + 1\right)^2$$
$$x + 5 = x + 2\sqrt{x} + 1$$
$$2\sqrt{x} = 4$$
$$\sqrt{x} = 2$$
$$x = 4$$

49. Solving the equation:
$$\sqrt{y+21} + \sqrt{y} = 7$$
$$\left(\sqrt{y+21}\right)^2 = \left(7 - \sqrt{y}\right)^2$$
$$y + 21 = 49 - 14\sqrt{y} + y$$
$$14\sqrt{y} = 28$$
$$\sqrt{y} = 2$$
$$y = 4$$

50. Solving the equation:
$$\sqrt{y+9} - \sqrt{y-6} = 3$$
$$\left(\sqrt{y+9}\right)^2 = \left(3 + \sqrt{y-6}\right)^2$$
$$y + 9 = 9 + 6\sqrt{y-6} + y - 6$$
$$6\sqrt{y-6} = 6$$
$$\sqrt{y-6} = 1$$
$$y - 6 = 1$$
$$y = 7$$

51. Solving for t:
$$16t^2 - 10t - h = 0$$
$$t = \frac{10 \pm \sqrt{100 + 64h}}{32} = \frac{10 \pm 2\sqrt{25 + 16h}}{32} = \frac{5 \pm \sqrt{25 + 16h}}{16}$$

52. Solving for t:
$$16t^2 - vt - 10 = 0$$
$$t = \frac{v \pm \sqrt{v^2 - 4(16)(-10)}}{32} = \frac{v \pm \sqrt{v^2 + 640}}{32}$$

53. Factoring the inequality:
$$x^2 - x - 2 < 0$$
$$(x - 2)(x + 1) < 0$$

Forming a sign chart:

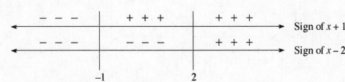

The solution set is $-1 < x < 2$. Graphing the solution set:

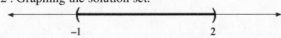

54. Factoring the inequality:
$$3x^2 - 14x + 8 \le 0$$
$$(3x - 2)(x - 4) \le 0$$

Forming a sign chart:

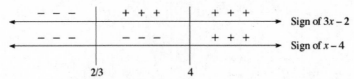

The solution set is $\frac{2}{3} \le x \le 4$. Graphing the solution set:

55. Factoring the inequality:
$$2x^2 + 5x - 12 \ge 0$$
$$(2x - 3)(x + 4) \ge 0$$

Forming a sign chart:

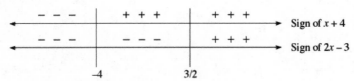

The solution set is $x \le -4$ or $x \ge \frac{3}{2}$. Graphing the solution set:

56. Forming a sign chart:

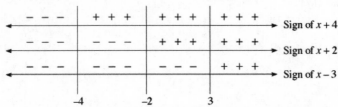

The solution set is $-4 < x < -2$ or $x > 3$. Graphing the solution set:

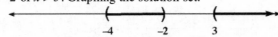

57. First complete the square: $y = x^2 - 6x + 8 = \left(x^2 - 6x + 9\right) + 8 - 9 = \left(x - 3\right)^2 - 1$

The x-intercepts are 2,4, and the vertex is $(3,-1)$. Graphing the parabola:

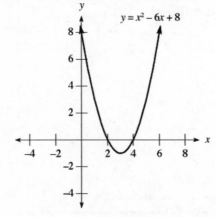

58.   The *x*-intercepts are ±2, and the vertex is (0,−4). Graphing the parabola:

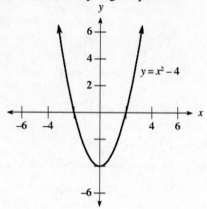

Chapter 9
Exponential and Logarithmic Functions

9.1 Exponential Functions

1. Evaluating: $g(0) = \left(\frac{1}{2}\right)^0 = 1$

3. Evaluating: $g(-1) = \left(\frac{1}{2}\right)^{-1} = 2$

5. Evaluating: $f(-3) = 3^{-3} = \frac{1}{27}$

7. Evaluating: $f(2) + g(-2) = 3^2 + \left(\frac{1}{2}\right)^{-2} = 9 + 4 = 13$

9. Graphing the function:

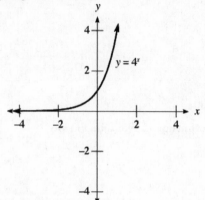

11. Graphing the function:

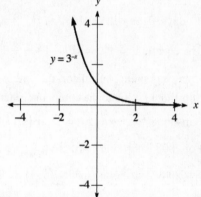

13. Graphing the function:

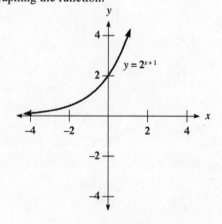

15. Graphing the function:

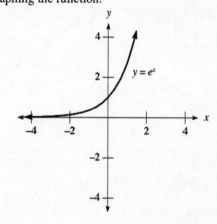

17.   Graphing the functions:

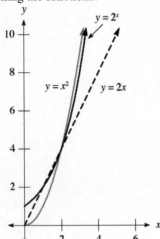

19.   Graphing the functions:

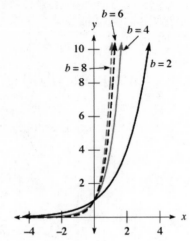

21.   The equation is: $h(n) = 6\left(\frac{2}{3}\right)^n$. Substituting $n = 5$: $h(5) = 6\left(\frac{2}{3}\right)^5 \approx 0.79$ feet

23.   After 8 days, there will be: $1400 \cdot 2^{-8/8} = 1400 \cdot \frac{1}{2} = 700$ micrograms

After 11 days, there will be: $1400 \cdot 2^{-11/8} \approx 539.8$ micrograms

25.   **a.**   The equation is $A(t) = 1200\left(1 + \frac{0.06}{4}\right)^{4t}$.   **b.**   Substitute $t = 8$: $A(8) = 1200\left(1 + \frac{0.06}{4}\right)^{32} \approx \$1,932.39$

  c.   Using a graphing calculator, the time is approximately 11.64 years.

  d.   Substitute $t = 8$ into the compound interest formula: $A(8) = 1200 e^{0.06 \times 8} \approx \$1,939.29$

27.   **a.**   Substitute $t = 20$: $E(20) = 78.16(1.11)^{20} \approx \$630$ billion. The estimate is $69 billion too low.

  b.   For 2008, substitute $t = 38$: $E(38) = 78.16(1.11)^{38} \approx \$4,123$ billion

  For 2009, substitute $t = 39$: $E(39) = 78.16(1.11)^{39} \approx \$4,577$ billion

  For 2010, substitute $t = 40$: $E(40) = 78.16(1.11)^{40} \approx \$5,080$ billion

29.   Graphing the function:

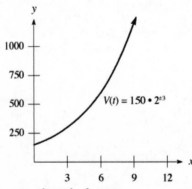

31.   The painting will be worth $600 after approximately 6 years.

33. **a.** Substitute $t = 3.5$: $V(5) = 450,000(1 - 0.30)^5 \approx \$129,138.48$

b. The domain is $\{t \mid 0 \leq t \leq 6\}$.

c. Sketching the graph:

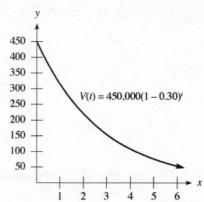

$V(t) = 450,000(1 - 0.30)^t$

d. The range is $\{V(t) \mid 52,942.05 \leq V(t) \leq 450,000\}$.

e. From the graph, the crane will be worth \$85,000 after approximately 4.7 years, or 4 years 8 months.

35. The domain is $\{1, 3, 4\}$ and the range is $\{1, 2, 4\}$. This is a function.

37. Find where the quantity inside the radical is non-negative:
$$3x + 1 \geq 0$$
$$3x \geq -1$$
$$x \geq -\frac{1}{3}$$
The domain is $\left\{ x \mid x \geq -\frac{1}{3} \right\}$.

39. Evaluating the function: $f(0) = 2(0)^2 - 18 = 0 - 18 = -18$

41. Simplifying the function: $\dfrac{g(x+h) - g(x)}{h} = \dfrac{2(x+h) - 6 - (2x - 6)}{h} = \dfrac{2x + 2h - 6 - 2x + 6}{h} = \dfrac{2h}{h} = 2$

43. Solving for y:
$$x = 2y - 3$$
$$2y = x + 3$$
$$y = \frac{x+3}{2}$$

45. Solving for y:
$$x = y^2 - 2$$
$$y^2 = x + 2$$
$$y = \pm\sqrt{x+2}$$

47. Solving for y:
$$x = \frac{y-4}{y-2}$$
$$x(y-2) = y - 4$$
$$xy - 2x = y - 4$$
$$xy - y = 2x - 4$$
$$y(x-1) = 2x - 4$$
$$y = \frac{2x-4}{x-1}$$

49. Solving for y:
$$x = \sqrt{y-3}$$
$$x^2 = y - 3$$
$$y = x^2 + 3$$

51.   Graphing the function:

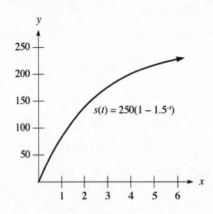

$$s(t) = 250(1 - 1.5^{-t})$$

9.2 The Inverse of a Function

1.   Let $y = f(x)$. Switch x and y and solve for y:

$$3y - 1 = x$$
$$3y = x + 1$$
$$y = \frac{x + 1}{3}$$

The inverse is $f^{-1}(x) = \dfrac{x + 1}{3}$.

3.   Let $y = f(x)$. Switch x and y and solve for y:

$$y^3 = x$$
$$y = \sqrt[3]{x}$$

The inverse is $f^{-1}(x) = \sqrt[3]{x}$.

5.   Let $y = f(x)$. Switch x and y and solve for y:

$$\frac{y - 3}{y - 1} = x$$
$$y - 3 = xy - x$$
$$y - xy = 3 - x$$
$$y(1 - x) = 3 - x$$
$$y = \frac{3 - x}{1 - x} = \frac{x - 3}{x - 1}$$

The inverse is $f^{-1}(x) = \dfrac{x - 3}{x - 1}$.

7.   Let $y = f(x)$. Switch x and y and solve for y:

$$\frac{y - 3}{4} = x$$
$$y - 3 = 4x$$
$$y = 4x + 3$$

The inverse is $f^{-1}(x) = 4x + 3$.

9.   Let $y = f(x)$. Switch x and y and solve for y:

$$\tfrac{1}{2}y - 3 = x$$
$$y - 6 = 2x$$
$$y = 2x + 6$$

The inverse is $f^{-1}(x) = 2x + 6$.

11.   Let $y = f(x)$. Switch x and y and solve for y:

$$\frac{2y + 1}{3y + 1} = x$$
$$2y + 1 = 3xy + x$$
$$2y - 3xy = x - 1$$
$$y(2 - 3x) = x - 1$$
$$y = \frac{x - 1}{2 - 3x} = \frac{1 - x}{3x - 2}$$

The inverse is $f^{-1}(x) = \dfrac{1 - x}{3x - 2}$.

13.   Finding the inverse:

$$2y - 1 = x$$
$$2y = x + 1$$
$$y = \frac{x + 1}{2}$$

The inverse is $y^{-1} = \frac{x+1}{2}$. Graphing each curve:

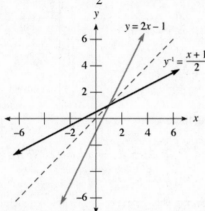

15.   Finding the inverse:

$$y^2 - 3 = x$$
$$y^2 = x + 3$$
$$y = \pm\sqrt{x + 3}$$

The inverse is $y^{-1} = \pm\sqrt{x+3}$. Graphing each curve:

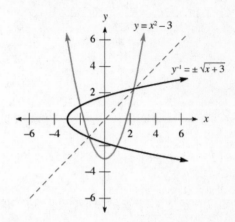

17.   Finding the inverse:

$$y^2 - 2y - 3 = x$$
$$y^2 - 2y + 1 = x + 3 + 1$$
$$(y - 1)^2 = x + 4$$
$$y - 1 = \pm\sqrt{x + 4}$$
$$y = 1 \pm \sqrt{x + 4}$$

The inverse is $y^{-1} = 1 \pm \sqrt{x+4}$. Graphing each curve:

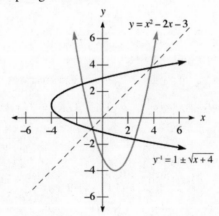

19. The inverse is $x = 3^y$. Graphing each curve:

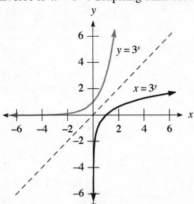

21. The inverse is $x = 4$. Graphing each curve:

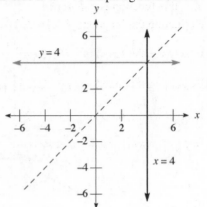

23. Finding the inverse:
$$\frac{1}{2}y^3 = x$$
$$y^3 = 2x$$
$$y = \sqrt[3]{2x}$$

The inverse is $y^{-1} = \sqrt[3]{2x}$. Graphing each curve:

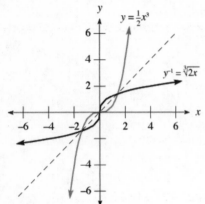

25. Finding the inverse:
$$\frac{1}{2}y + 2 = x$$
$$y + 4 = 2x$$
$$y = 2x - 4$$

The inverse is $y^{-1} = 2x - 4$. Graphing each curve:

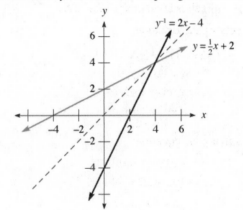

27. Finding the inverse:
$$\sqrt{y + 2} = x$$
$$y + 2 = x^2$$
$$y = x^2 - 2$$

The inverse is $y^{-1} = x^2 - 2, x \geq 0$. Graphing each curve:

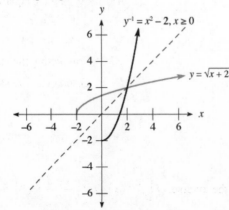

29.   **a.**   Yes, this function is one-to-one.   **b.**   No, this function is not one-to-one.
     c.   Yes, this function is one-to-one.

31.   **a.**   Evaluating the function: $f(2) = 3(2) - 2 = 6 - 2 = 4$

    b.   Evaluating the function: $f^{-1}(2) = \dfrac{2+2}{3} = \dfrac{4}{3}$

    c.   Evaluating the function: $f\left[f^{-1}(2)\right] = f\left(\frac{4}{3}\right) = 3\left(\frac{4}{3}\right) - 2 = 4 - 2 = 2$

    d.   Evaluating the function: $f^{-1}\left[f(2)\right] = f^{-1}(4) = \dfrac{4+2}{3} = \dfrac{6}{3} = 2$

33.   Let $y = f(x)$. Switch x and y and solve for y:

$$\frac{1}{y} = x$$
$$y = \frac{1}{x}$$

The inverse is $f^{-1}(x) = \dfrac{1}{x}$.

35.   **a.**   The value is −3.   **b.**   The value is −6.
    c.   The value is 2.   **d.**   The value is 3.
    e.   The value is −2.   **f.**   The value is 3.
    g.   Each is an inverse of the other.

37.   Solving the equation: **39.**   The number is 25, since $x^2 - 10x + 25 = (x-5)^2$.

$$(2x-1)^2 = 25$$
$$2x - 1 = \pm\sqrt{25}$$
$$2x - 1 = -5, 5$$
$$2x = -4, 6$$
$$x = -2, 3$$

41.   Solving the equation: **43.**   Solving the equation:

$$x^2 - 10x + 8 = 0$$
$$x^2 - 10x + 25 = -8 + 25$$
$$(x-5)^2 = 17$$
$$x - 5 = \pm\sqrt{17}$$
$$x = 5 \pm \sqrt{17}$$

$$3x^2 - 6x + 6 = 0$$
$$x^2 - 2x + 2 = 0$$
$$x^2 - 2x + 1 = -2 + 1$$
$$(x-1)^2 = -1$$
$$x - 1 = \pm i$$
$$x = 1 \pm i$$

45.   Simplifying: $3^{-2} = \dfrac{1}{3^2} = \dfrac{1}{9}$

47.   Solving the equation: **49.**   Solving the equation:
$$2 = 3x$$
$$x = \tfrac{2}{3}$$

$$4 = x^3$$
$$x = \sqrt[3]{4}$$

51.   Completing the statement: $8 = 2^3$ **53.**   Completing the statement: $10,000 = 10^4$

55.   Completing the statement: $81 = 3^4$ **57.**   Completing the statement: $6 = 6^1$

59.   Finding the inverse:
$$3y + 5 = x$$
$$3y = x - 5$$
$$y = \frac{x-5}{3}$$

So $f^{-1}(x) = \dfrac{x-5}{3}$. Now verifying the inverse: $f\left[f^{-1}(x)\right] = f\left(\dfrac{x-5}{3}\right) = 3\left(\dfrac{x-5}{3}\right) + 5 = x - 5 + 5 = x$

61. Finding the inverse:

$$y^3 + 1 = x$$
$$y^3 = x - 1$$
$$y = \sqrt[3]{x-1}$$

So $f^{-1}(x) = \sqrt[3]{x-1}$. Now verifying the inverse: $f\left[f^{-1}(x)\right] = f\left(\sqrt[3]{x-1}\right) = \left(\sqrt[3]{x-1}\right)^3 + 1 = x - 1 + 1 = x$

63. Finding the inverse:

$$\frac{y-4}{y-2} = x$$
$$y - 4 = xy - 2x$$
$$y - xy = 4 - 2x$$
$$y(1-x) = 4 - 2x$$
$$y = \frac{4-2x}{1-x} = \frac{2x-4}{x-1}$$

So $f^{-1}(x) = \dfrac{2x-4}{x-1}$. Now verifying the inverse:

$$f\left[f^{-1}(x)\right] = f\left(\frac{2x-4}{x-1}\right) = \frac{\dfrac{2x-4}{x-1} - 4}{\dfrac{2x-4}{x-1} - 2} = \frac{2x-4-4(x-1)}{2x-4-2(x-1)} = \frac{2x-4-4x+4}{2x-4-2x+2} = \frac{-2x}{-2} = x$$

65.
   a. From the graph: $f(0) = 1$
   b. From the graph: $f(1) = 2$
   c. From the graph: $f(2) = 5$
   d. From the graph: $f^{-1}(1) = 0$
   e. From the graph: $f^{-1}(2) = 1$
   f. From the graph: $f^{-1}(5) = 2$
   g. From the graph: $f^{-1}\left[f(2)\right] = 2$
   h. From the graph: $f\left[f^{-1}(5)\right] = 5$

9.3 Logarithms Are Exponents

1. Writing in logarithmic form: $\log_2 16 = 4$

3. Writing in logarithmic form: $\log_5 125 = 3$

5. Writing in logarithmic form: $\log_{10} 0.01 = -2$

7. Writing in logarithmic form: $\log_2 \frac{1}{32} = -5$

9. Writing in logarithmic form: $\log_{1/2} 8 = -3$

11. Writing in logarithmic form: $\log_3 27 = 3$

13. Writing in exponential form: $10^2 = 100$

15. Writing in exponential form: $2^6 = 64$

17. Writing in exponential form: $8^0 = 1$

19. Writing in exponential form: $10^{-3} = 0.001$

21. Writing in exponential form: $6^2 = 36$

23. Writing in exponential form: $5^{-2} = \frac{1}{25}$

25. Solving the equation:
$$\log_3 x = 2$$
$$x = 3^2 = 9$$

27. Solving the equation:
$$\log_5 x = -3$$
$$x = 5^{-3} = \frac{1}{125}$$

29. Solving the equation:
$$\log_2 16 = x$$
$$2^x = 16$$
$$x = 4$$

31. Solving the equation:
$$\log_8 2 = x$$
$$8^x = 2$$
$$x = \frac{1}{3}$$

33. Solving the equation:
$$\log_x 4 = 2$$
$$x^2 = 4$$
$$x = 2$$

35. Solving the equation:
$$\log_x 5 = 3$$
$$x^3 = 5$$
$$x = \sqrt[3]{5}$$

37.   Sketching the graph:

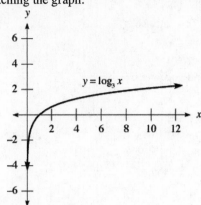

39.   Sketching the graph:

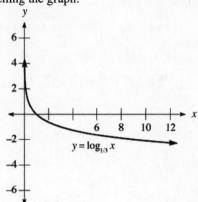

41.   Sketching the graph:

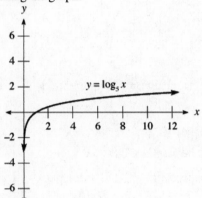

43.   Sketching the graph:

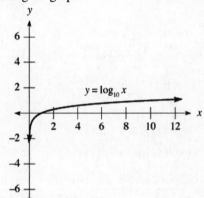

45.   Simplifying the logarithm:

$$x = \log_2 16$$
$$2^x = 16$$
$$x = 4$$

47.   Simplifying the logarithm:

$$x = \log_{25} 125$$
$$25^x = 125$$
$$5^{2x} = 5^3$$
$$2x = 3$$
$$x = \frac{3}{2}$$

49.   Simplifying the logarithm:

$$x = \log_{10} 1000$$
$$10^x = 1000$$
$$x = 3$$

51.   Simplifying the logarithm:

$$x = \log_3 3$$
$$3^x = 3$$
$$x = 1$$

53.   Simplifying the logarithm:

$$x = \log_5 1$$
$$5^x = 1$$
$$x = 0$$

55. First find $\log_6 6$:
$$x = \log_6 6$$
$$6^x = 6$$
$$x = 1$$
Now find $\log_3 1$:
$$x = \log_3 1$$
$$3^x = 1$$
$$x = 0$$

57. First find $\log_2 16$:
$$x = \log_2 16$$
$$2^x = 16$$
$$x = 4$$
Now find $\log_2 4$:
$$x = \log_2 4$$
$$2^x = 4$$
$$x = 2$$
Now find $\log_4 2$:
$$x = \log_4 2$$
$$4^x = 2$$
$$2^{2x} = 2$$
$$2x = 1$$
$$x = \tfrac{1}{2}$$

59. The pH is given by: $\text{pH} = -\log_{10}\left(10^{-7}\right) = -(-7) = 7$

61. The $\left[\text{H}^+\right]$ is given by:
$$-\log_{10}\left[\text{H}^+\right] = 6$$
$$\log_{10}\left[\text{H}^+\right] = -6$$
$$\left[\text{H}^+\right] = 10^{-6}$$

63. Using the relationship $M = \log_{10} T$:
$$M = \log_{10} 100$$
$$10^M = 100$$
$$M = 2$$

65. It is 10^8 times as large.

67. Completing the square: $x^2 + 10x + 25 = (x+5)^2$

69. Completing the square: $y^2 - 2y + 1 = (y-1)^2$

71. Solving the equation:
$$-y^2 = 9$$
$$y^2 = -9$$
$$y = \pm\sqrt{-9} = \pm 3i$$

73. Solving the equation:
$$-x^2 - 8 = -4$$
$$-x^2 = 4$$
$$x^2 = -4$$
$$x = \pm\sqrt{-4} = \pm 2i$$

75. Solving the equation:
$$2x^2 + 4x - 3 = 0$$
$$x = \frac{-4 \pm \sqrt{16+24}}{4} = \frac{-4 \pm \sqrt{40}}{4} = \frac{-4 \pm 2\sqrt{10}}{4} = \frac{-2 \pm \sqrt{10}}{2}$$

77. Solving the equation:
$$(2y-3)(2y-1) = -4$$
$$4y^2 - 8y + 3 = -4$$
$$4y^2 - 8y + 7 = 0$$
$$y = \frac{8 \pm \sqrt{64-112}}{8} = \frac{8 \pm \sqrt{-48}}{8} = \frac{8 \pm 4i\sqrt{3}}{8} = \frac{2 \pm i\sqrt{3}}{2}$$

79. Solving the equation:
$$t^3 - 125 = 0$$
$$(t-5)\left(t^2 + 5t + 25\right) = 0$$
$$t = 5,\ \frac{-5 \pm \sqrt{25-100}}{2} = \frac{-5 \pm \sqrt{-75}}{2} = \frac{-5 \pm 5i\sqrt{3}}{2}$$

81. Solving the equation:
$$4x^5 - 16x^4 = 20x^3$$
$$4x^5 - 16x^4 - 20x^3 = 0$$
$$4x^3\left(x^2 - 4x - 5\right) = 0$$
$$4x^3\left(x - 5\right)\left(x + 1\right) = 0$$
$$x = -1, 0, 5$$

83. Solving the equation:
$$\frac{1}{x-3} + \frac{1}{x+2} = 1$$
$$x + 2 + x - 3 = \left(x - 3\right)\left(x + 2\right)$$
$$2x - 1 = x^2 - x - 6$$
$$x^2 - 3x - 5 = 0$$
$$x = \frac{3 \pm \sqrt{9 + 20}}{2} = \frac{3 \pm \sqrt{29}}{2}$$

85. Simplifying; $8^{2/3} = \left(8^{1/3}\right)^2 = \left(\sqrt[3]{8}\right)^2 = 2^2 = 4$

87. Solving the equation:
$$\left(x + 2\right)\left(x\right) = 2^3$$
$$x^2 + 2x = 8$$
$$x^2 + 2x - 8 = 0$$
$$\left(x + 4\right)\left(x - 2\right) = 0$$
$$x = -4, 2$$

89. Solving the equation:
$$\frac{x-2}{x+1} = 9$$
$$x - 2 = 9\left(x + 1\right)$$
$$x - 2 = 9x + 9$$
$$-8x = 11$$
$$x = -\frac{11}{8}$$

91. Writing in exponential form: $2^3 = \left(x + 2\right)\left(x\right)$

93. Writing in exponential form: $3^4 = \frac{x-2}{x+1}$

95. **a.** Completing the table:

x	-1	0	1	2
$f(x)$	$\frac{1}{8}$	1	8	64

b. Completing the table:

x	$\frac{1}{8}$	1	8	64
$f^{-1}(x)$	-1	0	1	2

c. The equation is $f(x) = 8^x$.

d. The equation is $f^{-1}(x) = \log_8 x$.

9.4 Properties of Logarithms

1. Using properties of logarithms: $\log_3 4x = \log_3 4 + \log_3 x$

3. Using properties of logarithms: $\log_6 \frac{5}{x} = \log_6 5 - \log_6 x$

5. Using properties of logarithms: $\log_2 y^5 = 5\log_2 y$

7. Using properties of logarithms: $\log_9 \sqrt[3]{z} = \log_9 z^{1/3} = \frac{1}{3}\log_9 z$

9. Using properties of logarithms: $\log_6 x^2 y^4 = \log_6 x^2 + \log_6 y^4 = 2\log_6 x + 4\log_6 y$

11. Using properties of logarithms: $\log_5 \sqrt{x} \cdot y^4 = \log_5 x^{1/2} + \log_5 y^4 = \frac{1}{2}\log_5 x + 4\log_5 y$

13. Using properties of logarithms: $\log_b \frac{xy}{z} = \log_b xy - \log_b z = \log_b x + \log_b y - \log_b z$

15. Using properties of logarithms: $\log_{10} \frac{4}{xy} = \log_{10} 4 - \log_{10} xy = \log_{10} 4 - \log_{10} x - \log_{10} y$

17. Using properties of logarithms: $\log_{10} \frac{x^2 y}{\sqrt{z}} = \log_{10} x^2 + \log_{10} y - \log_{10} z^{1/2} = 2\log_{10} x + \log_{10} y - \frac{1}{2}\log_{10} z$

19.   Using properties of logarithms: $\log_{10} \dfrac{x^3 \sqrt{y}}{z^4} = \log_{10} x^3 + \log_{10} y^{1/2} - \log_{10} z^4 = 3\log_{10} x + \frac{1}{2}\log_{10} y - 4\log_{10} z$

21.   Using properties of logarithms:

$$\log_b \sqrt[3]{\dfrac{x^2 y}{z^4}} = \log_b \dfrac{x^{2/3} y^{1/3}}{z^{4/3}} = \log_b x^{2/3} + \log_b y^{1/3} - \log_b z^{4/3} = \tfrac{2}{3}\log_b x + \tfrac{1}{3}\log_b y - \tfrac{4}{3}\log_b z$$

23.   Writing as a single logarithm: $\log_b x + \log_b z = \log_b xz$

25.   Writing as a single logarithm: $2\log_3 x - 3\log_3 y = \log_3 x^2 - \log_3 y^3 = \log_3 \dfrac{x^2}{y^3}$

27.   Writing as a single logarithm: $\frac{1}{2}\log_{10} x + \frac{1}{3}\log_{10} y = \log_{10} x^{1/2} + \log_{10} y^{1/3} = \log_{10} \sqrt{x}\,\sqrt[3]{y}$

29.   Writing as a single logarithm: $3\log_2 x + \frac{1}{2}\log_2 y - \log_2 z = \log_2 x^3 + \log_2 y^{1/2} - \log_2 z = \log_2 \dfrac{x^3 \sqrt{y}}{z}$

31.   Writing as a single logarithm: $\frac{1}{2}\log_2 x - 3\log_2 y - 4\log_2 z = \log_2 x^{1/2} - \log_2 y^3 - \log_2 z^4 = \log_2 \dfrac{\sqrt{x}}{y^3 z^4}$

33.   Writing as a single logarithm:

$$\tfrac{3}{2}\log_{10} x - \tfrac{3}{4}\log_{10} y - \tfrac{4}{5}\log_{10} z = \log_{10} x^{3/2} - \log_{10} y^{3/4} - \log_{10} z^{4/5} = \log_{10} \dfrac{x^{3/2}}{y^{3/4} z^{4/5}}$$

35.   Solving the equation:

$$\log_2 x + \log_2 3 = 1$$
$$\log_2 3x = 1$$
$$3x = 2^1$$
$$3x = 2$$
$$x = \tfrac{2}{3}$$

37.   Solving the equation:

$$\log_3 x - \log_3 2 = 2$$
$$\log_3 \dfrac{x}{2} = 2$$
$$\dfrac{x}{2} = 3^2$$
$$\dfrac{x}{2} = 9$$
$$x = 18$$

39.   Solving the equation:

$$\log_3 x + \log_3 (x-2) = 1$$
$$\log_3 (x^2 - 2x) = 1$$
$$x^2 - 2x = 3^1$$
$$x^2 - 2x - 3 = 0$$
$$(x-3)(x+1) = 0$$
$$x = 3, -1$$

The solution is 3 (–1 does not check).

41.   Solving the equation:

$$\log_3 (x+3) - \log_3 (x-1) = 1$$
$$\log_3 \dfrac{x+3}{x-1} = 1$$
$$\dfrac{x+3}{x-1} = 3^1$$
$$x+3 = 3x - 3$$
$$-2x = -6$$
$$x = 3$$

43.   Solving the equation:

$$\log_2 x + \log_2 (x-2) = 3$$
$$\log_2 (x^2 - 2x) = 3$$
$$x^2 - 2x = 2^3$$
$$x^2 - 2x - 8 = 0$$
$$(x-4)(x+2) = 0$$
$$x = 4, -2$$

The solution is 4 (–2 does not check).

45.   Solving the equation:

$$\log_8 x + \log_8 (x-3) = \tfrac{2}{3}$$
$$\log_8 (x^2 - 3x) = \tfrac{2}{3}$$
$$x^2 - 3x = 8^{2/3}$$
$$x^2 - 3x - 4 = 0$$
$$(x-4)(x+1) = 0$$
$$x = 4, -1$$

The solution is 4 (–1 does not check).

47.   Solving the equation:

$$\log_5 \sqrt{x} + \log_5 \sqrt{6x+5} = 1$$
$$\log_5 \sqrt{6x^2 + 5x} = 1$$
$$\tfrac{1}{2}\log_5 \left(6x^2 + 5x\right) = 1$$
$$\log_5 \left(6x^2 + 5x\right) = 2$$
$$6x^2 + 5x = 5^2$$
$$6x^2 + 5x - 25 = 0$$
$$\left(3x - 5\right)\left(2x + 5\right) = 0$$
$$x = \tfrac{5}{3}, -\tfrac{5}{2}$$

The solution is $\tfrac{5}{3}$ ($-\tfrac{5}{2}$ does not check).

49.   Rewriting the formula:

$$D = 10\log_{10}\left(\frac{I}{I_0}\right)$$
$$D = 10\left(\log_{10} I - \log_{10} I_0\right)$$

51.   Solving for N:

$$N = \log_{10}\frac{100}{1} = \log_{10} 10^2 = 2$$

$$N = \log_{10} 100 - \log_{10} 1 = \log_{10} 10^2 = 2$$

So $N = 2$ in both cases.

53.   Dividing: $\dfrac{12x^2 + y^2}{36} = \dfrac{12x^2}{36} + \dfrac{y^2}{36} = \dfrac{x^2}{3} + \dfrac{y^2}{36}$

55.   Dividing: $\dfrac{25x^2 + 4y^2}{100} = \dfrac{25x^2}{100} + \dfrac{4y^2}{100} = \dfrac{x^2}{4} + \dfrac{y^2}{25}$

57.   Computing the discriminant: $D = \left(-5\right)^2 - 4\left(2\right)\left(4\right) = 25 - 32 = -7$. There are two complex solutions.

59.   Writing the equation:
$$\left(x + 3\right)\left(x - 5\right) = 0$$
$$x^2 - 2x - 15 = 0$$

61.   Writing the equation:
$$\left(3y - 2\right)\left(y - 3\right) = 0$$
$$3y^2 - 11y + 6 = 0$$

63.   Simplifying: $5^0 = 1$

65.   Simplifying: $\log_3 3 = \log_3 3^1 = 1$

67.   Simplifying: $\log_b b^4 = 4$

69.   Using a calculator: $10^{-5.6} \approx 2.5 \times 10^{-6}$

71.   Using a calculator: $\dfrac{2.00 \times 10^8}{3.96 \times 10^6} \approx 51$

9.5 Common Logarithms and Natural Logarithms

1.   Evaluating the logarithm: $\log 378 \approx 2.5775$

3.   Evaluating the logarithm: $\log 37.8 \approx 1.5775$

5.   Evaluating the logarithm: $\log 3{,}780 \approx 3.5775$

7.   Evaluating the logarithm: $\log 0.0378 \approx -1.4225$

9.   Evaluating the logarithm: $\log 37{,}800 \approx 4.5775$

11.   Evaluating the logarithm: $\log 600 \approx 2.7782$

13.   Evaluating the logarithm: $\log 2{,}010 \approx 3.3032$

15.   Evaluating the logarithm: $\log 0.00971 \approx -2.0128$

17.   Evaluating the logarithm: $\log 0.0314 \approx -1.5031$

19.   Evaluating the logarithm: $\log 0.399 \approx -0.3990$

21.   Solving for x:
$$\log x = 2.8802$$
$$x = 10^{2.8802} \approx 759$$

23.   Solving for x:
$$\log x = -2.1198$$
$$x = 10^{-2.1198} \approx 0.00759$$

25.   Solving for x:
$$\log x = 3.1553$$
$$x = 10^{3.1553} \approx 1{,}430$$

27.   Solving for x:
$$\log x = -5.3497$$
$$x = 10^{-5.3497} \approx 0.00000447$$

29.   Solving for x:
$$\log x = -7.0372$$
$$x = 10^{-7.0372} \approx 0.0000000918$$

31.   Solving for x:
$$\log x = 10$$
$$x = 10^{10}$$

33.   Solving for x:
$$\log x = -10$$
$$x = 10^{-10}$$

35.   Solving for x:
$$\log x = 20$$
$$x = 10^{20}$$

37. Solving for x:

$$\log x = -2$$
$$x = 10^{-2} = \frac{1}{100}$$

39. Solving for x:

$$\log x = \log_2 8$$
$$\log x = 3$$
$$x = 10^3 = 1,000$$

41. Simplifying the logarithm: $\ln e = \ln e^1 = 1$

43. Simplifying the logarithm: $\ln e^5 = 5$

45. Simplifying the logarithm: $\ln e^x = x$

47. Using properties of logarithms: $\ln 10e^{3t} = \ln 10 + \ln e^{3t} = \ln 10 + 3t$

49. Using properties of logarithms: $\ln Ae^{-2t} = \ln A + \ln e^{-2t} = \ln A - 2t$

51. Evaluating the logarithm: $\ln 15 = \ln(3 \bullet 5) = \ln 3 + \ln 5 = 1.0986 + 1.6094 = 2.7080$

53. Evaluating the logarithm: $\ln \frac{1}{3} = \ln 3^{-1} = -\ln 3 = -1.0986$

55. Evaluating the logarithm: $\ln 9 = \ln 3^2 = 2 \ln 3 = 2(1.0986) = 2.1972$

57. Evaluating the logarithm: $\ln 16 = \ln 2^4 = 4 \ln 2 = 4(0.6931) = 2.7724$

59. a. Using a calculator: $\dfrac{\log 25}{\log 15} \approx 1.1886$

    b. Using a calculator: $\log \frac{25}{15} \approx 0.2218$

61. a. Using a calculator: $\dfrac{\log 4}{\log 8} \approx 0.6667$

    b. Using a calculator: $\log \frac{4}{8} \approx -0.3010$

63. Computing the pH: $pH = -\log(6.50 \times 10^{-4}) \approx 3.19$

65. Finding the concentration:

$$4.75 = -\log\left[H^+\right]$$
$$-4.75 = \log\left[H^+\right]$$
$$\left[H^+\right] = 10^{-4.75} \approx 1.78 \times 10^{-5}$$

67. Finding the magnitude:

$$5.5 = \log T$$
$$T = 10^{5.5} \approx 3.16 \times 10^5$$

69. Finding the magnitude:

$$8.3 = \log T$$
$$T = 10^{8.3} \approx 2.00 \times 10^8$$

71. For the first earthquake:

$$\log T_1 = 6.5$$
$$T_1 = 10^{6.5}$$

For the second earthquake:

$$\log T_2 = 5.5$$
$$T_2 = 10^{5.5}$$

The ratio is $\dfrac{T_1}{T_2} = \dfrac{10^{6.5}}{10^{5.5}} = 10$ times stronger.

73. Completing the table:

Location	Date	Magnitude (M)	Shockwave (T)
Moresby Island	January 23	4.0	1.00×10^4
Vancouver Island	April 30	5.3	1.99×10^5
Quebec City	June 29	3.2	1.58×10^3
Mould Bay	November 13	5.2	1.58×10^5
St. Lawrence	December 14	3.7	5.01×10^3

75. Finding the rate of depreciation:

$$\log(1-r) = \tfrac{1}{5} \log \frac{4500}{9000}$$
$$\log(1-r) \approx -0.0602$$
$$1 - r \approx 10^{-0.0602}$$
$$r = 1 - 10^{-0.0602}$$
$$r \approx 0.129 = 12.9\%$$

77. Finding the rate of depreciation:

$$\log(1-r) = \tfrac{1}{5} \log \frac{5750}{7550}$$
$$\log(1-r) \approx -0.0237$$
$$1 - r \approx 10^{-0.0237}$$
$$r = 1 - 10^{-0.0237}$$
$$r \approx 0.053 = 5.3\%$$

79.   It appears to approach e. Completing the table:

x	$(1+x)^{1/x}$
1	2
0.5	2.25
0.1	2.5937
0.01	2.7048
0.001	2.7169
0.0001	2.7181
0.00001	2.7183

81.   Solving the equation:

$$(y+3)^2 + y^2 = 9$$
$$y^2 + 6y + 9 + y^2 = 9$$
$$2y^2 + 6y = 0$$
$$2y(y+3) = 0$$
$$y = -3, 0$$

83.   Solving the equation:

$$(x+3)^2 + 1^2 = 2$$
$$x^2 + 6x + 9 + 1 = 2$$
$$x^2 + 6x + 8 = 0$$
$$(x+4)(x+2) = 0$$
$$x = -4, -2$$

85.   Solving the equation:

$$x^4 - 2x^2 - 8 = 0$$
$$(x^2 - 4)(x^2 + 2) = 0$$
$$x^2 = 4, -2$$
$$x = \pm 2, \pm i\sqrt{2}$$

87.   Solving the equation:

$$2x - 5\sqrt{x} + 3 = 0$$
$$(2\sqrt{x} - 3)(\sqrt{x} - 1) = 0$$
$$\sqrt{x} = 1, \tfrac{3}{2}$$
$$x = 1, \tfrac{9}{4}$$

89.   Solving the equation:

$$5(2x+1) = 12$$
$$10x + 5 = 12$$
$$10x = 7$$
$$x = \tfrac{7}{10} = 0.7$$

91.   Using a calculator: $\dfrac{100,000}{32,000} = 3.125$

93.   Using a calculator: $\dfrac{1}{2}\left(\dfrac{-0.6931}{1.4289} + 3\right) \approx 1.2575$

95.   Rewriting the logarithm: $\log 1.05^t = t \log 1.05$

97.   Simplifying: $\ln e^{0.05t} = 0.05t$

9.6 Exponential Equations and Change of Base

1.   Solving the equation:

$$3^x = 5$$
$$\ln 3^x = \ln 5$$
$$x \ln 3 = \ln 5$$
$$x = \frac{\ln 5}{\ln 3} \approx 1.4650$$

3.   Solving the equation:

$$5^x = 3$$
$$\ln 5^x = \ln 3$$
$$x \ln 5 = \ln 3$$
$$x = \frac{\ln 3}{\ln 5} \approx 0.6826$$

5.   Solving the equation:

$$5^{-x} = 12$$
$$\ln 5^{-x} = \ln 12$$
$$-x \ln 5 = \ln 12$$
$$x = -\frac{\ln 12}{\ln 5} \approx -1.5440$$

7.   Solving the equation:

$$12^{-x} = 5$$
$$\ln 12^{-x} = \ln 5$$
$$-x \ln 12 = \ln 5$$
$$x = -\frac{\ln 5}{\ln 12} \approx -0.6477$$

9.   Solving the equation:

$$8^{x+1} = 4$$
$$2^{3x+3} = 2^2$$
$$3x + 3 = 2$$
$$3x = -1$$
$$x = -\tfrac{1}{3} \approx -0.3333$$

11.   Solving the equation:

$$4^{x-1} = 4$$
$$4^{x-1} = 4^1$$
$$x - 1 = 1$$
$$x = 2 = 2.0000$$

13. Solving the equation:
$$3^{2x+1} = 2$$
$$\ln 3^{2x+1} = \ln 2$$
$$(2x+1)\ln 3 = \ln 2$$
$$2x+1 = \frac{\ln 2}{\ln 3}$$
$$2x = \frac{\ln 2}{\ln 3} - 1$$
$$x = \tfrac{1}{2}\left(\frac{\ln 2}{\ln 3} - 1\right) \approx -0.1845$$

15. Solving the equation:
$$3^{1-2x} = 2$$
$$\ln 3^{1-2x} = \ln 2$$
$$(1-2x)\ln 3 = \ln 2$$
$$1-2x = \frac{\ln 2}{\ln 3}$$
$$-2x = \frac{\ln 2}{\ln 3} - 1$$
$$x = \tfrac{1}{2}\left(1 - \frac{\ln 2}{\ln 3}\right) \approx 0.1845$$

17. Solving the equation:
$$15^{3x-4} = 10$$
$$\ln 15^{3x-4} = \ln 10$$
$$(3x-4)\ln 15 = \ln 10$$
$$3x-4 = \frac{\ln 10}{\ln 15}$$
$$3x = \frac{\ln 10}{\ln 15} + 4$$
$$x = \tfrac{1}{3}\left(\frac{\ln 10}{\ln 15} + 4\right) \approx 1.6168$$

19. Solving the equation:
$$6^{5-2x} = 4$$
$$\ln 6^{5-2x} = \ln 4$$
$$(5-2x)\ln 6 = \ln 4$$
$$5-2x = \frac{\ln 4}{\ln 6}$$
$$-2x = \frac{\ln 4}{\ln 6} - 5$$
$$x = \tfrac{1}{2}\left(5 - \frac{\ln 4}{\ln 6}\right) \approx 2.1131$$

21. Evaluating the logarithm: $\log_8 16 = \dfrac{\log 16}{\log 8} \approx 1.3333$

23. Evaluating the logarithm: $\log_{16} 8 = \dfrac{\log 8}{\log 16} = 0.7500$

25. Evaluating the logarithm: $\log_7 15 = \dfrac{\log 15}{\log 7} \approx 1.3917$

27. Evaluating the logarithm: $\log_{15} 7 = \dfrac{\log 7}{\log 15} \approx 0.7186$

29. Evaluating the logarithm: $\log_8 240 = \dfrac{\log 240}{\log 8} \approx 2.6356$

31. Evaluating the logarithm: $\log_4 321 = \dfrac{\log 321}{\log 4} \approx 4.1632$

33. Evaluating the logarithm: $\ln 345 \approx 5.8435$

35. Evaluating the logarithm: $\ln 0.345 \approx -1.0642$

37. Evaluating the logarithm: $\ln 10 \approx 2.3026$

39. Evaluating the logarithm: $\ln 45,000 \approx 10.7144$

41. Using the compound interest formula:
$$500\left(1 + \frac{0.06}{2}\right)^{2t} = 1000$$
$$\left(1 + \frac{0.06}{2}\right)^{2t} = 2$$
$$\ln\left(1 + \frac{0.06}{2}\right)^{2t} = \ln 2$$
$$2t\ln\left(1 + \frac{0.06}{2}\right) = \ln 2$$
$$t = \frac{\ln 2}{2\ln\left(1 + \frac{0.06}{2}\right)} \approx 11.7$$

It will take 11.7 years.

43. Using the compound interest formula:
$$1000\left(1 + \frac{0.12}{6}\right)^{6t} = 3000$$
$$\left(1 + \frac{0.12}{6}\right)^{6t} = 3$$
$$\ln\left(1 + \frac{0.12}{6}\right)^{6t} = \ln 3$$
$$6t\ln\left(1 + \frac{0.12}{6}\right) = \ln 3$$
$$t = \frac{\ln 3}{6\ln\left(1 + \frac{0.12}{6}\right)} \approx 9.25$$

It will take 9.25 years.

45. Using the compound interest formula:

$$P\left(1+\frac{0.08}{4}\right)^{4t} = 2P$$

$$\left(1+\frac{0.08}{4}\right)^{4t} = 2$$

$$\ln\left(1+\frac{0.08}{4}\right)^{4t} = \ln 2$$

$$4t\ln\left(1+\frac{0.08}{4}\right) = \ln 2$$

$$t = \frac{\ln 2}{4\ln\left(1+\frac{0.08}{4}\right)} \approx 8.75$$

It will take 8.75 years.

47. Using the compound interest formula:

$$25\left(1+\frac{0.06}{2}\right)^{2t} = 75$$

$$\left(1+\frac{0.06}{2}\right)^{2t} = 3$$

$$\ln\left(1+\frac{0.06}{2}\right)^{2t} = \ln 3$$

$$2t\ln\left(1+\frac{0.06}{2}\right) = \ln 3$$

$$t = \frac{\ln 3}{2\ln\left(1+\frac{0.06}{2}\right)} \approx 18.58$$

It was invested 18.58 years ago.

49. Using the continuous interest formula:

$$500e^{0.06t} = 1000$$

$$e^{0.06t} = 2$$

$$0.06t = \ln 2$$

$$t = \frac{\ln 2}{0.06} \approx 11.55$$

It will take 11.55 years.

51. Using the continuous interest formula:

$$500e^{0.06t} = 1500$$

$$e^{0.06t} = 3$$

$$0.06t = \ln 3$$

$$t = \frac{\ln 3}{0.06} \approx 18.31$$

It will take 18.31 years.

53. Completing the square: $y = 2x^2 + 8x - 15 = 2\left(x^2 + 4x + 4\right) - 8 - 15 = 2\left(x+2\right)^2 - 23$. The lowest point is $(-2,-23)$.

55. Completing the square: $y = 12x - 4x^2 = -4\left(x^2 - 3x + \frac{9}{4}\right) + 9 = -4\left(x - \frac{3}{2}\right)^2 + 9$. The highest point is $\left(\frac{3}{2}, 9\right)$.

57. Completing the square: $y = 64t - 16t^2 = -16\left(t^2 - 4t + 4\right) + 64 = -16\left(t-2\right)^2 + 64$

The object reaches a maximum height after 2 seconds, and the maximum height is 64 feet.

59. Using the population model:

$$32000e^{0.05t} = 64000$$

$$e^{0.05t} = 2$$

$$0.05t = \ln 2$$

$$t = \frac{\ln 2}{0.05} \approx 13.9$$

The city will reach 64,000 toward the end of the year 2007.

61. Finding when $P(t) = 45,000$:

$$15,000e^{0.04t} = 45,000$$

$$e^{0.04t} = 3$$

$$0.04t = \ln 3$$

$$t = \frac{\ln 3}{0.04} \approx 27.5$$

It will take approximately 27.5 years.

63. Solving for t:

$$A = Pe^{rt}$$

$$e^{rt} = \frac{A}{P}$$

$$rt = \ln\frac{A}{P}$$

$$t = \frac{\ln A - \ln P}{r} = \frac{1}{r}\ln\frac{A}{P}$$

65. Solving for t:

$$A = P2^{-kt}$$

$$2^{-kt} = \frac{A}{P}$$

$$\ln 2^{-kt} = \ln \frac{A}{P}$$

$$-kt \ln 2 = \ln A - \ln P$$

$$t = \frac{\ln A - \ln P}{-k \ln 2} = \frac{\ln P - \ln A}{k \ln 2}$$

67. Solving for t:

$$A = P(1-r)^t$$

$$(1-r)^t = \frac{A}{P}$$

$$\ln(1-r)^t = \ln \frac{A}{P}$$

$$t \ln(1-r) = \ln A - \ln P$$

$$t = \frac{\ln A - \ln P}{\ln(1-r)}$$

Chapter 9 Review/Test

1. Evaluating the function: $f(4) = 2^4 = 16$

2. Evaluating the function: $f(-1) = 2^{-1} = \frac{1}{2}$

3. Evaluating the function: $g(2) = \left(\frac{1}{3}\right)^2 = \frac{1}{9}$

4. Evaluating the function: $f(2) - g(-2) = 2^2 - \left(\frac{1}{3}\right)^{-2} = 4 - 9 = -5$

5. Evaluating the function: $f(-1) + g(1) = 2^{-1} + \left(\frac{1}{3}\right)^1 = \frac{1}{2} + \frac{1}{3} = \frac{5}{6}$

6. Evaluating the function: $g(-1) + f(2) = \left(\frac{1}{3}\right)^{-1} + 2^2 = 3 + 4 = 7$

7. Graphing the function:

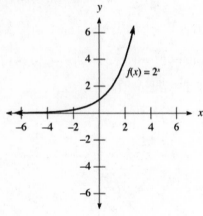

8. Graphing the function:

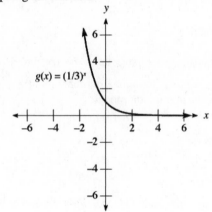

9. Finding the inverse:

$$2y + 1 = x$$
$$2y = x - 1$$
$$y = \frac{x - 1}{2}$$

The inverse is $y^{-1} = \frac{x - 1}{2}$. Sketching the graph:

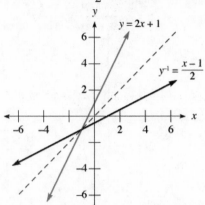

11. Finding the inverse:

$$2y + 3 = x$$
$$2y = x - 3$$
$$y = \frac{x - 3}{2}$$

The inverse is $f^{-1}(x) = \frac{x - 3}{2}$.

13. Finding the inverse:

$$\tfrac{1}{2} y + 2 = x$$
$$y + 4 = 2x$$
$$y = 2x - 4$$

The inverse is $f^{-1}(x) = 2x - 4$.

15. Writing in logarithmic form: $\log_3 81 = 4$

17. Writing in logarithmic form: $\log_{10} 0.01 = -2$

19. Writing in exponential form: $2^3 = 8$

21. Writing in exponential form: $4^{1/2} = 2$

23. Solving for x:

$$\log_5 x = 2$$
$$x = 5^2 = 25$$

10. Finding the inverse:

$$y^2 - 4 = x$$
$$y^2 = x + 4$$
$$y = \pm\sqrt{x + 4}$$

The inverse is $y^{-1} = \pm\sqrt{x + 4}$. Sketching the graph:

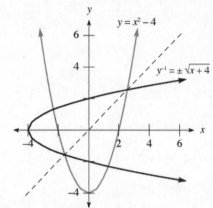

12. Finding the inverse:

$$y^2 - 1 = x$$
$$y^2 = x + 1$$
$$y = \pm\sqrt{x + 1}$$

The inverse is $y = \pm\sqrt{x + 1}$ (which is not a function).

14. Finding the inverse:

$$4 - 2y^2 = x$$
$$2y^2 = 4 - x$$
$$y^2 = \frac{4 - x}{2}$$
$$y = \pm\sqrt{\frac{4 - x}{2}}$$

The inverse is $y = \pm\sqrt{\frac{4 - x}{2}}$ (which is not a function).

16. Writing in logarithmic form: $\log_7 49 = 2$

18. Writing in logarithmic form: $\log_2 \frac{1}{8} = -3$

20. Writing in exponential form: $3^2 = 9$

22. Writing in exponential form: $4^1 = 4$

24. Solving for x:

$$\log_{16} 8 = x$$
$$16^x = 8$$
$$2^{4x} = 2^3$$
$$4x = 3$$
$$x = \frac{3}{4}$$

25. Solving for x:

$$\log_x 0.01 = -2$$
$$x^{-2} = 0.01$$
$$x^2 = 100$$
$$x = 10$$

26. Graphing the equation:

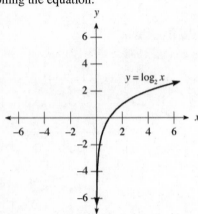

27. Graphing the equation:

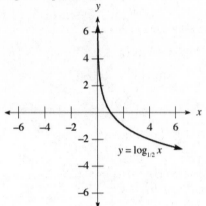

28. Simplifying the logarithm: $\log_4 16 = \log_4 4^2 = 2$

29. Simplifying the logarithm:

$$\log_{27} 9 = x$$
$$27^x = 9$$
$$3^{3x} = 3^2$$
$$3x = 2$$
$$x = \frac{2}{3}$$

30. Simplifying the logarithm: $\log_4 \left(\log_3 3 \right) = \log_4 1 = 0$

31. Expanding the logarithm: $\log_2 5x = \log_2 5 + \log_2 x$

32. Expanding the logarithm: $\log_{10} \dfrac{2x}{y} = \log_{10} 2x - \log_{10} y = \log_{10} 2 + \log_{10} x - \log_{10} y$

33. Expanding the logarithm: $\log_a \dfrac{y^3 \sqrt{x}}{z} = \log_a y^3 + \log_a x^{1/2} - \log_a z = 3\log_a y + \frac{1}{2}\log_a x - \log_a z$

34. Expanding the logarithm: $\log_{10} \dfrac{x^2}{y^3 z^4} = \log_{10} x^2 - \log_{10} y^3 - \log_{10} z^4 = 2\log_{10} x - 3\log_{10} y - 4\log_{10} z$

35. Writing as a single logarithm: $\log_2 x + \log_2 y = \log_2 xy$

36. Writing as a single logarithm: $\log_3 x - \log_3 4 = \log_3 \dfrac{x}{4}$

37. Writing as a single logarithm: $2\log_a 5 - \frac{1}{2}\log_a 9 = \log_a 5^2 - \log_a 9^{1/2} = \log_a 25 - \log_a 3 = \log_a \dfrac{25}{3}$

38. Writing as a single logarithm: $3\log_2 x + 2\log_2 y - 4\log_2 z = \log_2 x^3 + \log_2 y^2 - \log_2 z^4 = \log_2 \dfrac{x^3 y^2}{z^4}$

39. Solving the equation:

$$\log_2 x + \log_2 4 = 3$$
$$\log_2 4x = 3$$
$$4x = 2^3$$
$$4x = 8$$
$$x = 2$$

40. Solving the equation:

$$\log_2 x - \log_2 3 = 1$$
$$\log_2 \frac{x}{3} = 1$$
$$\frac{x}{3} = 2^1$$
$$\frac{x}{3} = 2$$
$$x = 6$$

41. Solving the equation:

$$\log_3 x + \log_3 (x-2) = 1$$
$$\log_3 (x^2 - 2x) = 1$$
$$x^2 - 2x = 3^1$$
$$x^2 - 2x - 3 = 0$$
$$(x-3)(x+1) = 0$$
$$x = 3, -1$$

The solution is 3 (–1 does not check).

42. Solving the equation:

$$\log_4 (x+1) - \log_4 (x-2) = 1$$
$$\log_4 \frac{x+1}{x-2} = 1$$
$$\frac{x+1}{x-2} = 4^1$$
$$x+1 = 4x - 8$$
$$-3x = -9$$
$$x = 3$$

43. Solving the equation:

$$\log_6 (x-1) + \log_6 x = 1$$
$$\log_6 (x^2 - x) = 1$$
$$x^2 - x = 6^1$$
$$x^2 - x - 6 = 0$$
$$(x-3)(x+2) = 0$$
$$x = 3, -2$$

The solution is 3 (–2 does not check).

44. Solving the equation:

$$\log_4 (x-3) + \log_4 x = 1$$
$$\log_4 (x^2 - 3x) = 1$$
$$x^2 - 3x = 4^1$$
$$x^2 - 3x - 4 = 0$$
$$(x-4)(x+1) = 0$$
$$x = 4, -1$$

The solution is 4 (–1 does not check).

45. Evaluating: $\log 346 \approx 2.5391$

46. Evaluating: $\log 0.713 \approx -0.1469$

47. Solving for x:
$$\log x = 3.9652$$
$$x = 10^{3.9652} \approx 9,230$$

48. Solving for x:
$$\log x = -1.6003$$
$$x = 10^{-1.6003} \approx 0.0251$$

49. Simplifying: $\ln e = \ln e^1 = 1$

50. Simplifying: $\ln 1 = \ln e^0 = 0$

51. Simplifying: $\ln e^2 = 2$

52. Simplifying: $\ln e^{-4} = -4$

53. Finding the pH: $pH = -\log\left(7.9 \times 10^{-3}\right) \approx 2.1$

54. Finding the pH: $pH = -\log\left(8.1 \times 10^{-6}\right) \approx 5.1$

55. Finding $\left[H^+\right]$:

$$2.7 = -\log\left[H^+\right]$$
$$-2.7 = \log\left[H^+\right]$$
$$\left[H^+\right] = 10^{-2.7} \approx 2.0 \times 10^{-3}$$

56. Finding $\left[H^+\right]$:

$$7.5 = -\log\left[H^+\right]$$
$$-7.5 = \log\left[H^+\right]$$
$$\left[H^+\right] = 10^{-7.5} \approx 3.2 \times 10^{-8}$$

57. Solving the equation:

$$4^x = 8$$
$$2^{2x} = 2^3$$
$$2x = 3$$
$$x = \frac{3}{2}$$

58. Solving the equation:

$$4^{3x+2} = 5$$
$$\ln 4^{3x+2} = \ln 5$$
$$(3x+2)\ln 4 = \ln 5$$
$$3x+2 = \frac{\ln 5}{\ln 4}$$
$$3x = \frac{\ln 5}{\ln 4} - 2$$
$$x = \frac{1}{3}\left(\frac{\ln 5}{\ln 4} - 2\right) \approx -0.28$$

59. Using a calculator: $\log_{16} 8 = \dfrac{\ln 8}{\ln 16} = 0.75$

60. Using a calculator: $\log_{12} 421 = \dfrac{\ln 421}{\ln 12} \approx 2.43$

61. Using the compound interest formula:

$$5000(1+0.16)^t = 10000$$
$$1.16^t = 2$$
$$\ln 1.16^t = \ln 2$$
$$t \ln 1.16 = \ln 2$$
$$t = \frac{\ln 2}{\ln 1.16} \approx 4.67$$

It will take approximately 4.67 years for the amount to double.

62. Using the compound interest formula:

$$10000\left(1+\frac{0.12}{6}\right)^{6t} = 30000$$
$$\left(1+\frac{0.12}{6}\right)^{6t} = 3$$
$$\ln\left(1+\frac{0.12}{6}\right)^{6t} = \ln 3$$
$$6t \ln 1.02 = \ln 3$$
$$t = \frac{\ln 3}{6\ln 1.02} \approx 9.25$$

It will take approximately 9.25 years for the amount to triple.

Chapter 10
Conic Sections

10.1 The Circle

1. Using the distance formula: $d = \sqrt{(6-3)^2 + (3-7)^2} = \sqrt{9+16} = \sqrt{25} = 5$

3. Using the distance formula: $d = \sqrt{(5-0)^2 + (0-9)^2} = \sqrt{25+81} = \sqrt{106}$

5. Using the distance formula: $d = \sqrt{(-2-3)^2 + (1+5)^2} = \sqrt{25+36} = \sqrt{61}$

7. Using the distance formula: $d = \sqrt{(-10+1)^2 + (5+2)^2} = \sqrt{81+49} = \sqrt{130}$

9. Solving the equation:
$$\sqrt{(x-1)^2 + (2-5)^2} = \sqrt{13}$$
$$(x-1)^2 + 9 = 13$$
$$(x-1)^2 = 4$$
$$x - 1 = \pm 2$$
$$x - 1 = -2, 2$$
$$x = -1, 3$$

11. Solving the equation:
$$\sqrt{(7-8)^2 + (y-3)^2} = 1$$
$$(y-3)^2 + 1 = 1$$
$$(y-3)^2 = 0$$
$$y - 3 = 0$$
$$y = 3$$

13. The equation is $(x-2)^2 + (y-3)^2 = 16$.

15. The equation is $(x-3)^2 + (y+2)^2 = 9$.

17. The equation is $(x+5)^2 + (y+1)^2 = 5$.

19. The equation is $x^2 + (y+5)^2 = 1$.

21. The equation is $x^2 + y^2 = 4$.

23. The center is $(0,0)$ and the radius is 2.

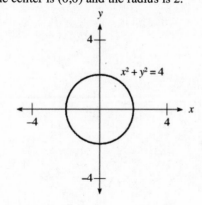

25. The center is $(1,3)$ and the radius is 5.

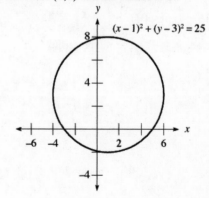

27. The center is $(-2,4)$ and the radius is $2\sqrt{2}$.

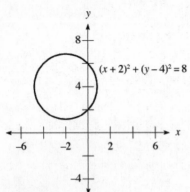

29. The center is $(-1,-1)$ and the radius is 1.

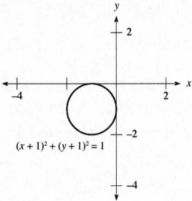

31. Completing the square:
$$x^2 + y^2 - 6y = 7$$
$$x^2 + \left(y^2 - 6y + 9\right) = 7 + 9$$
$$x^2 + \left(y - 3\right)^2 = 16$$
The center is $(0,3)$ and the radius is 4.

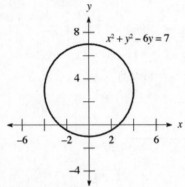

33. Completing the square:
$$x^2 + y^2 - 4x - 6y = -4$$
$$\left(x^2 - 4x + 4\right) + \left(y^2 - 6y + 9\right) = -4 + 4 + 9$$
$$\left(x - 2\right)^2 + \left(y - 3\right)^2 = 9$$
The center is $(-5,0)$ and the radius is 5.

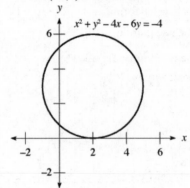

35. Completing the square:
$$x^2 + y^2 + 2x + y = \tfrac{11}{4}$$
$$\left(x^2 + 2x + 1\right) + \left(y^2 + y + \tfrac{1}{4}\right) = \tfrac{11}{4} + 1 + \tfrac{1}{4}$$
$$\left(x + 1\right)^2 + \left(y + \tfrac{1}{2}\right)^2 = 4$$
The center is $\left(-1, -\tfrac{1}{2}\right)$ and the radius is 2.

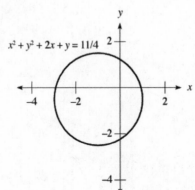

37. The equation is $\left(x - 3\right)^2 + \left(y - 4\right)^2 = 25$.

39. The equations are:

A: $\left(x-\frac{1}{2}\right)^2 + (y-1)^2 = \frac{1}{4}$

B: $(x-1)^2 + (y-1)^2 = 1$

C: $(x-2)^2 + (y-1)^2 = 4$

41. The equations are:

A: $(x+8)^2 + y^2 = 64$

B: $x^2 + y^2 = 64$

C: $(x-8)^2 + y^2 = 64$

43. The x-coordinate of the center is $x = 500$, the y-coordinate of the center is $12 + 120 = 132$, and the radius is 120. Thus the equation of the circle is $(x - 500)^2 + (y - 132)^2 = 120^2 = 14,400$.

45. Let $y = f(x)$. Switch x and y and solve for y:

$3^y = x$

$y = \log_3 x$

The inverse is $f^{-1}(x) = \log_3 x$.

47. Let $y = f(x)$. Switch x and y and solve for y:

$2y + 3 = x$

$2y = x - 3$

$y = \dfrac{x-3}{2}$

The inverse is $f^{-1}(x) = \dfrac{x-3}{2}$.

49. Let $y = f(x)$. Switch x and y and solve for y:

$\dfrac{y+3}{5} = x$

$y + 3 = 5x$

$y = 5x - 3$

The inverse is $f^{-1}(x) = 5x - 3$.

51. Solving the equation:

$y^2 = 9$

$y = \pm\sqrt{9} = \pm 3$

53. Solving the equation:

$-y^2 = 4$

$y^2 = -4$

$y = \pm\sqrt{-4} = \pm 2i$

55. Solving the equation:

$\dfrac{-x^2}{9} = 1$

$x^2 = -9$

$x = \pm\sqrt{-9} = \pm 3i$

57. Dividing: $\dfrac{4x^2 + 9y^2}{36} = \dfrac{4x^2}{36} + \dfrac{9y^2}{36} = \dfrac{x^2}{9} + \dfrac{y^2}{4}$

59. To find the x-intercept, let $y = 0$:

$3x - 4(0) = 12$

$3x - 0 = 12$

$3x = 12$

$x = 4$

To find the y-intercept, let $x = 0$:

$3(0) - 4y = 12$

$0 - 4y = 12$

$-4y = 12$

$y = -3$

61. Substituting $x = 3$:

$\dfrac{3^2}{25} + \dfrac{y^2}{9} = 1$

$\dfrac{9}{25} + \dfrac{y^2}{9} = 1$

$\dfrac{y^2}{9} = \dfrac{16}{25}$

$y^2 = \dfrac{144}{25}$

$y = \pm\sqrt{\dfrac{144}{25}} = \pm\dfrac{12}{5} = \pm 2.4$

63. The radius is 2, so the equation is $(x-2)^2 + (y-3)^2 = 4$.

65. The radius is 2, so the equation is $(x-2)^2 + (y-3)^2 = 4$.

67. Completing the square:

$$x^2 + y^2 - 6x + 8y = 144$$
$$\left(x^2 - 6x + 9\right) + \left(y^2 + 8y + 16\right) = 144 + 9 + 16$$
$$(x-3)^2 + (y+4)^2 = 169$$

The center is (3,–4), so the distance is: $d = \sqrt{(-3)^2 + 4^2} = \sqrt{9+16} = 5$

69. Completing the square:

$$x^2 + y^2 - 6x - 8y = 144$$
$$\left(x^2 - 6x + 9\right) + \left(y^2 - 8y + 16\right) = 144 + 9 + 16$$
$$(x-3)^2 + (y-4)^2 = 169$$

The center is (3,4), so the distance is: $d = \sqrt{3^2 + 4^2} = \sqrt{9+16} = 5$

71. The equation $y = \sqrt{9 - x^2}$ corresponds to the top half of the circle, and the equation $y = -\sqrt{9 - x^2}$ corresponds to the bottom half of the circle.

10.2 Ellipses and Hyperbolas

1. Graphing the ellipse:

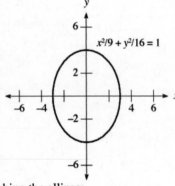

3. Graphing the ellipse:

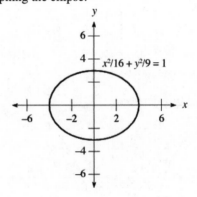

5. Graphing the ellipse:

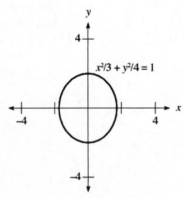

7. The standard form is $\dfrac{x^2}{25} + \dfrac{y^2}{4} = 1$. Graphing the ellipse:

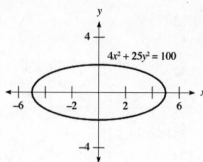

9. The standard form is $\dfrac{x^2}{16} + \dfrac{y^2}{2} = 1$. Graphing the ellipse:

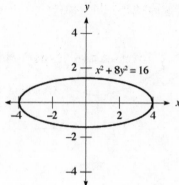

11. Graphing the hyperbola:

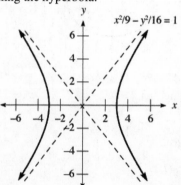

13. Graphing the hyperbola:

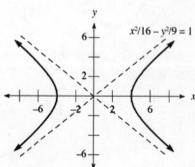

15. Graphing the hyperbola:

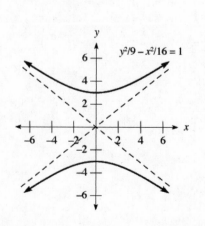

17. Graphing the hyperbola:

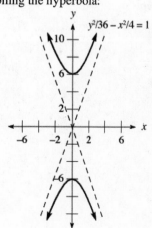

19. The standard form is $\dfrac{x^2}{4} - \dfrac{y^2}{1} = 1$. Graphing the hyperbola:

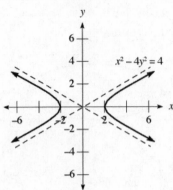

$x^2 - 4y^2 = 4$

21. The standard form is $\dfrac{y^2}{9} - \dfrac{x^2}{16} = 1$. Graphing the hyperbola:

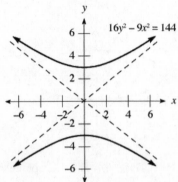

$16y^2 - 9x^2 = 144$

23. For the x-intercepts, set $y = 0$:
$$0.4x^2 = 3.6$$
$$x^2 = 9$$
$$x = \pm 3$$

For the y-intercepts, set $x = 0$:
$$0.9y^2 = 3.6$$
$$y^2 = 4$$
$$y = \pm 2$$

25. For the x-intercepts, set $y = 0$:
$$\frac{x^2}{0.04} = 1$$
$$x^2 = 0.04$$
$$x = \pm 0.2$$

For the y-intercepts, set $x = 0$:
$$-\frac{y^2}{0.09} = 1$$
$$y^2 = -0.09$$

There are no y-intercepts.

27. For the x-intercepts, set $y = 0$:
$$\frac{25x^2}{9} = 1$$
$$x^2 = \frac{9}{25}$$
$$x = \pm\frac{3}{5}$$

For the y-intercepts, set $x = 0$:
$$\frac{25y^2}{4} = 1$$
$$y^2 = \frac{4}{25}$$
$$y = \pm\frac{2}{5}$$

29.   Graphing the ellipse:

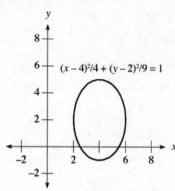

$(x-4)^2/4 + (y-2)^2/9 = 1$

31.   Completing the square:

$$4x^2 + y^2 - 4y - 12 = 0$$
$$4x^2 + (y^2 - 4y + 4) = 12 + 4$$
$$4x^2 + (y-2)^2 = 16$$
$$\frac{x^2}{4} + \frac{(y-2)^2}{16} = 1$$

Graphing the ellipse:

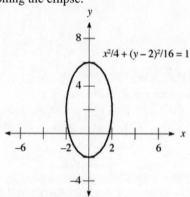

$x^2/4 + (y-2)^2/16 = 1$

33.   Completing the square:

$$x^2 + 9y^2 + 4x - 54y + 76 = 0$$
$$(x^2 + 4x + 4) + 9(y^2 - 6y + 9) = -76 + 4 + 81$$
$$(x+2)^2 + 9(y-3)^2 = 9$$
$$\frac{(x+2)^2}{9} + \frac{(y-3)^2}{1} = 1$$

Graphing the ellipse:

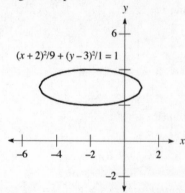

$(x+2)^2/9 + (y-3)^2/1 = 1$

35.   Graphing the hyperbola:

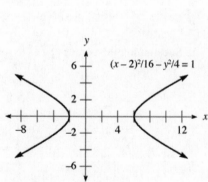

$(x-2)^2/16 - y^2/4 = 1$

37. Completing the square:
$$9y^2 - x^2 - 4x + 54y + 68 = 0$$
$$9(y^2 + 6y + 9) - (x^2 + 4x + 4) = -68 + 81 - 4$$
$$9(y + 3)^2 - (x + 2)^2 = 9$$
$$\frac{(y + 3)^2}{1} - \frac{(x + 2)^2}{9} = 1$$

Graphing the hyperbola:

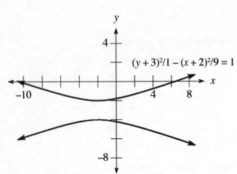

39. Completing the square:
$$4y^2 - 9x^2 - 16y + 72x - 164 = 0$$
$$4(y^2 - 4y + 4) - 9(x^2 - 8x + 16) = 164 + 16 - 144$$
$$4(y - 2)^2 - 9(x - 4)^2 = 36$$
$$\frac{(y - 2)^2}{9} - \frac{(x - 4)^2}{4} = 1$$

Graphing the hyperbola:

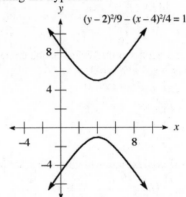

41. Substituting $x = 4$:
$$\frac{4^2}{25} + \frac{y^2}{9} = 1$$
$$\frac{y^2}{9} + \frac{16}{25} = 1$$
$$\frac{y^2}{9} = \frac{9}{25}$$
$$\frac{y}{3} = \pm\frac{3}{5}$$
$$y = \pm\frac{9}{5}$$

43. Substituting $x = 1.8$:
$$16(1.8)^2 + 9y^2 = 144$$
$$51.84 + 9y^2 = 144$$
$$9y^2 = 92.16$$
$$3y = \pm 9.6$$
$$y = \pm 3.2$$

45. The asymptotes are $y = \frac{3}{4}x$ and $y = -\frac{3}{4}x$.

47. The equation has the form $\dfrac{x^2}{16} + \dfrac{y^2}{b^2} = 1$. Substituting the point $\left(2, \sqrt{3}\right)$:
$$\frac{(2)^2}{16} + \frac{\left(\sqrt{3}\right)^2}{b^2} = 1$$
$$\frac{1}{4} + \frac{3}{b^2} = 1$$
$$\frac{3}{b^2} = \frac{3}{4}$$
$$b^2 = 4$$

The equation of the ellipse is $\dfrac{x^2}{16} + \dfrac{y^2}{4} = 1$.

49. The equation of the ellipse is $\dfrac{x^2}{20^2}+\dfrac{y^2}{10^2}=1$. Substituting $y=6$:

$$\dfrac{x^2}{20^2}+\dfrac{6^2}{10^2}=1$$
$$\dfrac{x^2}{400}+\dfrac{9}{25}=1$$
$$\dfrac{x^2}{400}=\dfrac{16}{25}$$
$$x^2=256$$
$$x=16$$

The man can walk within 16 feet of the center.

51. Substituting $a=4$ and $c=3$:
$$4^2=b^2+3^2$$
$$16=b^2+9$$
$$b^2=7$$
$$b=\sqrt{7}\approx 2.65$$

The width should be approximately $2(2.65)=5.3$ feet wide.

53. Evaluating: $f(4)=\dfrac{2}{4-2}=\dfrac{2}{2}=1$

55. Evaluating: $f\big[g(0)\big]=f\left(\dfrac{2}{0+2}\right)=f(1)=\dfrac{2}{1-2}=\dfrac{2}{-1}=-2$

57. Simplifying: $f(x)+g(x)=\dfrac{2}{x-2}+\dfrac{2}{x+2}=\dfrac{2}{x-2}\bullet\dfrac{x+2}{x+2}+\dfrac{2}{x+2}\bullet\dfrac{x-2}{x-2}=\dfrac{2x+4+2x-4}{(x-2)(x+2)}=\dfrac{4x}{(x-2)(x+2)}$

59. Since $4^2+0^2=16$ and $0^2+5^2=25$, while $0^2+0^2=0$, only $(0,0)$ is a solution.

61. Multiplying: $(2y+4)^2=(2y)^2+2(2y)(4)+4^2=4y^2+16y+16$

63. Solving for x:
$$x-2y=4$$
$$x=2y+4$$

65. Simplifying: $x^2-2\big(x^2-3\big)=x^2-2x^2+6=-x^2+6$

67. Factoring: $5y^2+16y+12=(5y+6)(y+2)$

69. Solving the equation:
$$y^2=4$$
$$y=\pm\sqrt{4}=\pm2$$

71. Solving the equation:
$$-x^2+6=2$$
$$-x^2=-4$$
$$x^2=4$$
$$x=\pm\sqrt{4}=\pm2$$

10.3 Second-Degree Inequalities and Nonlinear Systems

1. Graphing the inequality:

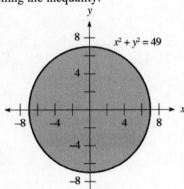

3. Graphing the inequality:

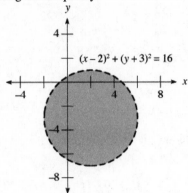

5. Graphing the inequality:

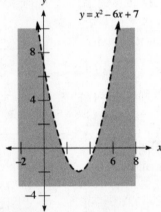

7. Graphing the inequality:

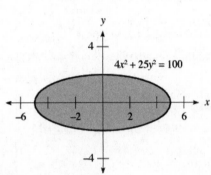

9. Solving the second equation for y yields $y = 3 - 2x$. Substituting into the first equation:

$$x^2 + \left(3 - 2x\right)^2 = 9$$
$$x^2 + 9 - 12x + 4x^2 = 9$$
$$5x^2 - 12x = 0$$
$$x\left(5x - 12\right) = 0$$
$$x = 0, \frac{12}{5}$$
$$y = 3, -\frac{9}{5}$$

The solutions are $\left(0, 3\right), \left(\frac{12}{5}, -\frac{9}{5}\right)$.

11. Solving the second equation for x yields $x = 8 - 2y$. Substituting into the first equation:

$$\left(8 - 2y\right)^2 + y^2 = 16$$
$$64 - 32y + 4y^2 + y^2 = 16$$
$$5y^2 - 32y + 48 = 0$$
$$\left(y - 4\right)\left(5y - 12\right) = 0$$
$$y = 4, \frac{12}{5}$$
$$x = 0, \frac{16}{5}$$

The solutions are $\left(0, 4\right), \left(\frac{16}{5}, \frac{12}{5}\right)$.

13.   Adding the two equations yields:

$$2x^2 = 50$$
$$x^2 = 25$$
$$x = -5, 5$$
$$y = 0$$

The solutions are $(-5,0), (5,0)$.

15.   Substituting into the first equation:

$$x^2 + \left(x^2 - 3\right)^2 = 9$$
$$x^2 + x^4 - 6x^2 + 9 = 9$$
$$x^4 - 5x^2 = 0$$
$$x^2\left(x^2 - 5\right) = 0$$
$$x = 0, -\sqrt{5}, \sqrt{5}$$
$$y = -3, 2, 2$$

The solutions are $\left(0,-3\right), \left(-\sqrt{5},2\right), \left(\sqrt{5},2\right)$.

17.   Substituting into the first equation:

$$x^2 + \left(x^2 - 4\right)^2 = 16$$
$$x^2 + x^4 - 8x^2 + 16 = 16$$
$$x^4 - 7x^2 = 0$$
$$x^2\left(x^2 - 7\right) = 0$$
$$x = 0, -\sqrt{7}, \sqrt{7}$$
$$y = -4, 3, 3$$

The solutions are $\left(0,-4\right), \left(-\sqrt{7},3\right), \left(\sqrt{7},3\right)$.

19.   Substituting into the first equation:

$$3x + 2\left(x^2 - 5\right) = 10$$
$$3x + 2x^2 - 10 = 10$$
$$2x^2 + 3x - 20 = 0$$
$$(x+4)(2x-5) = 0$$
$$x = -4, \tfrac{5}{2}$$
$$y = 11, \tfrac{5}{4}$$

The solutions are $(-4,11), \left(\tfrac{5}{2},\tfrac{5}{4}\right)$.

21.   Substituting into the first equation:

$$-x + 1 = x^2 + 2x - 3$$
$$x^2 + 3x - 4 = 0$$
$$(x+4)(x-1) = 0$$
$$x = -4, 1$$
$$y = 5, 0$$

The solutions are $(-4,5), (1,0)$.

23.   Substituting into the first equation:

$$x - 5 = x^2 - 6x + 5$$
$$x^2 - 7x + 10 = 0$$
$$(x-2)(x-5) = 0$$
$$x = 2, 5$$
$$y = -3, 0$$

The solutions are $(2,-3), (5,0)$.

25.   Adding the two equations yields:

$$8x^2 = 72$$
$$x^2 = 9$$
$$x = \pm 3$$
$$y = 0$$

The solutions are $(-3,0), (3,0)$.

27.   Solving the first equation for x yields $x = y + 4$. Substituting into the second equation:

$$(y+4)^2 + y^2 = 16$$
$$y^2 + 8y + 16 + y^2 = 16$$
$$2y^2 + 8y = 0$$
$$2y(y+4) = 0$$
$$y = 0, -4$$
$$x = 4, 0$$

The solutions are $(0,-4), (4,0)$.

29.   **a.**   Subtracting the two equations yields:

$$(x+8)^2 - x^2 = 0$$
$$x^2 + 16x + 64 - x^2 = 0$$
$$16x = -64$$
$$x = -4$$

Substituting to find y:

$$(-4)^2 + y^2 = 64$$
$$y^2 + 16 = 64$$
$$y^2 = 48$$
$$y = \pm\sqrt{48} = \pm 4\sqrt{3}$$

The intersection points are $\left(-4, -4\sqrt{3}\right)$ and $\left(-4, 4\sqrt{3}\right)$.

   b.   Subtracting the two equations yields:

$$x^2 - (x-8)^2 = 0$$
$$x^2 - x^2 + 16x - 64 = 0$$
$$16x = 64$$
$$x = 4$$

Substituting to find y:

$$4^2 + y^2 = 64$$
$$y^2 + 16 = 64$$
$$y^2 = 48$$
$$y = \pm\sqrt{48} = \pm 4\sqrt{3}$$

The intersection points are $\left(4, -4\sqrt{3}\right)$ and $\left(4, 4\sqrt{3}\right)$.

31.   Graphing the inequality:

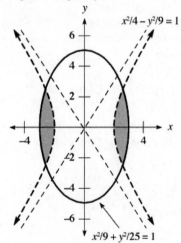

$y = x^2 - 1$

$x^2 + y^2 = 9$

33.   Graphing the inequality:

$x^2/4 - y^2/9 = 1$

$x^2/9 + y^2/25 = 1$

35. There is no intersection.

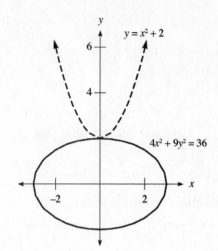

37. The system of inequalities is:

$$x^2 + y^2 < 16$$
$$y > 4 - \tfrac{1}{4}x^2$$

39. The system of equations is:

$$x^2 + y^2 = 89$$
$$x^2 - y^2 = 39$$

Adding the two equations yields:

$$2x^2 = 128$$
$$x^2 = 64$$
$$x = \pm 8$$
$$y = \pm 5$$

The numbers are either 8 and 5, 8 and –5, –8 and 5, or –8 and –5.

41. The system of equations is:

$$y = x^2 - 3$$
$$x + y = 9$$

Substituting into the second equation:

$$x + x^2 - 3 = 9$$
$$x^2 + x - 12 = 0$$
$$(x + 4)(x - 3) = 0$$
$$x = -4, 3$$
$$y = 13, 6$$

The numbers are either –4 and 13, or 3 and 6.

43. Solving the equation:

$$g(x) = 0$$
$$\left(x + \tfrac{2}{5}\right)^2 = 0$$
$$x + \tfrac{2}{5} = 0$$
$$x = -\tfrac{2}{5}$$

45. Solving the equation:

$$f(x) = g(x)$$
$$x^2 + 4x - 4 = 1$$
$$x^2 + 4x - 5 = 0$$
$$(x + 5)(x - 1) = 0$$
$$x = -5, 1$$

47.   Solving the equation:
$$f(x) = g(x)$$
$$x^2 + 4x - 4 = x - 6$$
$$x^2 + 3x + 2 = 0$$
$$(x+2)(x+1) = 0$$
$$x = -2, -1$$

Chapter 10 Review/Test

1.   Using the distance formula: $d = \sqrt{(-1-2)^2 + (5-6)^2} = \sqrt{9+1} = \sqrt{10}$

2.   Using the distance formula: $d = \sqrt{(1-3)^2 + (-1+4)^2} = \sqrt{4+9} = \sqrt{13}$

3.   Using the distance formula: $d = \sqrt{(-4-0)^2 + (0-3)^2} = \sqrt{16+9} = \sqrt{25} = 5$

4.   Using the distance formula: $d = \sqrt{(-3+3)^2 + (-2-7)^2} = \sqrt{0+81} = \sqrt{81} = 9$

5.   Solving the equation:

$$\sqrt{(x-2)^2 + (-1+4)^2} = 5$$
$$(x-2)^2 + 9 = 25$$
$$(x-2)^2 = 16$$
$$x-2 = \pm\sqrt{16}$$
$$x-2 = -4, 4$$
$$x = -2, 6$$

6.   Solving the equation:

$$\sqrt{(-3-3)^2 + (y+4)^2} = 10$$
$$(y+4)^2 + 36 = 100$$
$$(y+4)^2 = 64$$
$$y+4 = \pm\sqrt{64}$$
$$y+4 = -8, 8$$
$$y = -12, 4$$

7.   The equation is $(x-3)^2 + (y-1)^2 = 4$.

8.   The equation is $(x-3)^2 + (y+1)^2 = 16$.

9.   The equation is $(x+5)^2 + y^2 = 9$.

10.   The equation is $(x+3)^2 + (y-4)^2 = 18$.

11.   The equation is $x^2 + y^2 = 25$.

12.   The equation is $x^2 + y^2 = 9$.

13.   Finding the radius: $r = \sqrt{(-2-2)^2 + (3-0)^2} = \sqrt{16+9} = \sqrt{25} = 5$. The equation is $(x+2)^2 + (y-3)^2 = 25$.

14.   Finding the radius: $r = \sqrt{(-6)^2 + (8)^2} = \sqrt{36+64} = \sqrt{100} = 10$. The equation is $(x+6)^2 + (y-8)^2 = 100$.

15.   The center is (0,0) and the radius is 2.

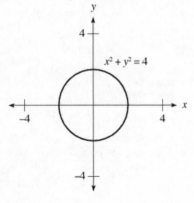

16.   The center is (3,−1) and the radius is 4.

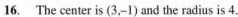
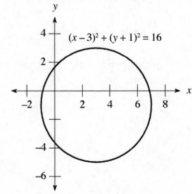

17.   Completing the square:

$$x^2 + y^2 - 6x + 4y = -4$$

$$\left(x^2 - 6x + 9\right) + \left(y^2 + 4y + 4\right) = -4 + 9 + 4$$

$$\left(x - 3\right)^2 + \left(y + 2\right)^2 = 9$$

The center is (3,–2) and the radius is 3.

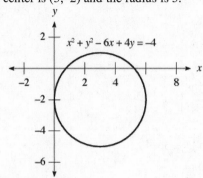

18.   Completing the square:

$$x^2 + y^2 + 4x - 2y = 4$$

$$\left(x^2 + 4x + 4\right) + \left(y^2 - 2y + 1\right) = 4 + 4 + 1$$

$$\left(x + 2\right)^2 + \left(y - 1\right)^2 = 9$$

The center is (–2,1) and the radius is 3.

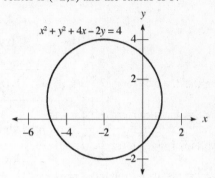

19.   Graphing the ellipse:

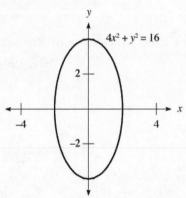

20.   The standard form is $\dfrac{x^2}{4} + \dfrac{y^2}{16} = 1$. Graphing the ellipse:

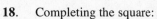

21.   Graphing the hyperbola:

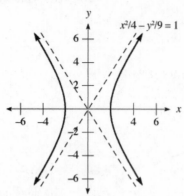

22.   The standard form is $\dfrac{x^2}{4} - \dfrac{y^2}{16} = 1$. Graphing the hyperbola:

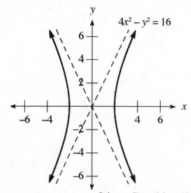

23.   Graphing the ellipse:

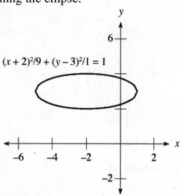

24.   Graphing the hyperbola:

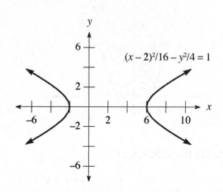

25.   Completing the square:
$$9y^2 - x^2 - 4x + 54y + 68 = 0$$
$$9\left(y^2 + 6y + 9\right) - \left(x^2 + 4x + 4\right) = -68 + 81 - 4$$
$$9\left(y + 3\right)^2 - \left(x + 2\right)^2 = 9$$
$$\frac{\left(y + 3\right)^2}{1} - \frac{\left(x + 2\right)^2}{9} = 1$$

26.   Completing the square:
$$9x^2 + 4y^2 - 72x - 16y + 124 = 0$$
$$9\left(x^2 - 8x + 16\right) + 4\left(y^2 - 4y + 4\right) = -124 + 144 + 16$$
$$9\left(x - 4\right)^2 + 4\left(y - 2\right)^2 = 36$$
$$\frac{\left(x - 4\right)^2}{4} + \frac{\left(y - 2\right)^2}{9} = 1$$

Graphing the hyperbola:

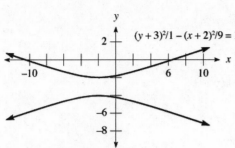

$(y+3)^2/1 - (x+2)^2/9 = 1$

Graphing the ellipse:

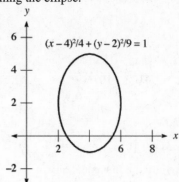

$(x-4)^2/4 + (y-2)^2/9 = 1$

27. Graphing the inequality:

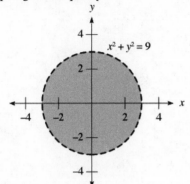

$x^2 + y^2 = 9$

28. Graphing the inequality:

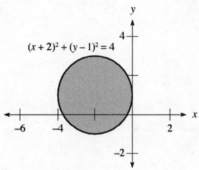

$(x+2)^2 + (y-1)^2 = 4$

29. Graphing the inequality:

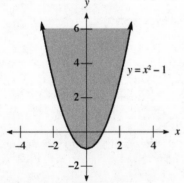

$y = x^2 - 1$

30. Graphing the inequality:

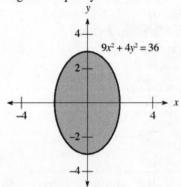

$9x^2 + 4y^2 = 36$

31. Graphing the solution set:

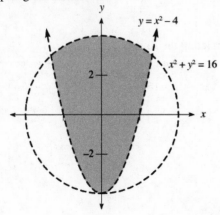

$y = x^2 - 4$

$x^2 + y^2 = 16$

32. Graphing the solution set:

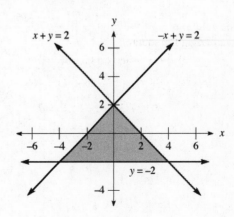

$x + y = 2$

$-x + y = 2$

$y = -2$

33.   Solving the second equation for y yields $y = 4 - 2x$. Substituting into the first equation:

$$x^2 + (4 - 2x)^2 = 16$$
$$x^2 + 16 - 16x + 4x^2 = 16$$
$$5x^2 - 16x = 0$$
$$x(5x - 16) = 0$$
$$x = 0, \frac{16}{5}$$
$$y = 4, -\frac{12}{5}$$

The solutions are $\left(0, 4\right), \left(\frac{16}{5}, -\frac{12}{5}\right)$.

34.   Substituting into the first equation:

$$x^2 + \left(x^2 - 2\right)^2 = 4$$
$$x^2 + x^4 - 4x^2 + 4 = 4$$
$$x^4 - 3x^2 = 0$$
$$x^2\left(x^2 - 3\right) = 0$$
$$x = 0, \pm\sqrt{3}$$
$$y = -2, 1$$

The solutions are $\left(0, -2\right), \left(-\sqrt{3}, 1\right), \left(\sqrt{3}, 1\right)$.

35.   Adding the two equations yields:

$$18x^2 = 72$$
$$x^2 = 4$$
$$x = \pm 2$$
$$y = 0$$

The solutions are $(-2, 0), (2, 0)$.

36.   Multiply the second equation by 2 and add it to the first equation:

$$2x^2 - 4y^2 = 8$$
$$2x^2 + 4y^2 = 20$$

Adding yields:

$$4x^2 = 28$$
$$x^2 = 7$$
$$x = \pm\sqrt{7}$$
$$y = \pm\sqrt{\frac{3}{2}} = \pm\frac{\sqrt{6}}{2}$$

The solutions are $\left(-\sqrt{7}, -\frac{\sqrt{6}}{2}\right), \left(-\sqrt{7}, \frac{\sqrt{6}}{2}\right), \left(\sqrt{7}, -\frac{\sqrt{6}}{2}\right), \left(\sqrt{7}, \frac{\sqrt{6}}{2}\right)$.

Chapter 11
Sequences and Series

11.1 Sequences

1. The first five terms are: $4, 7, 10, 13, 16$

3. The first five terms are: $3, 7, 11, 15, 19$

5. The first five terms are: $1, 2, 3, 4, 5$

7. The first five terms are: $4, 7, 12, 19, 28$

9. The first five terms are: $\frac{1}{4}, \frac{2}{5}, \frac{1}{2}, \frac{4}{7}, \frac{5}{8}$

11. The first five terms are: $1, \frac{1}{4}, \frac{1}{9}, \frac{1}{16}, \frac{1}{25}$

13. The first five terms are: $2, 4, 8, 16, 32$

15. The first five terms are: $2, \frac{3}{2}, \frac{4}{3}, \frac{5}{4}, \frac{6}{5}$

17. The first five terms are: $-2, 4, -8, 16, -32$

19. The first five terms are: $3, 5, 3, 5, 3$

21. The first five terms are: $1, -\frac{2}{3}, \frac{3}{5}, -\frac{4}{7}, \frac{5}{9}$

23. The first five terms are: $\frac{1}{2}, 1, \frac{9}{8}, 1, \frac{25}{32}$

25. The first five terms are: $3, -9, 27, -81, 243$

27. The first five terms are: $1, 5, 13, 29, 61$

29. The first five terms are: $2, 3, 5, 9, 17$

31. The first five terms are: $5, 11, 29, 83, 245$

33. The first five terms are: $4, 4, 4, 4, 4$

35. The general term is: $a_n = 4n$

37. The general term is: $a_n = n^2$

39. The general term is: $a_n = 2^{n+1}$

41. The general term is: $a_n = \dfrac{1}{2^{n+1}}$

43. The general term is: $a_n = 3n + 2$

45. The general term is: $a_n = -4n + 2$

47. The general term is: $a_n = (-2)^{n-1}$

49. The general term is: $a_n = \log_{n+1}(n+2)$

51. **a.** The sequence of salaries is: 28000, 29120, 30284.80, $31,496.19$, 32756.04

   b. The general term is: $a_n = 28000(1.04)^{n-1}$

53. **a.** The sequence of values is: 16 ft, 48 ft, 80 ft, 112 ft, 144 ft

   b. The sum of the values is 400 feet.   **c.** No, since the sum is less than 420 feet.

55. **a.** The sequence of angles is: $180°, 360°, 540°, 720°$ **b.** Substituting $n = 20$: $a_{20} = 180°(20 - 2) = 3,240°$

   c. The sum of the interior angles would be $0°$.

57. Solving for x:

$$\log_9 x = \tfrac{3}{2}$$
$$x = 9^{3/2} = 27$$

59. Simplifying: $\log_2 32 = \log_2 2^5 = 5$

61. Simplifying: $\log_3\left[\log_2 8\right] = \log_3\left[\log_2 2^3\right] = \log_3 3 = 1$

63. Simplifying: $-2 + 6 + 4 + 22 = 30$

65. Simplifying: $-8 + 16 - 32 + 64 = 40$

67. Simplifying: $(1 - 3) + (4 - 3) + (9 - 3) + (16 - 3) = -2 + 1 + 6 + 13 = 18$

69. Simplifying: $-\frac{1}{3} + \frac{1}{9} - \frac{1}{27} + \frac{1}{81} = -\frac{27}{81} + \frac{9}{81} - \frac{3}{81} + \frac{1}{81} = -\frac{20}{81}$ **71.** Simplifying: $\frac{1}{3} + \frac{1}{2} + \frac{3}{5} + \frac{2}{3} = \frac{10}{30} + \frac{15}{30} + \frac{18}{30} + \frac{20}{30} = \frac{63}{30} = \frac{21}{10}$

73. Computing the values: $a_{100} \approx 2.7048, a_{1,000} \approx 2.7169, a_{10,000} \approx 2.7181, a_{100,000} \approx 2.7183$

75. The first ten terms are: $1, 1, 2, 3, 5, 8, 13, 21, 34, 55$

11.2 Series

1. Expanding the sum: $\displaystyle\sum_{i=1}^{4}(2i+4) = 6+8+10+12 = 36$

3. Expanding the sum: $\displaystyle\sum_{i=2}^{3}(i^2-1) = 3+8 = 11$

5. Expanding the sum: $\displaystyle\sum_{i=1}^{4}(i^2-3) = -2+1+6+13 = 18$

7. Expanding the sum: $\displaystyle\sum_{i=1}^{4}\frac{i}{1+i} = \frac{1}{2}+\frac{2}{3}+\frac{3}{4}+\frac{4}{5} = \frac{30}{60}+\frac{40}{60}+\frac{45}{60}+\frac{48}{60} = \frac{163}{60}$

9. Expanding the sum: $\displaystyle\sum_{i=1}^{4}(-3)^i = -3+9-27+81 = 60$

11. Expanding the sum: $\displaystyle\sum_{i=3}^{6}(-2)^i = -8+16-32+64 = 40$

13. Expanding the sum: $\displaystyle\sum_{i=2}^{6}(-2)^i = 4-8+16-32+64 = 44$

15. Expanding the sum: $\displaystyle\sum_{i=1}^{5}\left(-\frac{1}{2}\right)^i = -\frac{1}{2}+\frac{1}{4}-\frac{1}{8}+\frac{1}{16}-\frac{1}{32} = -\frac{16}{32}+\frac{8}{32}-\frac{4}{32}+\frac{2}{32}-\frac{1}{32} = -\frac{11}{32}$

17. Expanding the sum: $\displaystyle\sum_{i=2}^{5}\frac{i-1}{i+1} = \frac{1}{3}+\frac{1}{2}+\frac{3}{5}+\frac{2}{3} = 1+\frac{5}{10}+\frac{6}{10} = \frac{21}{10}$

19. Expanding the sum: $\displaystyle\sum_{i=1}^{5}(x+i) = (x+1)+(x+2)+(x+3)+(x+4)+(x+5) = 5x+15$

21. Expanding the sum: $\displaystyle\sum_{i=1}^{4}(x-2)^i = (x-2)+(x-2)^2+(x-2)^3+(x-2)^4$

23. Expanding the sum: $\displaystyle\sum_{i=1}^{5}\frac{x+i}{x-1} = \frac{x+1}{x-1}+\frac{x+2}{x-1}+\frac{x+3}{x-1}+\frac{x+4}{x-1}+\frac{x+5}{x-1}$

25. Expanding the sum: $\displaystyle\sum_{i=3}^{8}(x+i)^i = (x+3)^3+(x+4)^4+(x+5)^5+(x+6)^6+(x+7)^7+(x+8)^8$

27. Expanding the sum: $\displaystyle\sum_{i=3}^{6}(x-2i)^{i+3} = (x-6)^6+(x-8)^7+(x-10)^8+(x-12)^9$

29. Writing with summation notation: $2+4+8+16 = \displaystyle\sum_{i=1}^{4}2^i$

31. Writing with summation notation: $4+8+16+32+64 = \displaystyle\sum_{i=2}^{6}2^i$

33. Writing with summation notation: $5+9+13+17+21 = \displaystyle\sum_{i=1}^{5}(4i+1)$

35. Writing with summation notation: $-4+8-16+32 = \displaystyle\sum_{i=2}^{5}-(-2)^i$

37. Writing with summation notation: $\frac{3}{4}+\frac{4}{5}+\frac{5}{6}+\frac{6}{7}+\frac{7}{8} = \displaystyle\sum_{i=3}^{7}\frac{i}{i+1}$

39. Writing with summation notation: $\frac{1}{3}+\frac{2}{5}+\frac{3}{7}+\frac{4}{9} = \displaystyle\sum_{i=1}^{4}\frac{i}{2i+1}$

41. Writing with summation notation: $(x-2)^6+(x-2)^7+(x-2)^8+(x-2)^9 = \displaystyle\sum_{i=6}^{9}(x-2)^i$

43. Writing with summation notation: $\left(1+\dfrac{1}{x}\right)^2+\left(1+\dfrac{2}{x}\right)^3+\left(1+\dfrac{3}{x}\right)^4+\left(1+\dfrac{4}{x}\right)^5=\displaystyle\sum_{i=1}^{4}\left(1+\dfrac{i}{x}\right)^{i+1}$

45. Writing with summation notation: $\dfrac{x}{x+3}+\dfrac{x}{x+4}+\dfrac{x}{x+5}=\displaystyle\sum_{i=3}^{5}\dfrac{x}{x+i}$

47. Writing with summation notation: $x^2(x+2)+x^3(x+3)+x^4(x+4)=\displaystyle\sum_{i=2}^{4}x^i(x+i)$

49. **a.** Writing as a series: $\frac{1}{3}=0.3+0.03+0.003+0.0003+\ldots$

    b. Writing as a series: $\frac{2}{9}=0.2+0.02+0.002+0.0002+\ldots$

    c. Writing as a series: $\frac{3}{11}=0.27+0.0027+0.000027+\ldots$

51. The sequence of values he falls is: 16, 48, 80, 112, 144, 176, 208
    During the seventh second he falls 208 feet, and the total he falls is 784 feet.

53. **a.** The series is $16+48+80+112+144$.   **b.** Writing in summation notation: $\displaystyle\sum_{i=1}^{5}(32i-16)$

55. Expanding the logarithm: $\log_2 x^3 y=\log_2 x^3+\log_2 y=3\log_2 x+\log_2 y$

57. Expanding the logarithm: $\log_{10}\dfrac{\sqrt[3]{x}}{y^2}=\log_{10}x^{1/3}-\log_{10}y^2=\dfrac{1}{3}\log_{10}x-2\log_{10}y$

59. Writing as a single logarithm: $\log_{10}x-\log_{10}y^2=\log_{10}\dfrac{x}{y^2}$

61. Writing as a single logarithm: $2\log_3 x-3\log_3 y-4\log_3 z=\log_3 x^2-\log_3 y^3-\log_3 z^4=\log_3\dfrac{x^2}{y^3z^4}$

63. Solving the equation:

$$\log_4 x-\log_4 5=2$$
$$\log_4\frac{x}{5}=2$$
$$\frac{x}{5}=4^2$$
$$\frac{x}{5}=16$$
$$x=80$$

The solution is 80.

65. Solving the equation:

$$\log_2 x+\log_2(x-7)=3$$
$$\log_2(x^2-7x)=3$$
$$x^2-7x=2^3$$
$$x^2-7x=8$$
$$x^2-7x-8=0$$
$$(x+1)(x-8)=0$$
$$x=-1,8$$

The solution is 8 (−1 does not check).

67. Simplifying: $2+9(8)=2+72=74$

69. Simplifying: $\frac{10}{2}\left(\frac{1}{2}+5\right)=5\left(\frac{1}{2}+\frac{10}{2}\right)=5\left(\frac{11}{2}\right)=\frac{55}{2}$

71. Simplifying: $3+(n-1)(2)=3+2n-2=2n+1$

73. Multiplying the first equation by −1:

$$-x-2y=-7$$
$$x+7y=17$$

Adding yields:

$$5y=10$$
$$y=2$$

Substituting into the first equation:

$$x+2(2)=7$$
$$x+4=7$$
$$x=3$$

The solution is (3,2).

75. Solving the equation for x:

$$\sum_{i=1}^{4}(x-i)=16$$

$$(x-1)+(x-2)+(x-3)+(x-4)=16$$

$$4x-10=16$$

$$4x=26$$

$$x=\tfrac{13}{2}$$

11.3 Arithmetic Sequences

1. The sequence is arithmetic: $d=1$

5. The sequence is arithmetic: $d=-5$

9. The sequence is arithmetic: $d=\tfrac{2}{3}$

3. The sequence is not arithmetic.

7. The sequence is not arithmetic.

11. Finding the general term: $a_n=3+(n-1)\cdot 4=3+4n-4=4n-1$. Therefore: $a_{24}=4\cdot 24-1=96-1=95$

13. Finding the required term: $a_{10}=6+(10-1)\cdot(-2)=6-18=-12$. Finding the sum: $S_{10}=\tfrac{10}{2}(6-12)=5(-6)=-30$

15. Writing out the equations:

$$a_6=a_1+5d \quad a_{12}=a_1+11d$$

$$17=a_1+5d \quad 29=a_1+11d$$

The system of equations is:

$$a_1+11d=29$$

$$a_1+5d=17$$

Subtracting yields:

$$6d=12$$

$$d=2$$

$$a_1=7$$

Finding the required term: $a_{30}=7+29\cdot 2=7+58=65$

17. Writing out the equations:

$$a_3=a_1+2d \quad a_8=a_1+7d$$

$$16=a_1+2d \quad 26=a_1+7d$$

The system of equations is:

$$a_1+7d=26$$

$$a_1+2d=16$$

Subtracting yields:

$$5d=10$$

$$d=2$$

$$a_1=12$$

Finding the required term: $a_{20}=12+19\cdot 2=12+38=50$. Finding the sum: $S_{20}=\tfrac{20}{2}(12+50)=10\cdot 62=620$

19. Finding the required term: $a_{20}=3+19\cdot 4=3+76=79$. Finding the sum: $S_{20}=\tfrac{20}{2}(3+79)=10\cdot 82=820$

21. Writing out the equations:
$$a_4 = a_1 + 3d \quad a_{10} = a_1 + 9d$$
$$14 = a_1 + 3d \quad 32 = a_1 + 9d$$
The system of equations is:
$$a_1 + 9d = 32$$
$$a_1 + 3d = 14$$
Subtracting yields:
$$6d = 18$$
$$d = 3$$
$$a_1 = 5$$
Finding the required term: $a_{40} = 5 + 39 \bullet 3 = 5 + 117 = 122$. Finding the sum: $S_{40} = \frac{40}{2}(5 + 122) = 20 \bullet 127 = 2540$

23. Using the summation formula:
$$S_6 = \frac{6}{2}(a_1 + a_6)$$
$$-12 = 3(a_1 - 17)$$
$$a_1 - 17 = -4$$
$$a_1 = 13$$
Now find d:
$$a_6 = 13 + 5 \bullet d$$
$$-17 = 13 + 5d$$
$$5d = -30$$
$$d = -6$$

25. Using $a_1 = 14$ and $d = -3$: $a_{85} = 14 + 84 \bullet (-3) = 14 - 252 = -238$

27. Using the summation formula:
$$S_{20} = \frac{20}{2}(a_1 + a_{20})$$
$$80 = 10(-4 + a_{20})$$
$$-4 + a_{20} = 8$$
$$a_{20} = 12$$
Now finding d:
$$a_{20} = a_1 + 19d$$
$$12 = -4 + 19d$$
$$16 = 19d$$
$$d = \frac{16}{19}$$
Finding the required term: $a_{39} = -4 + 38\left(\frac{16}{19}\right) = -4 + 32 = 28$

29. Using $a_1 = 5$ and $d = 4$: $a_{100} = a_1 + 99d = 5 + 99 \bullet 4 = 5 + 396 = 401$
Now finding the required sum: $S_{100} = \frac{100}{2}(5 + 401) = 50 \bullet 406 = 20,300$

31. Using $a_1 = 12$ and $d = -5$: $a_{35} = a_1 + 34d = 12 + 34 \bullet (-5) = 12 - 170 = -158$

33. Using $a_1 = \frac{1}{2}$ and $d = \frac{1}{2}$: $a_{10} = a_1 + 9d = \frac{1}{2} + 9 \bullet \frac{1}{2} = \frac{10}{2} = 5$. Finding the sum: $S_{10} = \frac{10}{2}\left(\frac{1}{2} + 5\right) = \frac{10}{2} \bullet \frac{11}{2} = \frac{55}{2}$

35. **a.** The first five terms are: $18,000, $14,700, $11,400, $8,100, $4,800
b. The common difference is –$3,300.
c. Constructing a line graph:

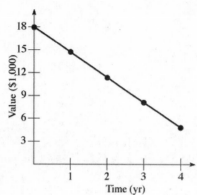

d. The value is approximately $9,750.
e. The recursive formula is: $a_0 = 18000; a_n = a_{n-1} - 3300$ for $n \geq 1$

37. **a.** The sequence of values is: 1500 ft, 1460 ft, 1420 ft, 1380 ft, 1340 ft, 1300 ft
b. It is arithmetic because the same amount is subtracted from each succeeding term.
c. The general term is: $a_n = 1500 + (n-1) \cdot (-40) = 1500 - 40n + 40 = 1540 - 40n$

39. **a.** The first 15 triangular numbers is: 1,3,6,10,15,21,28,36,45,55,66,78,91,105,120
b. The recursive formula is: $a_1 = 1; a_n = n + a_{n-1}$ for $n \geq 2$
c. It is not arithmetic because the same amount is not added to each term.

41. Finding the logarithm: $\log 576 \approx 2.7604$

43. Finding the logarithm: $\log 0.0576 \approx -1.2396$

45. Solving for x:
$$\log x = 2.6484$$
$$x = 10^{2.6484} \approx 445$$

47. Solving for x:
$$\log x = -7.3516$$
$$x = 10^{-7.3516} \approx 4.45 \times 10^{-8}$$

49. Simplifying: $\frac{1}{8}\left(\frac{1}{2}\right) = \frac{1}{16}$

51. Simplifying: $\frac{3\sqrt{3}}{3} = \sqrt{3}$

53. Simplifying: $2 \cdot 2^{n-1} = 2^{1+n-1} = 2^n$

55. Simplifying: $\frac{ar^6}{ar^3} = r^{6-3} = r^3$

57. Simplifying: $\frac{\frac{1}{5}}{1 - \frac{1}{2}} = \frac{\frac{1}{5}}{1 - \frac{1}{2}} \cdot \frac{10}{10} = \frac{2}{10 - 5} = \frac{2}{5}$

59. Simplifying: $\frac{-3\left[(-2)^8 - 1\right]}{-2 - 1} = \frac{-3(256 - 1)}{-2 - 1} = \frac{-3(255)}{-3} = 255$

61. Using the summation formulas:

$$S_7 = \frac{7}{2}(a_1 + a_7)$$
$$147 = \frac{7}{2}(a_1 + a_7)$$
$$a_1 + a_7 = 42$$
$$a_1 + (a_1 + 6d) = 42$$
$$2a_1 + 6d = 42$$
$$a_1 + 3d = 21$$

$$S_{13} = \frac{13}{2}(a_1 + a_{13})$$
$$429 = \frac{13}{2}(a_1 + a_{13})$$
$$a_1 + a_{13} = 66$$
$$a_1 + (a_1 + 12d) = 66$$
$$2a_1 + 12d = 66$$
$$a_1 + 6d = 33$$

So we have the system of equations:
$$a_1 + 6d = 33$$
$$a_1 + 3d = 21$$

Subtracting:
$$3d = 12$$
$$d = 4$$
$$a_1 = 33 - 6 \cdot 4 = 9$$

63. Solving the equation:
$$a_{15} - a_7 = -24$$
$$(a_1 + 14d) - (a_1 + 6d) = -24$$
$$a_1 + 14d - a_1 - 6d = -24$$
$$8d = -24$$
$$d = -3$$

65. Using $a_1 = 1$ and $d = \sqrt{2}$, find the term: $a_{50} = a_1 + 49d = 1 + 49\sqrt{2}$

Finding the sum: $S_{50} = \frac{50}{2}(a_1 + a_{50}) = 25(1 + 1 + 49\sqrt{2}) = 50 + 1225\sqrt{2}$

11.4 Geometric Sequences

1. The sequence is geometric: $r = 5$

3. The sequence is geometric: $r = \frac{1}{3}$

5. The sequence is not geometric.

7. The sequence is geometric: $r = -2$

9. The sequence is not geometric.

11. Finding the general term: $a_n = 4 \cdot 3^{n-1}$

13. Finding the term: $a_6 = -2\left(-\frac{1}{2}\right)^{6-1} = -2\left(-\frac{1}{2}\right)^5 = -2\left(-\frac{1}{32}\right) = \frac{1}{16}$

15. Finding the term: $a_{20} = 3(-1)^{20-1} = 3(-1)^{19} = -3$

17. Finding the sum: $S_{10} = \frac{10\left(2^{10} - 1\right)}{2 - 1} = 10 \cdot 1023 = 10,230$

19. Finding the sum: $S_{20} = \frac{1\left((-1)^{20} - 1\right)}{-1 - 1} = \frac{1 \cdot 0}{-2} = 0$

21. Using $a_1 = \frac{1}{5}$ and $r = \frac{1}{2}$, the term is: $a_8 = \frac{1}{5} \cdot \left(\frac{1}{2}\right)^{8-1} = \frac{1}{5} \cdot \left(\frac{1}{2}\right)^7 = \frac{1}{5} \cdot \frac{1}{128} = \frac{1}{640}$

23. Using $a_1 = -\frac{1}{2}$ and $r = \frac{1}{2}$, the sum is: $S_5 = \frac{-\frac{1}{2}\left(\left(\frac{1}{2}\right)^5 - 1\right)}{\frac{1}{2} - 1} = \frac{-\frac{1}{2}\left(\frac{1}{32} - 1\right)}{-\frac{1}{2}} = \frac{1}{32} - 1 = -\frac{31}{32}$

25. Using $a_1 = \sqrt{2}$ and $r = \sqrt{2}$, the term is: $a_{10} = \sqrt{2}\left(\sqrt{2}\right)^9 = \left(\sqrt{2}\right)^{10} = 2^5 = 32$

The sum is: $S_{10} = \frac{\sqrt{2}\left(\left(\sqrt{2}\right)^{10} - 1\right)}{\sqrt{2} - 1} = \frac{\sqrt{2}(32 - 1)}{\sqrt{2} - 1} = \frac{31\sqrt{2}}{\sqrt{2} - 1} \cdot \frac{\sqrt{2} + 1}{\sqrt{2} + 1} = \frac{62 + 31\sqrt{2}}{2 - 1} = 62 + 31\sqrt{2}$

27. Using $a_1 = 100$ and $r = 0.1$, the term is: $a_6 = 100(0.1)^5 = 10^2\left(10^{-5}\right) = 10^{-3} = \frac{1}{1000}$

The sum is: $S_6 = \frac{100\left((0.1)^6 - 1\right)}{0.1 - 1} = \frac{100\left(10^{-6} - 1\right)}{-0.9} = \frac{-99.9999}{-0.9} = 111.111$

29. Since $a_4 \cdot r \cdot r = a_6$, we have the equation:
$$a_4 r^2 = a_6$$
$$40 r^2 = 160$$
$$r^2 = 4$$
$$r = \pm 2$$

31. Since $a_1 = -3$ and $r = -2$, the values are:
$$a_8 = -3(-2)^7 = -3(-128) = 384$$
$$S_8 = \frac{-3\left((-2)^8 - 1\right)}{-2 - 1} = \frac{-3(256 - 1)}{-3} = 255$$

33.   Since $a_7 \cdot r \cdot r \cdot r = a_{10}$, we have the equation:

$$a_7 r^3 = a_{10}$$
$$13r^3 = 104$$
$$r^3 = 8$$
$$r = 2$$

35.   Using $a_1 = \frac{1}{2}$ and $r = \frac{1}{2}$ in the sum formula: $S = \dfrac{\frac{1}{2}}{1 - \frac{1}{2}} = \dfrac{\frac{1}{2}}{\frac{1}{2}} = 1$

37.   Using $a_1 = 4$ and $r = \frac{1}{2}$ in the sum formula: $S = \dfrac{4}{1 - \frac{1}{2}} = \dfrac{4}{\frac{1}{2}} = 8$

39.   Using $a_1 = 2$ and $r = \frac{1}{2}$ in the sum formula: $S = \dfrac{2}{1 - \frac{1}{2}} = \dfrac{2}{\frac{1}{2}} = 4$

41.   Using $a_1 = \frac{4}{3}$ and $r = -\frac{1}{2}$ in the sum formula: $S = \dfrac{\frac{4}{3}}{1 + \frac{1}{2}} = \dfrac{\frac{4}{3}}{\frac{3}{2}} = \frac{4}{3} \cdot \frac{2}{3} = \frac{8}{9}$

43.   Using $a_1 = \frac{2}{5}$ and $r = \frac{2}{5}$ in the sum formula: $S = \dfrac{\frac{2}{5}}{1 - \frac{2}{5}} = \dfrac{\frac{2}{5}}{\frac{3}{5}} = \frac{2}{5} \cdot \frac{5}{3} = \frac{2}{3}$

45.   Using $a_1 = \frac{3}{4}$ and $r = \frac{1}{3}$ in the sum formula: $S = \dfrac{\frac{3}{4}}{1 - \frac{1}{3}} = \dfrac{\frac{3}{4}}{\frac{2}{3}} = \frac{3}{4} \cdot \frac{3}{2} = \frac{9}{8}$

47.   Interpreting the decimal as an infinite sum with $a_1 = 0.4$ and $r = 0.1$: $S = \dfrac{0.4}{1 - 0.1} = \dfrac{0.4}{0.9} \cdot \dfrac{10}{10} = \frac{4}{9}$

49.   Interpreting the decimal as an infinite sum with $a_1 = 0.27$ and $r = 0.01$: $S = \dfrac{0.27}{1 - 0.01} = \dfrac{0.27}{0.99} \cdot \dfrac{100}{100} = \frac{27}{99} = \frac{3}{11}$

51.   **a.**   The first five terms are: \$450,000, \$315,000, \$220,500, \$154,350, \$108,045

     b.   The common ratio is 0.7.

     c.   Constructing a line graph:

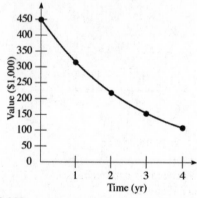

     d.   The value is approximately \$90,000.

     e.   The recursive formula is: $a_0 = 450000; a_n = 0.7a_{n-1}$ for $n \geq 1$

53.   **a.**   Using $a_1 = \frac{1}{3}$ and $r = \frac{1}{3}$ in the sum formula: $S = \dfrac{\frac{1}{3}}{1 - \frac{1}{3}} = \dfrac{\frac{1}{3}}{\frac{2}{3}} = \frac{1}{2}$

     b.   Finding the sum: $S_6 = \dfrac{\frac{1}{3}\left(\left(\frac{1}{3}\right)^6 - 1\right)}{\frac{1}{3} - 1} = \dfrac{\frac{1}{3}\left(\frac{1}{729} - 1\right)}{-\frac{2}{3}} = \dfrac{\frac{1}{3}\left(-\frac{728}{729}\right)}{-\frac{2}{3}} = -\frac{1}{2}\left(-\frac{728}{729}\right) = \frac{364}{729}$

     c.   Finding the difference of these two answers: $S - S_6 = \frac{1}{2} - \frac{364}{729} = \frac{729}{1458} - \frac{728}{1458} = \frac{1}{1458}$

55. There are two infinite series for these heights. For the amount the ball falls, use $a_1 = 20$ and $r = \frac{7}{8}$:

$$S_{\text{fall}} = \frac{20}{1 - \frac{7}{8}} = \frac{20}{\frac{1}{8}} = 160$$

For the amount the ball rises, use $a_1 = \frac{7}{8}(20) = \frac{35}{2}$ and $r = \frac{7}{8}$: $S_{\text{rise}} = \frac{\frac{35}{2}}{1 - \frac{7}{8}} = \frac{\frac{35}{2}}{\frac{1}{8}} = \frac{35}{2} \cdot 8 = 140$

The total distance traveled is then 160 feet + 140 feet = 300 feet

57. **a.** The general term is: $a_n = 15\left(\frac{4}{5}\right)^{n-1}$ **b.** Finding the sum: $S = \frac{15}{1 - \frac{4}{5}} = \frac{15}{\frac{1}{5}} = 75$ feet

59. Evaluating the logarithm: $\log_4 20 = \frac{\log 20}{\log 4} \approx 2.16$ **61.** Evaluating the logarithm: $\ln 576 \approx 6.36$

63. Solving for t:

$$A = 10e^{5t}$$

$$\frac{A}{10} = e^{5t}$$

$$5t = \ln\left(\frac{A}{10}\right)$$

$$t = \frac{\ln A - \ln 10}{5}$$

65. Expanding: $(x + y)^0 = 1$ **67.** Expanding: $(x + y)^2 = (x + y)(x + y) = x^2 + 2xy + y^2$

69. Simplifying: $\frac{6 \cdot 5 \cdot 4 \cdot 3 \cdot 2 \cdot 1}{(2 \cdot 1)(4 \cdot 3 \cdot 2 \cdot 1)} = \frac{6 \cdot 5}{2 \cdot 1} = \frac{30}{2} = 15$

71. Interpreting the decimal as an infinite sum with $a_1 = 0.63$ and $r = 0.01$: $S = \frac{0.63}{1 - 0.01} = \frac{0.63}{0.99} = \frac{63}{99} = \frac{7}{11}$

73. Finding the common ratio:

$$S = \frac{a_1}{1 - r}$$

$$6 = \frac{4}{1 - r}$$

$$6 - 6r = 4$$

$$-6r = -2$$

$$r = \frac{1}{3}$$

75. **a.** The areas are: stage 2: $\frac{3}{4}$; stage 3: $\frac{9}{16}$; stage 4: $\frac{27}{64}$

b. They form a geometric sequence.

c. The area is 0, since the sequence of areas is approaching 0.

d. The perimeters form an increasing sequence.

11.5 The Binomial Expansion

1.   Using the binomial formula:
$$(x+2)^4 = \binom{4}{0}x^4 + \binom{4}{1}x^3(2) + \binom{4}{2}x^2(2)^2 + \binom{4}{3}x(2)^3 + \binom{4}{4}(2)^4$$
$$= x^4 + 4 \cdot 2x^3 + 6 \cdot 4x^2 + 4 \cdot 8x + 16$$
$$= x^4 + 8x^3 + 24x^2 + 32x + 16$$

3.   Using the binomial formula:
$$(x+y)^6 = \binom{6}{0}x^6 + \binom{6}{1}x^5y + \binom{6}{2}x^4y^2 + \binom{6}{3}x^3y^3 + \binom{6}{4}x^2y^4 + \binom{6}{5}xy^5 + \binom{6}{6}y^6$$
$$= x^6 + 6x^5y + 15x^4y^2 + 20x^3y^3 + 15x^2y^4 + 6xy^5 + y^6$$

5.   Using the binomial formula:
$$(2x+1)^5 = \binom{5}{0}(2x)^5 + \binom{5}{1}(2x)^4(1) + \binom{5}{2}(2x)^3(1)^2 + \binom{5}{3}(2x)^2(1)^3 + \binom{5}{4}(2x)(1)^4 + \binom{5}{5}(1)^5$$
$$= 32x^5 + 5 \cdot 16x^4 + 10 \cdot 8x^3 + 10 \cdot 4x^2 + 5 \cdot 2x + 1$$
$$= 32x^5 + 80x^4 + 80x^3 + 40x^2 + 10x + 1$$

7.   Using the binomial formula:
$$(x-2y)^5 = \binom{5}{0}x^5 + \binom{5}{1}x^4(-2y) + \binom{5}{2}x^3(-2y)^2 + \binom{5}{3}x^2(-2y)^3 + \binom{5}{4}x(-2y)^4 + \binom{5}{5}(-2y)^5$$
$$= x^5 - 5 \cdot 2x^4y + 10 \cdot 4x^3y^2 - 10 \cdot 8x^2y^3 + 5 \cdot 16xy^4 - 32y^5$$
$$= x^5 - 10x^4y + 40x^3y^2 - 80x^2y^3 + 80xy^4 - 32y^5$$

9.   Using the binomial formula:
$$(3x-2)^4 = \binom{4}{0}(3x)^4 + \binom{4}{1}(3x)^3(-2) + \binom{4}{2}(3x)^2(-2)^2 + \binom{4}{3}(3x)(-2)^3 + \binom{4}{4}(-2)^4$$
$$= 81x^4 - 4 \cdot 54x^3 + 6 \cdot 36x^2 - 4 \cdot 24x + 16$$
$$= 81x^4 - 216x^3 + 216x^2 - 96x + 16$$

11.   Using the binomial formula:
$$(4x-3y)^3 = \binom{3}{0}(4x)^3 + \binom{3}{1}(4x)^2(-3y) + \binom{3}{2}(4x)(-3y)^2 + \binom{3}{3}(-3y)^3$$
$$= 64x^3 - 3 \cdot 48x^2y + 3 \cdot 36xy^2 - 27y^3$$
$$= 64x^3 - 144x^2y + 108xy^2 - 27y^3$$

13.   Using the binomial formula:
$$(x^2+2)^4 = \binom{4}{0}(x^2)^4 + \binom{4}{1}(x^2)^3(2) + \binom{4}{2}(x^2)^2(2)^2 + \binom{4}{3}(x^2)(2)^3 + \binom{4}{4}(2)^4$$
$$= x^8 + 4 \cdot 2x^6 + 6 \cdot 4x^4 + 4 \cdot 8x^2 + 16$$
$$= x^8 + 8x^6 + 24x^4 + 32x^2 + 16$$

15.   Using the binomial formula:
$$(x^2+y^2)^3 = \binom{3}{0}(x^2)^3 + \binom{3}{1}(x^2)^2(y^2) + \binom{3}{2}(x^2)(y^2)^2 + \binom{3}{3}(y^2)^3 = x^6 + 3x^4y^2 + 3x^2y^4 + y^6$$

17.   Using the binomial formula:
$$(2x+3y)^4 = \binom{4}{0}(2x)^4 + \binom{4}{1}(2x)^3(3y) + \binom{4}{2}(2x)^2(3y)^2 + \binom{4}{3}(2x)(3y)^3 + \binom{4}{4}(3y)^4$$
$$= 16x^4 + 4 \cdot 24x^3y + 6 \cdot 36x^2y^2 + 4 \cdot 54xy^3 + 81y^4$$
$$= 16x^4 + 96x^3y + 216x^2y^2 + 216xy^3 + 81y^4$$

19. Using the binomial formula:

$$\left(\frac{x}{2}+\frac{y}{3}\right)^3 = \binom{3}{0}\left(\frac{x}{2}\right)^3+\binom{3}{1}\left(\frac{x}{2}\right)^2\left(\frac{y}{3}\right)+\binom{3}{2}\left(\frac{x}{2}\right)\left(\frac{y}{3}\right)^2+\binom{3}{3}\left(\frac{y}{3}\right)^3$$

$$=\frac{x^3}{8}+3\cdot\frac{x^2 y}{12}+3\cdot\frac{xy^2}{18}+\frac{y^3}{27}$$

$$=\frac{x^3}{8}+\frac{x^2 y}{4}+\frac{xy^2}{6}+\frac{y^3}{27}$$

21. Using the binomial formula:

$$\left(\frac{x}{2}-4\right)^3 = \binom{3}{0}\left(\frac{x}{2}\right)^3+\binom{3}{1}\left(\frac{x}{2}\right)^2(-4)+\binom{3}{2}\left(\frac{x}{2}\right)(-4)^2+\binom{3}{3}(-4)^3$$

$$=\frac{x^3}{8}-3\cdot x^2+3\cdot 8x-64$$

$$=\frac{x^3}{8}-3x^2+24x-64$$

23. Using the binomial formula:

$$\left(\frac{x}{3}+\frac{y}{2}\right)^4 = \binom{4}{0}\left(\frac{x}{3}\right)^4+\binom{4}{1}\left(\frac{x}{3}\right)^3\left(\frac{y}{2}\right)+\binom{4}{2}\left(\frac{x}{3}\right)^2\left(\frac{y}{2}\right)^2+\binom{4}{3}\left(\frac{x}{3}\right)\left(\frac{y}{2}\right)^3+\binom{4}{4}\left(\frac{y}{2}\right)^4$$

$$=\frac{x^4}{81}+4\cdot\frac{x^3 y}{54}+6\cdot\frac{x^2 y^2}{36}+4\cdot\frac{xy^3}{24}+\frac{y^4}{16}$$

$$=\frac{x^4}{81}+\frac{2x^3 y}{27}+\frac{x^2 y^2}{6}+\frac{xy^3}{6}+\frac{y^4}{16}$$

25. Writing the first four terms:

$$\binom{9}{0}x^9+\binom{9}{1}x^8(2)+\binom{9}{2}x^7(2)^2+\binom{9}{3}x^6(2)^3$$

$$=x^9+9\cdot 2x^8+36\cdot 4x^7+84\cdot 8x^6$$

$$=x^9+18x^8+144x^7+672x^6$$

27. Writing the first four terms:

$$\binom{10}{0}x^{10}+\binom{10}{1}x^9(-y)+\binom{10}{2}x^8(-y)^2+\binom{10}{3}x^7(-y)^3=x^{10}-10x^9 y+45x^8 y^2-120x^7 y^3$$

29. Writing the first four terms:

$$\binom{25}{0}x^{25}+\binom{25}{1}x^{24}(3)+\binom{25}{2}x^{23}(3)^2+\binom{25}{3}x^{22}(3)^3$$

$$=x^{25}+25\cdot 3x^{24}+300\cdot 9x^{23}+2300\cdot 27x^{22}$$

$$=x^{25}+75x^{24}+2700x^{23}+62100x^{22}$$

31. Writing the first four terms:

$$\binom{60}{0}x^{60}+\binom{60}{1}x^{59}(-2)+\binom{60}{2}x^{58}(-2)^2+\binom{60}{3}x^{57}(-2)^3$$

$$=x^{60}-60\cdot 2x^{59}+1770\cdot 4x^{58}-34220\cdot 8x^{57}$$

$$=x^{60}-120x^{59}+7080x^{58}-273760x^{57}$$

33. Writing the first four terms:

$$\binom{18}{0}x^{18}+\binom{18}{1}x^{17}(-y)+\binom{18}{2}x^{16}(-y)^2+\binom{18}{3}x^{15}(-y)^3=x^{18}-18x^{17}y+153x^{16}y^2-816x^{15}y^3$$

35. Writing the first three terms: $\binom{15}{0}x^{15}+\binom{15}{1}x^{14}(1)+\binom{15}{2}x^{13}(1)^2=x^{15}+15x^{14}+105x^{13}$

37. Writing the first three terms: $\binom{12}{0}x^{12}+\binom{12}{1}x^{11}(-y)+\binom{12}{2}x^{10}(-y)^2 = x^{12}-12x^{11}y+66x^{10}y^2$

39. Writing the first three terms:

$$\binom{20}{0}x^{20}+\binom{20}{1}x^{19}(2)+\binom{20}{2}x^{18}(2)^2 = x^{20}+20\cdot 2x^{19}+190\cdot 4x^{18} = x^{20}+40x^{19}+760x^{18}$$

41. Writing the first two terms: $\binom{100}{0}x^{100}+\binom{100}{1}x^{99}(2) = x^{100}+100\cdot 2x^{99} = x^{100}+200x^{99}$

43. Writing the first two terms: $\binom{50}{0}x^{50}+\binom{50}{1}x^{49}y = x^{50}+50x^{49}y$

45. Finding the required term: $\binom{12}{8}(2x)^4(3y)^8 = 495\cdot 2^4\cdot 3^8 x^4 y^8 = 51,963,120 x^4 y^8$

47. Finding the required term: $\binom{10}{4}x^6(-2)^4 = 210\cdot 16x^6 = 3360x^6$

49. Finding the required term: $\binom{12}{5}x^7(-2)^5 = -792\cdot 32x^7 = -25344x^7$

51. Finding the required term: $\binom{25}{2}x^{23}(-3y)^2 = 300\cdot 9x^{23}y^2 = 2700x^{23}y^2$

53. Finding the required term: $\binom{20}{11}(2x)^9(5y)^{11} = \dfrac{20!}{11!9!}(2x)^9(5y)^{11}$

55. Writing the first three terms:

$$\binom{10}{0}(x^2y)^{10}+\binom{10}{1}(x^2y)^9(-3)+\binom{10}{2}(x^2y)^8(-3)^2$$

$$= x^{20}y^{10}-10\cdot 3x^{18}y^9+45\cdot 9x^{16}y^8$$

$$= x^{20}y^{10}-30x^{18}y^9+405x^{16}y^8$$

57. Finding the third term: $\binom{7}{2}\left(\tfrac{1}{2}\right)^5\left(\tfrac{1}{2}\right)^2 = 21\cdot\tfrac{1}{128} = \tfrac{21}{128}$

59. Solving the equation:

$$5^x = 7$$
$$\log 5^x = \log 7$$
$$x\log 5 = \log 7$$
$$x = \frac{\log 7}{\log 5}\approx 1.21$$

61. Solving the equation:

$$8^{2x+1} = 16$$
$$2^{6x+3} = 2^4$$
$$6x+3 = 4$$
$$6x = 1$$
$$x = \tfrac{1}{6}\approx 0.17$$

63. Using the compound interest formula:

$$400\left(1+\frac{0.10}{4}\right)^{4t} = 800$$

$$(1.025)^{4t} = 2$$

$$\ln(1.025)^{4t} = \ln 2$$

$$4t\ln 1.025 = \ln 2$$

$$t = \frac{\ln 2}{4\ln 1.025}\approx 7.02$$

It will take 7.02 years.

65. They both equal 56.

67. They both equal 125,970.

69. Smaller triangles (Serpinski triangles) begin to emerge.

Chapter 11 Review/Test

1. Writing the first four terms: 7,9,11,13
2. Writing the first four terms: 1,4,7,10
3. Writing the first four terms: 0,3,8,15
4. Writing the first four terms: $\frac{4}{3},\frac{5}{4},\frac{6}{5},\frac{7}{6}$
5. Writing the first four terms: 4,16,64,256
6. Writing the first four terms: $\frac{1}{4},\frac{1}{16},\frac{1}{64},\frac{1}{256}$
7. The general term is: $a_n = 3n - 1$
8. The general term is: $a_n = 2n - 5$
9. The general term is: $a_n = n^4$
10. The general term is: $a_n = n^2 + 1$
11. The general term is: $a_n = \left(\frac{1}{2}\right)^n = 2^{-n}$
12. The general term is: $a_n = \frac{n+1}{n^2}$
13. Expanding the sum: $\sum_{i=1}^{4}(2i+3) = 5+7+9+11 = 32$
14. Expanding the sum: $\sum_{i=1}^{3}(2i^2 - 1) = 1+7+17 = 25$
15. Expanding the sum: $\sum_{i=2}^{3}\frac{i^2}{i+2} = 1+\frac{9}{5} = \frac{14}{5}$
16. Expanding the sum: $\sum_{i=1}^{4}(-2)^{i-1} = 1-2+4-8 = -5$
17. Expanding the sum: $\sum_{i=3}^{5}\left(4i+i^2\right) = 21+32+45 = 98$
18. Expanding the sum: $\sum_{i=4}^{6}\frac{i+2}{i} = \frac{3}{2}+\frac{7}{5}+\frac{4}{3} = \frac{45}{30}+\frac{42}{30}+\frac{40}{30} = \frac{127}{30}$
19. Writing in summation notation: $\sum_{i=1}^{4}3i$
20. Writing in summation notation: $\sum_{i=1}^{4}(4i-1)$
21. Writing in summation notation: $\sum_{i=1}^{5}(2i+3)$
22. Writing in summation notation: $\sum_{i=2}^{4}i^2$
23. Writing in summation notation: $\sum_{i=1}^{4}\frac{1}{i+2}$
24. Writing in summation notation: $\sum_{i=1}^{5}\frac{i}{3^i}$
25. Writing in summation notation: $\sum_{i=1}^{3}(x-2i)$
26. Writing in summation notation: $\sum_{i=1}^{4}\frac{x}{x+i}$
27. The sequence is geometric.
28. The sequence is arithmetic.
29. The sequence is arithmetic.
30. The sequence is neither.
31. The sequence is geometric.
32. The sequence is geometric.
33. The sequence is arithmetic.
34. The sequence is neither.
35. Finding the general term: $a_n = 2+(n-1)3 = 2+3n-3 = 3n-1$. Now finding a_{20}: $a_{20} = 3(20)-1 = 59$
36. Finding the general term: $a_n = 5+(n-1)(-3) = 5-3n+3 = 8-3n$. Now finding a_{16}: $a_{16} = 8-3(16) = -40$
37. Finding a_{10}: $a_{10} = -2+9\cdot4 = 34$. Now finding S_{10}: $S_{10} = \frac{10}{2}(-2+34) = 5\cdot32 = 160$
38. Finding a_{16}: $a_{16} = 3+15\cdot5 = 78$. Now finding S_{16}: $S_{16} = \frac{16}{2}(3+78) = 8\cdot81 = 648$
39. First write the equations:
$$a_5 = a_1 + 4d \quad a_8 = a_1 + 7d$$
$$21 = a_1 + 4d \quad 33 = a_1 + 7d$$
We have the system of equations:
$$a_1 + 7d = 33$$
$$a_1 + 4d = 21$$
Subtracting yields:
$$3d = 12$$
$$d = 4$$
$$a_1 = 33 - 28 = 5$$
Now finding a_{10}: $a_{10} = 5+9\cdot4 = 41$

40.   First write the equations:

$$a_3 = a_1 + 2d \quad a_7 = a_1 + 6d$$

$$14 = a_1 + 2d \quad 26 = a_1 + 6d$$

We have the system of equations:

$$a_1 + 6d = 26$$

$$a_1 + 2d = 14$$

Subtracting yields:

$$4d = 12$$

$$d = 3$$

$$a_1 = 26 - 18 = 8$$

Now finding $a_9 : a_9 = 8 + 8 \bullet 3 = 32$. Finding $S_9 : S_9 = \frac{9}{2}(8 + 32) = \frac{9}{2} \bullet 40 = 180$

41.   First write the equations:

$$a_4 = a_1 + 3d \quad a_8 = a_1 + 7d$$

$$-10 = a_1 + 3d \quad -18 = a_1 + 7d$$

We have the system of equations:

$$a_1 + 7d = -18$$

$$a_1 + 3d = -10$$

Subtracting yields:

$$4d = -8$$

$$d = -2$$

$$a_1 = -18 + 14 = -4$$

Now finding $a_{20} : a_{20} = -4 + 19 \bullet (-2) = -42$. Finding $S_{20} : S_{20} = \frac{20}{2}(-4 - 42) = 10 \bullet (-46) = -460$

42.   Using $a_1 = 3$ and $d = 4$, find $a_{100} : a_{100} = 3 + 99 \bullet 4 = 399$

Now finding the sum: $S_{100} = \frac{100}{2}(3 + 399) = 50 \bullet 402 = 20,100$

43.   Using $a_1 = 100$ and $d = -5 : a_{40} = 100 + 39 \bullet (-5) = 100 - 195 = -95$

44.   The general term is: $a_n = 3(2)^{n-1}$. Now finding $a_{20} : a_{20} = 3(2)^{19} = 1,572,864$

45.   The general term is: $a_n = 5(-2)^{n-1}$. Now finding $a_{16} : a_{16} = 5(-2)^{15} = -163,840$

46.   The general term is: $a_n = 4\left(\frac{1}{2}\right)^{n-1}$. Now finding $a_{10} : a_{10} = 4\left(\frac{1}{2}\right)^9 = \frac{1}{128}$

47.   Finding the sum: $S = \dfrac{-2}{1 - \frac{1}{3}} = \dfrac{-2}{\frac{2}{3}} = -3$ **48.**   Finding the sum: $S = \dfrac{4}{1 - \frac{1}{2}} = \dfrac{4}{\frac{1}{2}} = 8$

49.   Since $a_3 \bullet r = a_4$, we have:

$$12r = 24$$

$$r = 2$$

Finding the first term:

$$a_3 = a_1 r^2$$

$$12 = a_1 \bullet 2^2$$

$$12 = 4a_1$$

$$a_1 = 3$$

Now finding $a_6 : a_6 = a_1 r^5 = 3 \bullet 2^5 = 96$

50.   Using $a_1 = 3$ and $r = \sqrt{3} : a_{10} = a_1 r^9 = 3 \bullet \left(\sqrt{3}\right)^9 = 3 \bullet 81\sqrt{3} = 243\sqrt{3}$

51. Using the binomial formula:

$$(x-2)^4 = \binom{4}{0}x^4 + \binom{4}{1}x^3(-2) + \binom{4}{2}x^2(-2)^2 + \binom{4}{3}x(-2)^3 + \binom{4}{4}(-2)^4$$

$$= x^4 - 4 \cdot 2x^3 + 6 \cdot 4x^2 - 4 \cdot 8x + 16$$

$$= x^4 - 8x^3 + 24x^2 - 32x + 16$$

52. Using the binomial formula:

$$(2x+3)^4 = \binom{4}{0}(2x)^4 + \binom{4}{1}(2x)^3(3) + \binom{4}{2}(2x)^2(3)^2 + \binom{4}{3}(2x)(3)^3 + \binom{4}{4}(3)^4$$

$$= 16x^4 + 4 \cdot 24x^3 + 6 \cdot 36x^2 + 4 \cdot 54x + 81$$

$$= 16x^4 + 96x^3 + 216x^2 + 216x + 81$$

53. Using the binomial formula:

$$(3x+2y)^3 = \binom{3}{0}(3x)^3 + \binom{3}{1}(3x)^2(2y) + \binom{3}{2}(3x)(2y)^2 + \binom{3}{3}(2y)^3$$

$$= 27x^3 + 3 \cdot 18x^2y + 3 \cdot 12xy^2 + 8y^3$$

$$= 27x^3 + 54x^2y + 36xy^2 + 8y^3$$

54. Using the binomial formula:

$$(x^2-2)^5 = \binom{5}{0}(x^2)^5 + \binom{5}{1}(x^2)^4(-2) + \binom{5}{2}(x^2)^3(-2)^2 + \binom{5}{3}(x^2)^2(-2)^3 + \binom{5}{4}(x^2)(-2)^4 + \binom{5}{5}(-2)^5$$

$$= x^{10} - 5 \cdot 2x^8 + 10 \cdot 4x^6 - 10 \cdot 8x^4 + 5 \cdot 16x^2 - 32$$

$$= x^{10} - 10x^8 + 40x^6 - 80x^4 + 80x^2 - 32$$

55. Using the binomial formula:

$$\left(\frac{x}{2}+3\right)^4 = \binom{4}{0}\left(\frac{x}{2}\right)^4 + \binom{4}{1}\left(\frac{x}{2}\right)^3(3) + \binom{4}{2}\left(\frac{x}{2}\right)^2(3)^2 + \binom{4}{3}\left(\frac{x}{2}\right)(3)^3 + \binom{4}{4}(3)^4$$

$$= \tfrac{1}{16}x^4 + 4 \cdot \tfrac{3}{8}x^3 + 6 \cdot \tfrac{9}{4}x^2 + 4 \cdot \tfrac{27}{2}x + 81$$

$$= \tfrac{1}{16}x^4 + \tfrac{3}{2}x^3 + \tfrac{27}{2}x^2 + 54x + 81$$

56. Using the binomial formula:

$$\left(\frac{x}{3}-\frac{y}{2}\right)^3 = \binom{3}{0}\left(\frac{x}{3}\right)^3 + \binom{3}{1}\left(\frac{x}{3}\right)^2\left(-\frac{y}{2}\right) + \binom{3}{2}\left(\frac{x}{3}\right)\left(-\frac{y}{2}\right)^2 + \binom{3}{3}\left(-\frac{y}{2}\right)^3$$

$$= \tfrac{1}{27}x^3 - 3 \cdot \tfrac{1}{18}x^2y + 3 \cdot \tfrac{1}{12}xy^2 - \tfrac{1}{8}y^3$$

$$= \tfrac{1}{27}x^3 - \tfrac{1}{6}x^2y + \tfrac{1}{4}xy^2 - \tfrac{1}{8}y^3$$

57. Writing the first three terms:

$$\binom{10}{0}x^{10} + \binom{10}{1}x^9(3y) + \binom{10}{2}x^8(3y)^2 = x^{10} + 10 \cdot 3x^9y + 45 \cdot 9x^8y^2 = x^{10} + 30x^9y + 405x^8y^2$$

58. Writing the first three terms:

$$\binom{9}{0}x^9 + \binom{9}{1}x^8(-3y) + \binom{9}{2}x^7(-3y)^2 = x^9 - 9 \cdot 3x^8y + 36 \cdot 9x^7y^2 = x^9 - 27x^8y + 324x^7y^2$$

59. Writing the first three terms:

$$\binom{11}{0}x^{11} + \binom{11}{1}x^{10}y + \binom{11}{2}x^9y^2 = x^{11} + 11x^{10}y + 55x^9y^2$$

60. Writing the first three terms:

$$\binom{12}{0}x^{12} + \binom{12}{1}x^{11}(-2y) + \binom{12}{2}x^{10}(-2y)^2 = x^{12} - 12 \cdot 2x^{11}y + 66 \cdot 4x^{10}y^2 = x^{12} - 24x^{11}y + 264x^{10}y^2$$

61. Writing the first two terms: $\begin{pmatrix} 16 \\ 0 \end{pmatrix} x^{16} + \begin{pmatrix} 16 \\ 1 \end{pmatrix} x^{15}(-2y) = x^{16} - 16 \cdot 2x^{15}y = x^{16} - 32x^{15}y$

62. Writing the first two terms: $\begin{pmatrix} 32 \\ 0 \end{pmatrix} x^{32} + \begin{pmatrix} 32 \\ 1 \end{pmatrix} x^{31}(2y) = x^{32} + 32 \cdot 2x^{31}y = x^{32} + 64x^{31}y$

63. Writing the first two terms: $\begin{pmatrix} 50 \\ 0 \end{pmatrix} x^{50} + \begin{pmatrix} 50 \\ 1 \end{pmatrix} x^{49}(-1) = x^{50} - 50x^{49}$

64. Writing the first two terms: $\begin{pmatrix} 150 \\ 0 \end{pmatrix} x^{150} + \begin{pmatrix} 150 \\ 1 \end{pmatrix} x^{149}y = x^{150} + 150x^{149}y$

65. Finding the sixth term: $\begin{pmatrix} 10 \\ 5 \end{pmatrix} x^{5}(-3)^{5} = 252 \cdot (-243)x^{5} = -61,236x^{5}$

66. Finding the fourth term: $\begin{pmatrix} 9 \\ 3 \end{pmatrix}(2x)^{6}(1)^{3} = 84 \cdot 64x^{6} = 5376x^{6}$